# Post-Fire Management and Restoration of Southern European Forests

# Managing Forest Ecosystems

Volume 24

*Series Editors:*

## Klaus von Gadow

*Georg-August-University,*
*Göttingen, Germany*

## Timo Pukkala

*University of Joensuu,*
*Joensuu, Finland*

and

## Margarida Tomé

*Instituto Superior de Agronomía,*
*Lisbon, Portugal*

**Aims & Scope:**

Well-managed forests and woodlands are a renewable resource, producing essential raw material with minimum waste and energy use. Rich in habitat and species diversity, forests may contribute to increased ecosystem stability. They can absorb the effects of unwanted deposition and other disturbances and protect neighbouring ecosystems by maintaining stable nutrient and energy cycles and by preventing soil degradation and erosion. They provide much-needed recreation and their continued existence contributes to stabilizing rural communities.

Forests are managed for timber production and species, habitat and process conservation. A subtle shift from *multiple-use management to ecosystems management* is being observed and the new ecological perspective of *multi-functional forest management* is based on the principles of ecosystem diversity, stability and elasticity, and the dynamic equilibrium of primary and secondary production.

Making full use of new technology is one of the challenges facing forest management today. Resource information must be obtained with a limited budget. This requires better timing of resource assessment activities and improved use of multiple data sources. Sound ecosystems management, like any other management activity, relies on effective forecasting and operational control.

The aim of the book series **Managing Forest Ecosystems** is to present state-of-the-art research results relating to the practice of forest management. Contributions are solicited from prominent authors. Each reference book, monograph or proceedings volume will be focused to deal with a specific context. Typical issues of the series are: resource assessment techniques, evaluating sustainability for even-aged and uneven-aged forests, multi-objective management, predicting forest development, optimizing forest management, biodiversity management and monitoring, risk assessment and economic analysis.

For further volumes:
http://www.springer.com/series/6247

Francisco Moreira • Margarita Arianoutsou
Piermaria Corona • Jorge De las Heras
Editors

# Post-Fire Management and Restoration of Southern European Forests

Springer

*Editors*

Francisco Moreira
Institute of Agronomy
Centre of Applied Ecology
Technical University of Lisbon
Tapada da Ajuda, 1349-017 Lisbon
Portugal
fmoreira@isa.utl.pt

Piermaria Corona
Department for Innovation in Biological
Agro-Food and Forest Systems
University of Tuscia
Via San Camillo de Lellis Snc
01100 Viterbo
Italy
piermaria.corona@unitus.it

Margarita Arianoutsou
Department of Ecology and Systematics
Faculty of Biology
School of Sciences
National and Kapodistrian University
of Athens
Panepistimiopolis, Ilisia
15784 Athens
Greece
marianou@biol.uoa.gr

Jorge De las Heras
Plant Production and Agronomy
Technology
University of Castilla-La Mancha
Campus Universitario
02071 Albacete
Spain
Jorge.heras@uclm.es

ISSN 1568-1319          e-ISSN 1568-1319
ISBN 978-94-007-2207-1          e-ISBN 978-94-007-2208-8
DOI 10.1007/978-94-007-2208-8
Springer Dordrecht Heidelberg London New York

Library of Congress Control Number: 2011941755

*Cover picture legend*: Typical scene in the State of Durango where forests are managed by communities known as Ejidos: management is by selective tree removal, clear-felling is not allowed. Animals (ganado) are part of the multiple use system practiced there. (Photo by K.v. Gadow, autumn 2009)

Printed on acid-free paper

Springer is part of Springer Science+Business Media (www.springer.com)

# Foreword

In spite of the huge areas affected by wildfires every year in Europe, the post-fire management of burned areas has been given much less attention than fire management. Even from a scientific research perspective, priority has been given to fire management, with the most relevant EU project in recent years, FIRE PARADOX (http://www.fireparadox.org/), focusing on the important issue of the use of fire as a tool to combating wildfires.

However, the high relevance of post-fire management has been recognized in several political contexts, such as the EU Forest Action Plan, which in its Key Action 9 (Enhance the protection of EU forests) asks for the support for the restoration of forests damaged by natural disasters and fire, or a FAO review on international cooperation in fire management, which identifies post-fire rehabilitation and management of invasive species as priority topics for cooperative action (FAO 2006). Though relevant scientific knowledge on these specific topics is available in some countries, mainly of Southern Europe, research innovations have not been sufficiently transferred to the management practice yet, especially at the European level. The EU project EUFIRELAB (http://www.eufirelab.org) was a significant starting point to this, by making a preliminary assessment of the state-of-the-art concerning post-fire management techniques and restoration efforts, but much remains to be done. Ongoing initiatives consist of networks of research institutions, such as the PHOENIX Project Centre (www.phoenixefi.org/) created in the frame of the European Forest Institute, and the COST Action FP0701 "Post-Fire Forest Management in Southern Europe" (http://uaeco.biol.uoa.gr/cost/), established in the frame of the European Science Foundation, aiming at gathering research efforts on post-fire management topics and at identifying the best management practices for post-fire ecosystem rehabilitation in Europe. This book has been written in the framework of the PHOENIX and COST initiatives.

Why is this book needed? Firstly, because post-fire management and restoration of burned areas is a relevant topic that has not received much attention in Europe. Secondly, because there is a lack of information on restoration approaches and assessment of currently applied post-fire management practices and techniques for post-fire ecosystem rehabilitation in Europe. Finally, because information on the legal and social implications of wildfires is scarce, and so are the methods for the economic assessment of their impacts.

The main objective of this book is to assemble and disseminate scientifically based decision criteria for post-fire forest management and restoration, by reviewing the results of previous and ongoing scientific research. It aims at transferring this scientific knowledge into management practices, by bridging the gap between science and practice in post-fire management. It covers a wide range of spatial scales, from stand to landscape level planning, thus the main target users of this book include not only forest managers but also landscape managers and policy makers at national and regional levels. In fact, several practices for restoration are implemented on a large landscape scale and go beyond forest policies, stretching to agricultural and socio-economic policies. Nevertheless, the book uses a scientific language and approach that makes it suitable also for a broader audience including scientists, university lecturers and graduate students. It is primarily based on a review of the most relevant scientific literature, but it also includes previously unpublished scientific information, with the aim of offering a timely synthesis and novel elements for guiding research and monitoring programs, management guidelines and policy, concerning the restoration of burned lands. The short term expected result is to increase the scientific basis for undertaking appropriate post-fire management practices in Europe, whereas the long term expected result is to improve our ability to effectively restore burned areas and reduce fire hazard in European forests and landscapes. Although focused on Southern Europe, where fire hazard is currently higher, the outcome is also highly relevant for central and northern European countries as well, as climate change is increasing fire hazard in these regions.

The book begins with an introduction, where key questions and concepts related to post-fire management are placed. Chapter 2 deals with the relationships between recent land-cover changes and fire regime in Europe. The economic, legal and social aspects of post-fire management are considered in Chap. 3. Chap. 4 provides an overview of the fire hazard of different forest types. Chapter 5 addresses the main questions related to the post-fire management, common to all forest types. The last chapters (6–12) present distinctive post-fire management issues related to different forest types in Europe. Following the classification of forest types made by the European Environmental Agency, we focused on relevant fire-affected forest types (serotinous pine forests, non-serotinous pine forests, cork oak forests, other Mediterranean broadleaved forests, forests of exotic species, newly fire affected threatened forest types) and shrublands. These chapters are roughly similar in structure, each including the ecological context, the current post-fire management practices, the main management alternatives, and some relevant post-fire management case studies.

We are grateful to external reviewers, Dimitrakopoulos Panayiotis (University of the Aegean), Francisco Rego (Technical University of Lisbon), Filipe Catry (Technical University of Lisbon), Giovanni Bovio (University of Turin), for their valuable contribution for the improvement of the quality of the book.

This publication is supported by COST.

The editors

# COST: European Cooperation in the Field of Scientific and Technical Research

COST – the acronym for European Cooperation in Science and Technology- is the oldest and widest European intergovernmental network for cooperation in research. Established by the Ministerial Conference in November 1971, COST is presently used by the scientific communities of 36 European countries to cooperate in common research projects supported by national funds.

The funds provided by COST – less than 1% of the total value of the projects – support the COST cooperation networks (COST Actions) through which, with EUR 30 million per year, more than 30,000 European scientists are involved in research having a total value which exceeds EUR 2 billion per year. This is the financial worth of the European added value which COST achieves.

A "bottom up approach" (the initiative of launching a COST Action comes from the European scientists themselves), "à la carte participation" (only countries interested in the Action participate), "equality of access" (participation is open also to the scientific communities of countries not belonging to the European Union) and "flexible structure" (easy implementation and light management of the research initiatives) are the main characteristics of COST.

As precursor of advanced multidisciplinary research COST has a very important role for the realisation of the European Research Area (ERA) anticipating and complementing the activities of the Framework Programmes, constituting a "bridge" towards the scientific communities of emerging countries, increasing the mobility of researchers across Europe and fostering the establishment of "Networks of Excellence" in many key scientific domains such as: Biomedicine and Molecular Biosciences; Food and Agriculture; Forests, their Products and Services; Materials, Physical and Nanosciences; Chemistry and Molecular Sciences and Technologies; Earth System Science and Environmental Management; Information and Communication Technologies; Transport and Urban Development; Individuals, Societies, Cultures and Health. It covers basic and more applied research and also addresses issues of prenormative nature or of societal importance.

Web: http://www.cost.eu

# Contents

# Chapter 1
# Setting the Scene for Post-Fire Management

**Francisco Moreira, Margarita Arianoutsou, V. Ramón Vallejo,
Jorge de las Heras, Piermaria Corona, Gavriil Xanthopoulos,
Paulo Fernandes, and Kostas Papageorgiou**

## 1.1 Introduction

Every year, around 45,000 wildfires occur in Europe, burning an area of 0.5 million hectares (San-Miguel and Camia 2009). Between 1995 and 2004, more than four million hectares were burned in the Mediterranean region alone, corresponding to an area larger than the Netherlands. In addition to social and environmental impacts, wildfires also produce considerable economic damages due to: (i) the huge amount of resources spent in fire suppression and prevention; (ii) the loss of commercial value of damaged wood products; (iii) the costs related to loss of public non-market services (i.e., biodiversity protection, water cycle regulation, supply of recreational areas, soil protection, carbon sequestration, etc.).

F. Moreira (✉)
Centre of Applied Ecology, Institute of Agronomy, Technical University of Lisbon,
Lisbon, Portugal
e-mail: fmoreira@isa.utl.pt

M. Arianoutsou
Faculty of Biology, Department of Ecology and Systematics, School of Sciences,
National and Kapodistrian University of Athens, Athens, Greece

V.R. Vallejo
Fundacion CEAM, Parque Tecnologico, Paterna, Spain

J. de las Heras
Escuela Tecnica Superior de Ingenieros Agronomos, University of Castilla-La Mancha,
Albacete, Spain

P. Corona
Department for Innovation in Biological, Agro-Food and Forest Systems,
University of Tuscia, Viterbo, Italy

F. Moreira et al. (eds.), *Post-Fire Management and Restoration of Southern European
Forests*, Managing Forest Ecosystems 24, DOI 10.1007/978-94-007-2208-8_1,
© Springer Science+Business Media B.V. 2012

The post-fire management of burned areas has been given much less attention than fire suppression and prevention in Europe and elsewhere. However, important questions raise public concern and require scientifically-based knowledge: how can we accurately evaluate fire damages in economic terms? What are the most suitable short-term intervention techniques to minimise soil erosion and runoff? How should burned trees be managed? What is the best approach to long-term planning for the rehabilitation of burned areas? Along side the damage they incur, wildfires can also be regarded as an opportunity to plan and establish less flammable and more resilient forests and landscapes in recently burned areas. What information is available on these topics and how should administrations and stakeholders react after large fires? These questions are relevant not only in a southern European perspective, where wildfires are more frequent, but all over Europe. In fact, climate change and land-use trends are expected to increase fire incidence in Central and Northern Europe (Lindner et al. 2010), and new geographical areas (and forest ecosystems) where wildfires were infrequent are likely to become more fire-prone. Thus, further knowledge is needed on how to manage the millions of hectares burned in Europe, including planning of post-fire management, short-term intervention techniques to minimise soil erosion and runoff, and longer-term ecosystem recovery and restoration.

## 1.2 Wildfires in Europe

Fire is an integral part of many terrestrial biomes including the Mediterranean ones, but is also a major factor of disturbance (Pausas et al. 2008). Natural fire regimes have been increasingly changed by man for many thousands of years, so that in many regions of the world human-caused fires have become more frequent than natural sources of ignition (Goldammer and Crutzen 1993). During the last decades, an increase in the number of fires and the area burned is observed (Flannigan et al. 2009; Moreno et al. 1998; Piñol et al. 1998). In the Southern European Mediterranean countries, the major driving forces behind this change in the fire regime are land abandonment and afforestation of former agricultural land, leading to fuel accumulation and landscape-level connectivity of flammable patches.

G. Xanthopoulos
National Agricultural Research Foundation, Institute of Mediterranean Forest Ecosystems and Forest Products Technology, Athens, Greece

P. Fernandes
Centre for the Research and Technology of Agro-Environmental and Biological Sciences, Universidade de Trás-os-Montes e Alto Douro, Vila Real, Portugal

K. Papageorgiou
Department of Forests, Forest Fire Protection Office, Nicosia, Cyprus

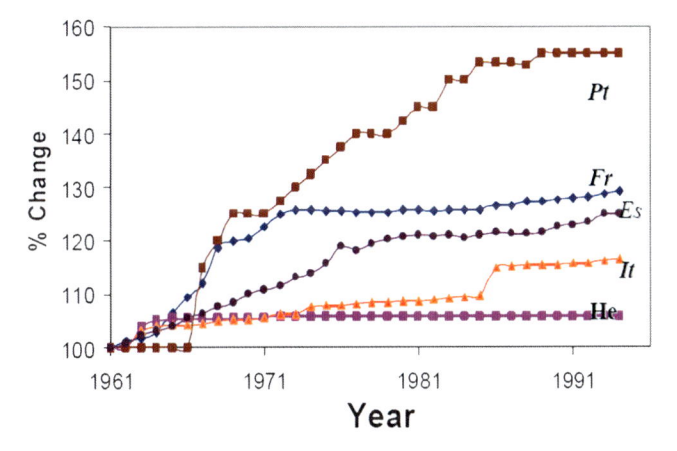

**Fig. 1.1** Change (%) in forest area in some Mediterranean countries of Europe during the second half of the twentieth century. *Pt* Portugal, *Fr* France, *Es* Spain, *It* Italy, *He* Hellas (Greece). (Source: FAOSTAT)

By the second half of the twentieth century, the process of extensive land use was halted and reversed in most regions of Europe. Rural exodus (Hill et al. 2008), mechanization of agriculture, reduced livestock grazing, afforestation in many marginal areas and changes in life styles caused important changes in land-use and land-cover patterns (Arianoutsou 2001; Lehouerou 1992; MacDonald et al. 2000; Moreira et al. 2001, 2009a, 2011). Poorly vegetated landscapes gave rise to others which gradually accumulated vegetation biomass, with growing trees or shrubs, and in which active management became occasional. This tendency to increase forest and non-managed land, that is, of increasing fuel loads in the landscape, has been observed in most countries (Fig. 1.1).

In addition to land cover changes, the influence of climatic changes upon shifts in fire regime cannot be ruled out (Pausas 2004; Pausas et al. 2008; Piñol et al. 1998). In fact, climatic extremes, that are more frequently observed nowadays (Founda and Giannakopoulos 2009; Tolika et al. 2009), may be more critical than fuel accumulation in controlling fire behaviour (Cumming 2001). Pausas and Fernandéz-Muñoz (2011) suggest that fire regime changes in the western Mediterranean were different before the 1970s, where fires were mostly fuel-limited, from the present, where they are mostly drought-driven. However, the effects of climate change on increased fire occurrence are not always obvious. For example, in Eastern Iberian Peninsula, Pausas (2004) found a positive correlation between summer rainfall and the area burned two years later, suggesting that this rainfall increases fuel loads for the subsequent fire seasons. In this perspective, warmer and drier summers could decrease fuel loads and, thus, fire hazard, which is supported by process-based simulations of fire activity under future climates (Thonicke et al. 2010).

Although forest fire statistics are incomplete for the first part of the twentieth century, available data indicate that wildfires were not important until the middle of

**Table 1.1** Number of fires and burned area in the five Southern States in the last 30 years

| Number of fires | Portugal | Spain | France | Italy | Greece[a] | Total |
|---|---|---|---|---|---|---|
| 2009 | 26,119 | 15,391 | 4,800 | 5,422 | 1,063 | 52,795 |
| % of total in 2009 | 49 | 29 | 9 | 10 | 2 | 100 |
| Average 1980–1989 | 7,381 | 9,515 | 4,910 | 11,575 | 1,264 | 34,645 |
| Average 1990–1999 | 22,250 | 18,152 | 5,538 | 11,164 | 1,748 | 58,851 |
| Average 2000–2009 | 24,949 | 18,337 | 4,406 | 7,259 | 1,569 | 56,645 |
| Average 1980–2009 | 18,194 | 15,335 | 4,951 | 9,999 | 1,569 | 50,047 |
| Total (1980–2009) | 545,805 | 452,848 | 148,531 | 299,977 | 47,058 | 1,501,409 |
| Burned areas (ha) | Portugal | Spain | France | Italy | Greece[a] | Total |
| 2009 | 87,416 | 110,783 | 17,000 | 73,355 | 35,342 | 323,896 |
| % of total in 2009 | 27 | 34 | 5 | 23 | 11 | 100 |
| Average 1980–1989 | 73,484 | 244,788 | 39,157 | 147,150 | 52,417 | 556,995 |
| Average 1990–1999 | 102,203 | 161,319 | 22,735 | 118,573 | 44,108 | 448,938 |
| Average 2000–2009 | 150,101 | 125,239 | 22,342 | 83,878 | 49,238 | 430,798 |
| Average 1980–2009 | 108,956 | 177,115 | 28,078 | 116,534 | 48,587 | 478,910 |
| Total (1980–2009) | 3,257,886 | 5,313,457 | 842,332 | 3,496,005 | 1,457,624 | 14,367,304 |

[a]Provisional data for 2009
Source: EFFIS report 2009

the century, at least in forested areas, which were the ones for which statistics were compiled. By the late 1960s wildfires started to occur at an increasing rate in the Euro-Mediterranean countries (Alexandrian and Esnault 1998). The number of fires has continued rising, although part of this trend is due to a change in the compilation of statistics (Moreno 2010). Additionally, an increase in ignition sources due to changes in socioeconomic and land-use cannot be excluded (Vélez 2009). Therefore, while a climate effect cannot be ruled out, certainly other factors came into play (Moreno 2010). Area burned, which is less sensitive to compilation procedures, increased during the 1970s and into the 1980s, by which time Spain and Italy had reached maximum values. Greece and Portugal followed with some delay (Moreno et al. 1998). This increase occurred while fire suppression was being strengthened in all the Euro-Mediterranean countries.

Around the Mediterranean most fires (≈95%) are caused by humans, either by accidents or intentionally (EFFIS 2009). During the last decades the area burned in Spain, France and Italy has decreased, but that is not the case for Greece and, much less so, for other Balkan countries and Portugal, where the area burned per year has markedly increased. During 2009, fires in these five countries burned a total area of 323,896 ha (Table 1.1), which, although almost doubling the area burned in 2008, is still below the average for the last 29 years. The number of fires that occurred (52,795) is also slightly below the average of the last two decades. Since the area of each country is different, and the area at risk within each country is also different, comparisons among countries are not straightforward. Fires also occur in other parts of Central and northern Europe, although with a much lower significance (EFFIS 2009).

Large fires represent a small fraction of the total number of fires, but are responsible for a large percentage of the total land area burned in the Mediterranean basin (Diaz-Delgado et al. 2004; Bermudez et al. 2009). Large fires are relatively new in the recent history of the Mediterranean Basin (Lloret and Marí 2001). The recent exceptional fire-seasons (e.g. 1978/79 and 1994 in Spain, 1998, 2000 and 2007 in Greece, 2005 in Portugal, 2003 throughout Europe) contributed to highlight the importance of large fires in the Euro-Mediterranean (e.g. Oliveras et al. 2009; Pausas 2004; Piñol et al. 1998; Xanthopoulos 2007).

A more detailed analysis of current fire regimes is made in Chap. 2.

## 1.3  What to Do After Fire?

The traditional strategy for the management of burned areas, and other degraded lands, in the Mediterranean region was based on afforestation or reforestation with conifers, particularly since the nineteenth century. Massive plantations covering millions of hectares were carried out in many Mediterranean countries, mostly using pines, and reforestation rates have been further promoted by EU agricultural policies that aimed to convert marginal agricultural land into forested areas (Pausas et al. 2004; Vallejo 2005a). These plantations provided jobs in rural areas, and aimed to increase forest productivity, protect watersheds and in some cases stabilize coastal dune systems. This strategy also assumed that the restoration of degraded areas involved a first stage in which a pioneer conifer was used, followed by the posterior introduction of late-successional hardwoods (Pausas et al. 2004). This traditional view ended up having a very low level of application, due to the cost of implementing it. In addition, changes in fire regime since the last decades of the twentieth century strongly compromised the effectiveness of this strategy.

Nowadays, the range of alternatives, in terms of management objectives for burned areas and techniques available for restoration, is much wider, and the usual political response of "planting trees in 5,000 ha if 5,000 ha were burned" is a simplistic approach no longer justified. Previously, restoring a burned area was equivalent to carrying out a reforestation or afforestation. But in fact, depending on the local conditions and objectives for the burned areas, these are often not the best management alternatives.

### 1.3.1  Major Questions and Some Answers

After a wildfire, forest managers and stakeholders face a series of questions that may not have an easy answer: should afforestations or reforestations be carried out? If so, in the whole area or just in part of it? Or is it better doing nothing? And if action is decided, when should it be taken? Using which techniques? Planting or seeding? Or wait for natural regeneration? But, more important than all the previous

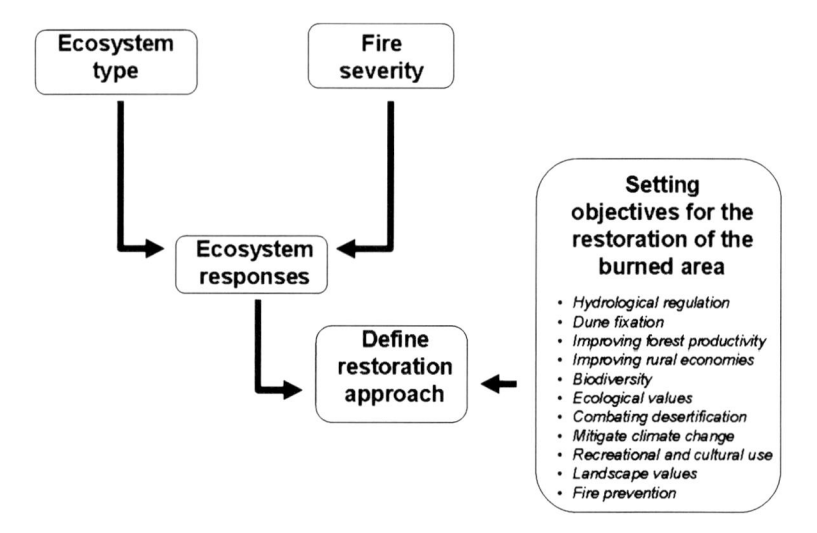

**Fig. 1.2** The definition of a restoration approach for a burned area depends on the expected ecosystem responses, which will be determined by ecosystem type and by fire severity, and on the objectives determined for the burned area. These are mostly set at on a local scale

questions, the key question is probably: "for which purpose?". Which are the objectives defined for the burned area and its management?

The answers to these questions depend on two fundamental topics: (i) our capacity to predict how affected ecosystems will react to fire; (ii) the definition of management objectives for the burned area (Vallejo 2005b). Both will determine the restoration approach and techniques that can be used (Fig. 1.2).

Ecosystem responses to fire are dependent on the regeneration capacity of its plant species. But predicting how plant communities will respond to fire is also dependent on the characteristics of the fire itself. Even for the same vegetation type different response patterns are expected whether a fire is quite intense and severe, or of low intensity and severity (e.g. Bond and van Wilgen 1996; Belligham and Sparrow 2000; Moreira et al. 2009b).

The management objectives for a burned area may be quite variable depending on the local situation. "Traditional" objectives included soil erosion prevention, water regulation, or increase forest productivity, but these have been replaced by new objectives such as biodiversity conservation, carbon storage, enhancing landscape values or reducing wildfire hazard (Fig. 1.2). These objectives are mostly local and can be quite variable from place to place, depending on the severity of impacts, the geographic and climate context, and socio-economic and cultural drivers. In the case of woodland restoration in the Mediterranean, Vallejo et al. (2006) suggested that the main priorities should be soil and water conservation, improving the resistance and resilience of vegetation to fire, increasing mature forests, promoting biodiversity and fostering the re-introduction of key species that might have disappeared.

## 1.3.2 Key Concepts in Restoration Ecology

These concepts have been addressed in several publications and books (e.g. Society for Ecological Restoration International Science and Policy Working Group 2004), thus we review them shortly with a focus on burned area restoration.

### 1.3.2.1 Restoration, Rehabilitation and Replacement

*Restoration* is the "process of assisting the recovery of an ecosystem that has been degraded, damaged or destroyed" (SER 2004). This definition is applicable to native forest ecosystems that were degraded or destroyed. Restoration aims to return an ecosystem to its historical condition, although setting this base reference is often difficult in a Mediterranean context where human management has been shaping landscapes for thousands of years. However, in the case of burned area management, our goals may not include restoration at all, in particular if we aim to change the ecosystem type that was burned (e.g. because it had no conservation value, or if we intend to reduce the fuel load in a particular location, independently of the previous land cover).

*Rehabilitation* shares with restoration a fundamental focus on historical or pre-existing ecosystems as a reference, but the two activities differ in their goals and strategies. Rehabilitation emphasizes the reparation of ecosystem processes, productivity and services, but not necessarily the re-establishment of the pre-existing biotic integrity in terms of species composition and community structure.

In *replacement*, or re-allocation (Aronson et al. 1993), the objective is to build up a new, productive ecosystem, often simpler than the original.

Forest restoration is a global concept that may have different degrees and intensities of management intervention, depending on the degradation stage of the forest and the specific management objectives considered. In the past, forest restoration has been mostly interpreted as planting trees, that is, afforestation or reforestation. Nowadays, this view is being replaced by a more holistic one, considering a wide set of restoration alternatives and approaches.

### 1.3.2.2 Active and Indirect Restoration

*Active restoration* uses techniques including plantations and direct seeding. These are relatively expensive tools for restoration, as they require site preparation, equipment, man-power, seedlings from nurseries, transport to the area, fertilizers, tree shelters, etc. (e.g. Moreira et al. 2009c). The survival of planted seedlings is quite variable, and often quite low in the case of broadleaved trees (Pausas et al. 2004). Direct seeding usually has lower costs compared to tree planting (Lamb and Gilmour 2003; Mansourian et al. 2005), but often only a low proportion of seeds is able to germinate and thrive (e.g. Pausas et al. 2004).

*Indirect restoration* implies the use of natural regeneration, and it can be either passive or assisted. *Passive restoration* is based on protecting the area from further disturbances and let ecological succession work (Lamb and Gilmour 2003). In burned areas regeneration may occur from seeds (e.g. Pausas et al. 2004), from resprouting of burned trees and stumps (mostly basal resprouting) (e.g. Espelta et al. 2003) or resprouting of burned shrubs or herbs. Tree resprouts, in particular, have significant advantages over seedlings or planted trees because they have an established root system which may confer higher probability of survival and better growth (e.g. Moreira et al. 2009c; Simões and Marques 2007). Further stages in natural regeneration management imply *assisted restoration* and may involve thinning, the selection of shoots in coppices, and the control of unwanted vegetation or protection from grazing animals (e.g. Lamb and Gilmour 2003; Moreira and Vallejo 2009; Vallejo et al. 2006; Whisenant 2005). The use of indirect restoration has been often neglected by managers and policy makers, and some regional and national governments have even subsidised active restoration in burned areas where natural regeneration was occurring.

Mediterranean-type ecosystems are highly resilient to fire when dominated by shrub and tree species that have the ability to resprout or produce seedlings after fire. Thus, these traits should be used in post-fire restoration, mainly through assisting natural regeneration that will likely result in less costly interventions and higher rate of vegetation recovery (Moreira and Vallejo 2009).

## 1.4 A Framework for Planning Post-Fire Restoration

In this section we describe a framework that can be used in post-fire management and restoration. It is based on five major steps (Fig. 1.3).

### 1.4.1 Predicting Vulnerable Areas

Even before fires occur, forest and landscape managers have the tools to map area vulnerability to wildfires, and to identify priority areas for fire prevention and intervention after fire. The key data to build these maps include soil information, topography (slope in particular), vegetation type, and also the location of values-at-risk (infrastructures, buildings, valuable ecosystems). One example is the work done by Alloza and Vallejo (2006) for the region of Valencia (over two million hectares). Using a Geographic Information System (GIS), these authors have mapped vulnerable areas based on the joint evaluation of the potential regeneration capacity of the vegetation and of the degradation risk (the environmental factors that condition the regeneration potential). Regeneration capacity was based on the combination of autosuccession potential (the ability to recover the pre-fire vegetation type) and the rate of plant recovery, which determines how quickly vegetation recovers to protect

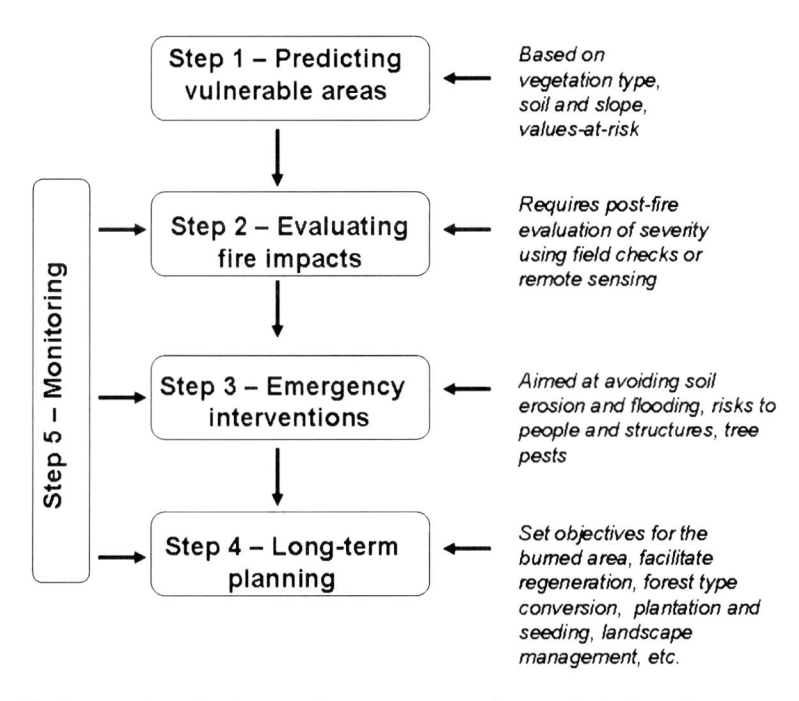

**Fig. 1.3** Framework to planning post-fire management and restoration in burned areas

the soil and decrease excessive erosion and runoff risk, for different vegetation types. The degradation risk was estimated based on the erosion potential (based on the Universal Soil Loss Equation) plus the drought risk (an estimator of the length of the dry period). At the end, the combination of the regeneration capacity and degradation risk yielded a map of ecosystem vulnerability that enabled the identification of priority areas for post-fire intervention, in the event of wildfires. These authors have found that 14% of the Valencia region is highly vulnerable in relation to forest fires, i.e. in the event of a forest fire some degradation of the vegetation cover can be expected in the short term. In Cape Sounion National Park, Central Greece, ecological and landscape data have been also integrated with decision-support techniques in a Geographic Information Systems (GIS) framework to evaluate the risk of loosing post-fire resilience in *Pinus halepensis* forests (Arianoutsou et al. 2011). Criteria related to the significance of several indicators (bio-indicators: plant woody cover, pine density and geo-indicators: fire history, parent rock material and slope) were incorporated within a weights coefficient and then integrated into a multicriteria rule that was used to map the risk of loosing resilience. This map is useful for identifying 'risk hotspots' where post-fire management measures should have priority.

This type of approach can be used all over Europe, in order to identify vulnerable areas and prioritise fire prevention and post-fire intervention actions after wildfires.

### 1.4.2   Evaluating Fire Impacts

Even if a given area was identified as highly vulnerable to wildfires, post-fire soil erosion and degradation might be negligible if a wildfire had low severity. The impact of a fire upon a site will depend on fire characteristics, and the key variable to evaluate how a given ecosystem will respond to fire, fire severity (Fig. 1.2), can obviously be assessed only after a wildfire has occurred. It is, therefore, important to evaluate severity levels as soon as possible after fire, and this can be done either by field inspections, by using remote sensing at high resolution (e.g. satellite images), or a combination of both. The use of remote sensing to evaluate fire severity will only be effective if the information is available for the forest managers in the short-term. Additionally, this technique has some inaccuracies (Lentile et al. 2006), for example in forest areas remotely sensed burn severity is often highly correlated with fire effects on the tree canopy but exhibits low correlation with ground and soil severity. Often, it is more practical to do field checks in the first couple of weeks after the fire, to visually evaluate fire severity. Several guidelines exist (e.g. USDI 2003) that can be used to quickly evaluate fire severity and identify areas for emergency interventions.

### 1.4.3   Emergency Interventions

These emergency interventions, sometimes called first-aid rehabilitation, aim to stabilise the affected area, prevent degradation processes and minimise risks for people (Robichaud et al. 2000). They may aim at soil protection to avoid erosion and decrease water runoff and risk of flooding, to decrease risks to people and property (e.g. hazard from falling burned trees), or to the prevention of tree pests and diseases. They should be undertaken as soon as possible, at most a few months after the fire, and preferably before the first autumn rains when in the Mediterranean region.

### 1.4.4   Long-Term Planning

This is related to setting the objectives for the burned area and the actions needed to accomplish these objectives. Depending on the situation, it may include facilitating the natural regeneration, carry out conversion to other forest types, afforestation or reforestation, landscape management to promote specific land covers, etc.

### 1.4.5   The Importance of Monitoring and Evaluation

Restoration projects are traditionally weak in monitoring and evaluation. This limits the opportunities to learn from past successes and failures (Vallejo 2005a). A properly planned restoration project or management action should attempt to fulfil clearly stated objectives. Assessing whether objectives were fulfilled, or how far we are from attaining these objectives, is only possible through monitoring and evaluation. What cannot be measured and monitored in an objective and unbiased way cannot be effectively managed (Corona et al. 2003). Thus, objectives, performance standards, and protocols for monitoring and for data assessment should be incorporated into restoration schemes prior to the start of a project.

Post-fire monitoring and evaluation is essential to gain understanding of forest ecosystems successional pathways after fire and, accordingly, to plan appropriate restoration actions. It will also allow re-directing restoration actions in an adaptive management context.

According to SER (2004), there are three strategies for conducting an evaluation: direct comparison, attribute analysis, and trajectory analysis. In direct comparison, selected parameters are measured in the reference and restoration sites. One example is to assess the plant composition of a post-fire recovering forest and compare it with that of an adjacent unburned plot of the same forest type. In attribute analysis, a set of desirable characteristics for the project result are defined at the beginning and the measured parameters are compared with this set. In trajectory analysis, data are collected periodically and trends examined to confirm that the project is following the intended trajectory.

Independently of the followed strategy, two major points should always be taken into account. The first is the knowledge of the *baseline situation*. As an example, consider how to assess the success of a given action in increasing soil cover if we did not measure it at the start? This implies assessing the current status of our parameter of interest before the start of the management action or experiment. The second point concerns the *monitoring of untreated control plots*, which provide the only way of evaluating the net effect of our action. For example, the meaning of a 50% increase in soil cover after a given treatment is totally different according to whether this increase in an untreated plot, during the same time period, was 5% (our treatment was highly effective, compared to the control) or 45% (denoting low effectiveness of the treatment, just 5% higher than the control).

Finally, the social impacts of any restoration project should also be taken into account in the evaluation of its outcomes as, beside the ecological objectives, there are always more or less explicit social and economic objectives and implications. Bautista and Alloza (2009) have recently developed a protocol for evaluating Mediterranean forest restoration projects considering both biophysical and socio-economic criteria. See also Table 3.5 (chapter 3).

## 1.5  Spatial Scales for Management and Restoration: From the Forest Stand to Land Use Planning

Post-fire management and restoration actions can be taken at a variety of spatial scales, from the forest stand level to landscape regional planning (Fig. 1.4).

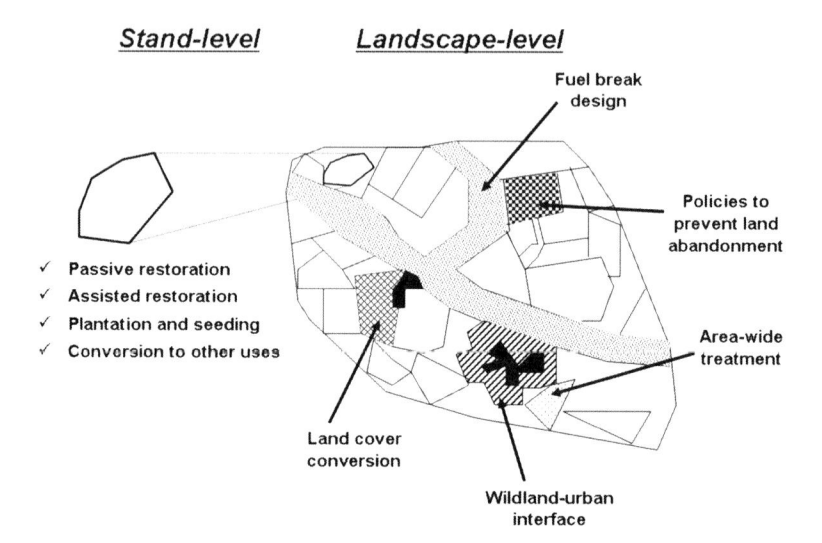

**Fig. 1.4** Spatial scales to carry out post-fire management and restoration. At landscape level, it may include the definition of landscape-scale fuel breaks or area-wide treatments, implementation of policies for preventing land abandonment or keeping specific land covers, land cover conversion to less fire-prone cover, specific regulations for the wildland-urban interfaces. At stand level, different options include passive restoration, assisted restoration, plantation and seeding (of the same forest species or converting to another tree species), or conversion to other non-forest uses

## 1.5.1 Stand-Level Silvicultural Practices

The planning and implementation of restoration actions in burned forest areas need to be spatially framed according to the stand-level variability in terms of fire affected forest types and levels of fire severity. In fact, restoration strategies should be put into place depending on fire damage levels and post-fire ecology of affected forest types, notably the post-fire regeneration strategies (seeders, resprouters). Current post-fire management practices and recommendations are proposed in the Chaps. 6–12 for the relevant fire-affected forest types in Europe.

First of all, it is recommended to accurately plan the post-fire harvest of trees, if needed. The pros and cons of salvage logging have been addressed in several studies (e.g. Lindenmayer and Noss 2006). Once emergency interventions have been completed, there are basically three possible options for the restoration of forests impacted by fire.

Indirect restoration techniques are applicable where forest stand functional/structural damage is limited and resilience is high. The first option is the management of burned forest (or shrubland) areas through natural regeneration (passive restoration). The exclusion or restriction of some land use activities (e.g. livestock grazing) for some years and the implementation of measures to prevent further degrading events (e.g. additional fires) is generally enough to ensure satisfactory post-fire regeneration. Management thus involves careful monitoring of the dynamics of recovery and verifying their coherence with the management objectives set in terms of forest structure and composition. The second option of indirect restoration is assisted restoration, with the application of appropriate silvicultural techniques to support post-fire natural regeneration in order to promote faster achievement of the mature stages of development. However, it has to be decided when to apply restoration interventions: soon after the fire or wait until the natural regeneration is established? Early silvicultural interventions aim at creating favourable conditions for the establishment of natural regeneration, while latter interventions are intended to support the development of already established regeneration. Different approaches can be used: (i) to favour vegetative regeneration through stump undercutting and selection of shoots, namely when fire affects wooded lands dominated by resprouting species like broadleaved evergreen or thermophilous deciduous oaks (coppices); (ii) to support regeneration of new individuals through pruning and thinning to stimulate seeding, by giving more light and potential growing space to dominant branches and crowns of dominant standing trees, whose seed production is generally higher than the other trees, and subsequent localised cleaning to favour seedling survival and growth, and, when necessary, localized planting to restore tree cover.

The third option is active restoration, through active seeding and plantation in stand-replacing fires. For a number of reasons (e.g. high fire severity, post-fire ecology of the affected vegetation, juvenile forest stand) natural regeneration may not guarantee the self-restoration of the fire affected forest types, mainly in the case of pine stands. In these cases seeding or planting are suitable techniques to ensure long-term restoration, and may include the use of a wide variety of species, depending on the objectives. For example, if forest conversion, e.g. to a less fire-prone composition, is an objective, plant species that did not exist before the fire may be planted or sown.

One last alternative of management is the conversion to other non-forest uses, usually within the scope of planning at a larger spatial scale (see Sect. 1.5.2).

As mentioned, several tree forest species, particularly Mediterranean ones, have high intrinsic resilience to fire, and therefore are potentially capable of guaranteeing an efficient natural recovery after fire. Yet, the post-fire woodland restoration methods currently practised in Europe mostly follow conventional practices and administrative regulations that do not always take into consideration the fire ecology of affected forest types, as will be illustrated in the specific chapters (see Chaps. 6–12); thus they are often not the most suitable to facilitate the natural ability of the vegetation to return to the pre-disturbance stage, through the autosuccessional process.

## *1.5.2   Landscape Management Planning*

Planning at the landscape-scale aims to reduce fire hazard in order to produce landscapes that are less fire-prone (Fig. 1.4).

At landscape level, wildfire initiation and spread result from the interaction among ignition sources, weather, topography and land cover (e.g. Mermoz et al. 2005; Rothermel 1983). From a management perspective, land cover (related to vegetation composition and flammability) is the only variable that can be manipulated by man. In general, agricultural areas and deciduous broadleaved forests are the least fire-prone land cover, whereas shrublands and pine woodlands are the most fire-prone in the Mediterranean basin (e.g. Moreira et al. 2009a). A straightforward application of this knowledge lies in the definition of landscape-scale fuel breaks and area-wide (block) treatments, whose main objective is to reduce fuel loads or change the spatial arrangement of fuels (i.e. the landscape structure), so that when a wildfire ignites in a managed landscape, it spreads more slowly, burns with less intensity and severity, and is less costly to suppress. Thus, the main objectives of landscape fuel treatments are to break up the continuity of hazardous fuels across a landscape, and to reduce the intensity of wildfires, providing broad zones within which fire-fighters can conduct suppression operations more safely and effectively. In addition, these areas can also be used to provide other types of benefits (e.g. habitat diversity, landscape scenery, protection of heritage sites) (Agee et al. 2000; Cumming 2001; Rigolot 2002; Weatherspoon and Skinner 1996).

The management implications of understanding the landscape-fire relationships are not restricted to fuel break or block treatment designs. The definition of land use management rules and the design and implementation of policies to achieve specific landscape objectives, ranging from forest to agricultural, rural development or urban policies (mainly in the wildland-urban interface), all contribute to "making" landscapes with lower fire hazard. For example, the "rural exodus syndrome" (Hill et al. 2008) is causing a widespread increase in vegetation biomass over large areas of the Mediterranean Europe, mainly in mountain areas, and a subsequent increase in fire hazard. Population decline, agricultural and pastoral land abandonment (and the subsequent natural regeneration of forests), and policies promoting forest cover, particularly in former agricultural land, are the main driving forces of this process. This trend can only be counteracted effectively through policies enabling the improvement of the socio-economic conditions of people living in rural areas, promoting new immigration to these regions, and implementing rural development policies that foster activities contributing to reduce fire hazard, such as agriculture and livestock grazing. These policies are mainly related to agricultural, rural development and economic issues, rather than forest management.

In the absence of other fuel treatment drivers, prescribed fire or controlled occupational burning (used by shepherds to renew pastures) can be useful tools to reduce fire hazard. In this perspective, energy policies supporting the environmentally compatible use of renewable bioenergy potential from agriculture and forestry products (agricultural waste, crop mix for biomass production, complementary fellings and residues from silvicultural activities and/or fuel treatments) may also contribute to decrease fire hazard, while providing job opportunities to rural populations.

At regional/local levels, these policies should be implemented mainly in areas where landscape-level planning to reduce fire hazard identified priority sites for fuel treatments, such as wildland-urban interface areas surrounding villages. In fact, one of the most serious consequences of the abandonment of traditional practices is that villages in mountain areas, traditionally surrounded by a belt of farmland that acted as a landscape fuel break, nowadays have forests and shrublands in the vicinity of houses and other infrastructures, which greatly increase fire hazard.

Under moderate to severe fire weather conditions, fuel management should be focused on increasing the fire suppression options and effectiveness by limiting fire ignition and fire spread in strategic locations. However, under exceptional fire weather conditions, fire may propagate throughout a whole landscape regardless of land cover and fuel management. Prevention strategies should therefore include self-protection options to limit fire intensity and damages on both ecosystems and human assets, under the assumption that fire fighting will not always be possible in these circumstances. This implies a major shift in the way fire management is seen, and a cross sectorial approach integrating agricultural, forest and urban planning policies. Currently, the unbalanced fire management being practiced in Europe, with too much resources being allocated to pre-suppression and suppression actions compared to poor fuel management measures (e.g. Fernandes 2008), is increasingly questioned. Simultaneously, the "learning to live with fire" objective is increasingly shared (Birot and Rigolot 2009), by recognising that fire cannot be excluded from the Mediterranean environment (Rego et al. 2010). Under this objective, fuel management is not only devoted to limiting wildfire spread, but also to lower fire impacts on human resources and assets.

Land ownership may be a problem when implementing landscape-scale management. In situations where most of the land is private and property size is small, it may be difficult to coordinate the management of different land owners in order to achieve a spatial dimension that suits management objectives. In countries or regions where fragmented private property prevails, the only way of assuring effective management is to promote coordinated action among land and forest owners. In Portugal, for example, the government is promoting the association of land owners with contiguous properties, so that a joint management plant is implemented to achieve this objective.

## 1.6  Key Messages

- The restoration of a burned area is not just a matter of how to carry out reforestations. The post-fire management approaches and techniques that can be used are quite variable and depend on (1) our capacity to predict how affected ecosystems will react to fire and (2) the definition of management objectives for the burned area. It is also important to adopt of an adaptive management approach which systematically integrates results of previous interventions to iteratively improve and accommodate change by learning from the outcomes of experimented practices;
- The management objectives for a burned area are mostly local and can be quite variable from place to place, depending on the severity of impacts, the geographic

and climate context, and the socio-economic and cultural context. But the main priorities should always be soil and water conservation;

- Ecosystems dominated by shrub and tree species that have the ability to resprout or produce seedlings after fire are usually highly resilient to fire. These traits should be used in post-fire restoration, mainly by assisting natural regeneration that will likely result in less costly interventions and higher rate of vegetation recovery;
- We suggest a framework for planning post-fire management and restoration based on five steps: (1) identifying vulnerable areas, (2) evaluating fire impacts, (3) carrying out emergency interventions, (4) long-term planning, and (5) monitoring;
- Spatial scales for post-fire management range from the forest stand to landscape management. At the landscape level, besides fuel break or area-wide treatment designs, the implementation of policies to achieve specific land management objectives, ranging from forest to agricultural, rural development or urban policies (mainly in the wildland-urban interface) is essential to promote landscapes which are less fire prone;
- The unbalanced fire management being practiced in Europe, with too much resources being allocated to pre-suppression and suppression actions compared to poor fuel management measures, needs to be changed to a greater focus on fuel management. Adoption of correct post-fire management practices is the first step towards adequate fuel management to decrease the damage caused by subsequent fires.

**Acknowledgments** We thank Francisco Rego for reviewing this chapter. Many of the ideas expressed here originated at the PHOENIX project centre of the European Forest Institute and were further developed under the scope of COST Action FP0701 "Post-fire forest management in Southern Europe".

# References

Agee JK, Bahrob B, Finney MA, Omid PN, Sapsise DB, Skinnerf CN, Wagtendonkg J, Weatherspoon P (2000) The use of shaded fuelbreaks in landscape fire management. For Ecol Manag 127:55–66

Alexandrian D, Esnault F (1998) Políticas públicas que afectan a los incendios forestales en la cuenca del Mediterráneo. In: FAO (ed) Reunión sobre Políticas Públicas que Afectan a los Incendios Forestales. FAO, Rome

Allosa JA, Vallejo R (2006) Restoration of burned areas in forest management plans. In: Kepner WG, Rubio JL, Mouat DA, Pedrazzini F (eds) Desertification in the Mediterranean region: a security issue. Springer, The Netherlands

Arianoutsou M (2001) Landscape changes in Mediterranean ecosystems of Greece: implications for fire and biodiversity issues. J Medit Ecol 2:165–178

Arianoutsou M, Koukoulas S, Kazanis D (2011) Evaluating post-fire forest resilience using GIS and multi-criteria analysis: an example from Cape Sounion National Park, Greece. Environ Manag 47:384–397. doi:10.1007/s00267-011-9614-7

Aronson J, Floret C, Le Floc'h E, Ovalle C, Pontanier R (1993) Restoration and rehabilitation of degraded ecosystems in arid and semi-arid lands. I. A view from the south. Restor Ecol 1:8–17

Bautista S, Alloza J (2009) Evaluation of forest restoration projects. In: Bautista S, Aronson J, Vallejo VR (eds) Land restoration to combat desertification. CEAM, Valencia

Bellingham PJ, Sparrow AD (2000) Resprouting as a life history strategy in woody plant communities. Oikos 89:409–416

Bermudez Z, Mendes J, Pereira JMC, Turkman KF, Vasconcelos MJP (2009) Spatial and temporal extremes of wildfire sizes in Portugal. Int J Wildland Fire 18:983–991

Birot Y, Rigolot E, (2009) The need for strategy anticipating climate … and other changes. In: Birot Y (ed) Living with wildfire: what science can tell us, EFI Discussion Paper 15, EFI, Joensuu

Bond WJ, van Wilgen BW (1996) Fire and plants. Chapman and Hall, London

Corona P, Koehl M, Marchetti M (eds) (2003) Advances in forest inventory for sustainable forest management and biodiversity monitoring. Kluwer, Dordrecht. ISBN 1-4020-1715-4

Cumming SG (2001) Forest type and wildfire in the Alberta boreal mixedwood: What do fires burn? Ecol Applic 11:97–110

Diaz-Delgado R, Lloret F, Pons X (2004) Spatial patterns of fire occurrence in Catalonia, NE, Spain. Landsc Ecol 7:731–745

EFFIS (2009) Forest fires in Europe. Report 10. EUR 24502 EN – 2010. JRC Scientific and Technical Reports, European Commission, Directorate General Environment

Espelta JM, Retana J, Habrouk A (2003) Resprouting patterns after fire and response to stool cleaning of two coexisting Mediterranean oaks with contrasting leaf habits on two different sites. For Ecol Manag 179:401–414

Fernandes P (2008) Forest fires in Galicia (Spain): the outcome of unbalanced fire management. J For Econ 14:155–157

Flannigan MD, Krawchuk MA, De Groot WJ, Wotton BM, Gowman LM (2009) Implications of changing climate for global wildland fire. Int J Wildland Fire 18:483–507

Founda D, Giannakopoulos C (2009) The exceptionally hot summer of 2007 in Athens, Greece – a typical summer in the future climate? Glob Planet Change 67:227–236

Goldammer JC, Crutzen PJ (1993) Fire in the environment: scientific rationale and summary of results of the Dahlem Workshop. In: Goldammer JG, Crutzen PJ (eds) Fire in the environment: the ecological, atmospheric and climatic importance of vegetation fire. John Wiley and Sons, Toronto

Hill J, Stellmes M, Udelhoven T, Röder A, Sommer S (2008) Mediterranean desertification and land degradation: Mapping related land use change syndromes based on satellite observations. Glob Planet Change 64:146–157

Lamb D, Guilmour D (2003) Rehabilitation and restoration of degraded forests. IUCN, Gland, Switzerland and Cambridge, UK and WWF, Gland, Switzerland

Lehouerou HN (1992) Climatic-change and desertization. Impact Sci Soc 42.183–201

Lentile LB, Holden ZA, Smith AMS, Falkowski MJ, Hudak AT, Morgan P, Lewis SA, Gessler PE, Benson NC (2006) Remote sensing techniques to assess active fire characteristics and post-fire effects. Int J Wildland Fire 15:319–345

Lindenmayer D, Noss R (2006) Salvage logging, ecosystem processes, and biodiversity conservation. Conserv Biol 20:949–958

Lindner M, Maroschek M, Netherer S, Kremer A, Barbati A, Garcia-Gonzalo J, Seidl R, Delzon S, Corona P, Kolstrom M, Lexer MJ, Marchetti M (2010) Climate change impacts, adaptive capacity, and vulnerability of European forest ecosystems. For Ecol Manag 259:698–709

Lloret F, Marí G (2001) A comparison of the medieval and the current fire regimes in managed pine forests of Catalonia (NE Spain). For Ecol Manag 141:155–163

MacDonald D, Crabtree JR, Wiesinger G, Dax T, Stamou N, Fleury P, Lazpita JG, Gibon A (2000) Agricultural abandonment in mountain areas of Europe: environmental consequences and policy response. J Environ Manag 59:47–69

Mansourian S, Lamb D, Gilmour D (2005) Overview of technical approaches to restoring tree cover at site level. In: Mansourian S, Vallauri D, Dudley N (eds) Forest restoration in landscapes. Beyond planting trees. Springer, New York

Mermoz M, Kitzberger T, Veblen TT (2005) Landscape influences on occurrence and spread of wildfires in Patagonian forests and shrublands. Ecology 86:2705–2715

Moreira F, Vallejo R (2009) What to do after fire? Post-fire restoration. In: Birot Y (ed) Living with wildfires: what science can tell us, EFI Discussion Paper 15, EFI, Joensuu

Moreira F, Rego FC, Ferreira PG (2001) Temporal (1958–1995) pattern of change in a cultural landscape of northwestern Portugal: implications for fire occurrence. Landsc Ecol 16:557–567

Moreira F, Vaz P, Catry F, Silva JS (2009a) Regional variations in wildfire susceptibility of land-cover types in Portugal: implications for landscape management to minimize fire hazard. Int J Wildland Fire 18:563–574

Moreira F, Catry F, Duarte I, Acácio V, Silva JS (2009b) A conceptual model of sprouting responses in relation to fire damage: an example with cork oak (Quercus suber L.) trees in Southern Portugal. Plant Ecol 201:77–85

Moreira F, Viedma O, Arianoustou M, Curt T, Koutsias N, Rigolot E, Barbati A, Corona P, Vaz P, Xanthopoulos G, Mouillot F, Bilgili E (2011) Landscape-wildfire interactions in Southern Europe. Implications for landscape management. J Env Manag 92:2389–2402

Moreira F, Catry F, Lopes T, Bugalho M, Rego F (2009c) Comparing survival and size of resprouts and planted trees for post-fire forest restoration in central Portugal. Ecol Eng 35:870–873

Moreno JM (2010) Climate change, wildland fires and biodiversity in Europe. Document prepared for The Council of Europe Convention of the Conservation of European Wildlife and Natural Habitats, Strasbourg

Moreno JM, Vásquez A, Veléz R (1998) Recent history of forest fires in Spain. In: Moreno JM (ed) Large forest fires. Backhuys Publishers, Leiden

Oliveras I, Gracia M, Moré G, Retana J (2009) Factors influencing the pattern of fire severities in a large wildfire under extreme meteorological conditions in the Mediterranean basin. Int J Wildland Fire 18:755–764

Pausas JG (2004) Changes in fire and climate in the eastern Iberian Peninsula (Mediterranean Basin). Clim Chang 63:337–350

Pausas JG, Fernandez-Muñoz S (2011) Fire regime changes in the western Mediterranean basin: from fuel-limited to drought-driven fire regime. Clim Chang. doi:10.1007/s10584-011-0060-6

Pausas JG, Blade C, Valdecantos A, Seva JP, Fuentes D, Alloza JA, Vilagrosa A et al (2004) Pines and oaks in the restoration of Mediterranean landscapes of Spain: new perspectives for an old practice – a review. Plant Ecol 171:209–220

Pausas JG, Llovet J, Rodrigo A, Vallejo R (2008) Are wildfires a disaster in the Mediterranean basin? – A review. Int J Wildland Fire 17:713–723

Piñol J, Terradas J, Lloret F (1998) Climate warming and wildfire hazard and wildfire occurrence in coastal eastern Spain. Clim Chang 38:345–357

Rego F, Rigolot E, Fernandes P, Montiel C, Silva JS (2010) Towards integrated fire management. European Forest Institute Policy Brief 4

Rigolot E (2002) Du plan départemental à la coupure de combustible. Guide méthodologique et pratique. In: Réseau Coupures de Combustible n° 6, Éditions de la Cárdère, Morières, France

Robichaud PR, Bayers JL, Neary DG (2000) Evaluating the effectiveness of post-fire rehabilitation treatments. General Technical Report RMRS-GTR-63, Rocky Mountain Research Station, Fort Collins

Rothermel R (1983) How to predict the spread and intensity of forest and range fires. USDA, Forest Service, Intermountain Forest and Range Experiment Station, General Technical Report INT-143, Ogden, UT

San-Miguel J, Camia A (2009). Forest fires at a glance: facts, figures and trends in the EU. In: Birot Y (ed) Living with wildfires: what science can tell us. A contribution to the science-policy dialogue, EFI Discussion Paper 15, EFI, Joensuu

SER – Society for Ecological Restoration International Science & Policy Working Group (2004) The SER international primer on ecological restoration. Society for Ecological Restoration International. www.ser.org &Tucson

Simões C, Marques M (2007) The role of sprouts in the restoration of Atlantic rainforest in Southern Brazil. Restor Ecol 15:53–59

Thonicke K, Rammig A, Gumpenberger M (2010) Changes in managed fires and wildfires under climate and land use change and the role of prescribed burning to reduce fire hazard under future climate conditions. Deliverable D4.2-1c / D4.2-4 of the Integrated Project FIRE PARADOX, Project no. FP6-018505, European Commission

Tolika K, Maheras P, Tegoulias I (2009) Extreme temperatures in Greece during 2007: could this be a "return to the future"? Geophys Res Lett 36:L10813. doi:10.1029/2009GL038538

USDI (2003) Fire monitoring handbook. Fire Management Program Center, National Interagency Fire Center, Boise

Vallejo R (2005a) Restoring Mediterranean forests. In: Mansourian S, Vallauri D, Dudley N (eds) Forest restoration in landscapes. Beyond planting trees. Springer, New York

Vallejo R (2005b) Identification des besoins et évaluation des techniques de restauration post-feu. Forêt Méditerr 26:217–224

Vallejo R, Aronson J, Pausas J, Cortina J (2006) Restoration of Mediterranean woodlands. In: van Andel J, Aronson J (eds) Restoration ecology: the new frontier. Blackwell Science, Oxford

Vélez R (2009) The causing factors: a focus on economic and social driving forces. In: Birot Y (ed) Living with wildfires: What science can tell us? EFI discussion paper 15, EFI, Joensuu

Weatherspoon CP, Skinner CN (1996) Landscape-level strategies for forest fuel management. In: Sierra Nevada Ecosystem Project: final report to Congress, II: assessments, scientific basis for management options, University of California, Centers for Water and Wildland Resources, Water Resources Center Report No. 37

Whisenant S (2005) Managing and directing natural succession. In: Mansourian S, Vallauri D, Dudley N (eds) Forest restoration in landscapes. Beyond planting trees. Springer, New York

Xanthopoulos G (2007) Olympic flames. Wildfire 16:10–18

# Chapter 2
# Land Cover Change and Fire Regime in the European Mediterranean Region

Jesús San-Miguel-Ayanz, Marcos Rodrigues, Sandra Santos de Oliveira, Claudia Kemper Pacheco, Francisco Moreira, Beatriz Duguy, and Andrea Camia

## 2.1  Introduction

Although fire is an integral component of Mediterranean ecosystems, the dynamics of fire regimes in Southern Europe is driven mainly by human factors. In fact, humans are responsible for over 95% of the fires taking place in this region (San-Miguel Ayanz and Camia 2009). Traditional usage of fire in agricultural and cattle raising practices in the region is one of the main causes of forest fires. Demographic changes related to the abandonment of rural areas are also related to increased fire hazard. Fuel accumulation due to the lack of forest management practices in the region leads to uncontrolled forest fires. Although, overall, the rural population in Southern Europe has decreased, peaks of high population density in recreational wildland areas during holiday periods increased fire ignition in

J. San-Miguel-Ayanz (✉) • A. Camia
Institute for Environment and Sustainability, Joint Research Centre, European Commission,
Ispra, Italy
e-mail: jesus.san-miguel@jrc.ec.europa.eu

M. Rodrigues
Institute for Environment and Sustainability, Joint Research Centre, European Commission,
Ispra, Italy

Grupo GEOFOREST, Departamento de Geografía y Ordenación del Territorio,
Universidad de Zaragoza, Zaragoza, Spain

S.S. de Oliveira
Institute for Environment and Sustainability, Joint Research Centre, European Commission,
Ispra, Italy

Centre of Forestry Studies, Institute of Agronomy, Technical University of Lisbon,
Lisbon, Portugal

F. Moreira et al. (eds.), *Post-Fire Management and Restoration of Southern European Forests*, Managing Forest Ecosystems 24, DOI 10.1007/978-94-007-2208-8_2,
© Springer Science+Business Media B.V. 2012

summer months. This is further enhanced by the expansion of urban areas into wildland areas. This effect, which is due to either the expansion of cities or the construction of secondary houses in rural areas, has lead to an extended Wildland Urban Interface (WUI) in the region. The difficult fire management of the extensive WUI in Southern Europe has been the cause of catastrophic fires such as those in Portugal in 2003 or Greece in 2007.

Land cover is a fundamental component of fire dynamics. It influences all the phases of the fire, from ignition to fire behavior and post-fire restoration. The analysis of land cover changes in the last decades is tackled in the first section of this chapter. This is followed by the analysis of fire regimes in the region, both in terms of number of fires and burned areas. The last two sections of the chapter are dedicated to an in depth analysis of land cover changes in areas affected by fires and the effects of fire on land cover dynamics.

## 2.2   Overview of Land Cover Changes in Europe

Land cover changes are related to fire hazard through changes in fuel load which, along with topography and weather, are the main drivers of fire intensity and rate of spread (Fernandes 2009; Moreira et al. 2009; Rothermel 1983). Thus, increased fire hazard is expected where land cover changes promote an increase in plant biomass (fuel load) while decreased fire hazard is linked to changes associated with the removal of biomass. The CORINE Land Cover database (http://www.eea.europa. eu/data-and-maps) was used to analyze the changes in land cover in southern Europe between 1990 and 2006. The analysis was carried out in 4 out of the 5 European Mediterranean countries that are most significantly affected by forest fires, i.e. Portugal, Spain, France and Italy. Greece was excluded from this analysis due to the lack of CORINE 2006 data for this country.

C.K. Pacheco
DIBAF, University of Tuscia, Viterbo, Italy

Institute of Agro-Envinronmental and Forest Biology (IBAF),
National Research Council (CNR), Rome, Italy

F. Moreira
Centre of Applied Eccology, Institute of Agronomy, Technical University of Lisbon,
Lisbon, Portugal

B. Duguy
Facultat de Biologia, Departament de Biologia Vegetal, Universitat de Barcelona,
Barcelona, Spain

CORINE provides a thematic legend of 44 land cover classes grouped in three hierarchical levels. These land cover classes were grouped into six general categories: Urban, Artificial, Agriculture, Forest, Shrubland, and No-vegetation. The analysis of transition of areas among these categories was carried out.

The largest land cover change observed was the transition from forests to shrublands (over three million hectares) (Table 2.1), which could be interpreted as forest degradation due to several causes (e.g. logging, fire, drought). Forests have also been replaced by urban, agricultural, artificial areas and areas with no vegetation. There were also significant areas of shrublands that have been replaced by agricultural areas (over one million hectares), areas with no vegetation, artificial and urban areas. All these changes contributed to decrease fire hazard.

On the other hand, a significant proportion of shrublands have become forests (over two million hectares) and the transition of former agricultural areas to forests (over 800,000 ha) and shrublands (over one million hectares) has also been significant. The transition of areas with no vegetation to shrublands was also relevant (over 450,000 ha). These changes are probably the consequence of secondary succession in shrublands and abandoned agricultural fields, along with afforestation programs promoted by EU agricultural and forest policies during the study period.

A large number of regional studies have also provided evidence of increased fire hazard in the Mediterranean areas in the last decades, mainly due to the increased cover of forests and shrublands in areas with former lower fuel loads. For example, Van Doorn and Baker (2007), in a region of southern Portugal, registered a 75% decline in the area of agricultural fields during the period 1985–2000, and an increase in shrublands and forest plantations. Similarly, Falcucci et al. (2007) measured a 74% increase in forest cover in Italy during the period 1960–2000, and a 20% decrease in agricultural areas.

The balance between land cover changes promoting an increase in fire hazard (summing 4.9 million hectares) and the ones decreasing it (5.4 million hectares) would suggest that southern Europe has become less fire prone in the period 1990–2006. These results are in line with those presented in the analysis of land cover changes during the period 1990–2000 in 24 European countries by Feranec et al. (2010). These authors also found that the establishment of forests by planting or natural regeneration provoked a significant proportion of land cover transitions corresponding to an increased fire hazard, as they result in an increase in fuel load at the landscape level. The authors concluded that afforestation was the most prominent land cover flow across all Europe, during this time period, particularly in Portugal, Spain and France. In contrast, three flow types – deforestation, intensification of agriculture and urbanization/ industrialization – included several transitions associated to decreased fire hazard. The greatest losses in forest have been observed in Spain, France and Portugal, mainly because of disturbances such as fire and wind. In the countries of Southern Europe, intensification of agriculture was more prevalent in Spain, whereas urbanization processes were more extensive in Spain, France, Italy and Portugal (Feranec et al. 2010).

**Table 2.1** Changes (in hectares) among land cover classes from 1990 to 2006 in Mediterranean Europe (Portugal, Spain, France and Italy)

| | To 2006 | Urban | Artificial | Agricultural | Forest | Shrubland | No vegetation |
|---|---|---|---|---|---|---|---|
| | Urban | 11,012,514 | 75,795 | 197,427 | **11,957** | **5,897** | 723 |
| From 1990 | Artificial | 70,972 | 2,826,169 | 107,870 | **21,353** | **62,476** | 7,292 |
| | Agricultural | 1,220,019 | 765,940 | 180,616,767 | **815,487** | **1,188,775** | 25,636 |
| | Forest | *43,099* | *101,413* | *583,320* | 85,029,222 | *3,192,178* | *82,721* |
| | Shrubland | *44,331* | *99,291* | *1,071,732* | **2,299,768** | 26,980,642 | *199,171* |
| | No vegetation | 3,362 | 12,183 | 57,788 | **43,162** | **468,750** | 4,120,242 |

*Note*: Changes associated to significant increases in fire hazard are signaled in bold, whereas changes leading to major decreases in fire hazard are in italics

## 2.3   Overview of Changes in Number of Fires and Burned Area in the European Mediterranean Countries

The Mediterranean region of Europe is strongly affected by forest fires. According to European Statistics (EC 2010), from 1980 until 2009 fires have burned an average of circa 478,900 ha of land per year in the five Southern European countries most affected by fire (Portugal, Spain, France, Italy and Greece). Data on the number of fires and burned area in this region have been collected since the 1980s by each country and compiled in the European Fire Database (Camia et al. 2010). The analysis of the spatial and temporal trends of fires is crucial to understand the underlying causes of the fires and their environmental and socio-economic impacts, assuming a key role in fire prevention and management. The purpose of this section is to analyze the spatial and temporal trends of fire frequency (number of fires) and burned area size, two essential components of the fire regime of an area.

The analysis of the number of fires, total burned area and average fire size was carried out at different spatial levels:

- At regional (supranational) level, considering the Euro-Mediterranean region as a whole, with the purpose of characterizing its fire regimes, known to be markedly different from the rest of Europe. The region under study, shortly referred to as EUMed in what follows, comprises Portugal, Spain, France, Italy and Greece;
- At country level, by analyzing the data of each country individually in order to assess differences between countries that may depend on national settings and policies;
- At province level (NUTS3), to investigate the potential influence of local environmental and socio-economic conditions.

Temporal trends were analyzed separately for the whole study period (1980–2009) and for the last 10 years (2000–2009). These trends were compared using the Mann–Kendall test, a non-parametric statistical test used to identify trends in time series data (Kendall 1975). In addition, seasonal trends were also characterized both at regional and country levels, by examining separately the months corresponding to the main fire season (June to October) and the other months.

### 2.3.1   Overall Trends for the EUMed Region

The general trend for the whole study period was a slight increase in the number of fires (Fig. 2.1), even though annual fluctuations are evident. In the 1990s a substantial increase was observed, while in the last 10 years (since 2000), the number of fires decreased, except for the years 2003 and 2005. The increase observed in the 1990s can be partly due to the changes in the reporting systems in the countries, mostly driven by EC regulations. Other reasons for the rise in the number of fires

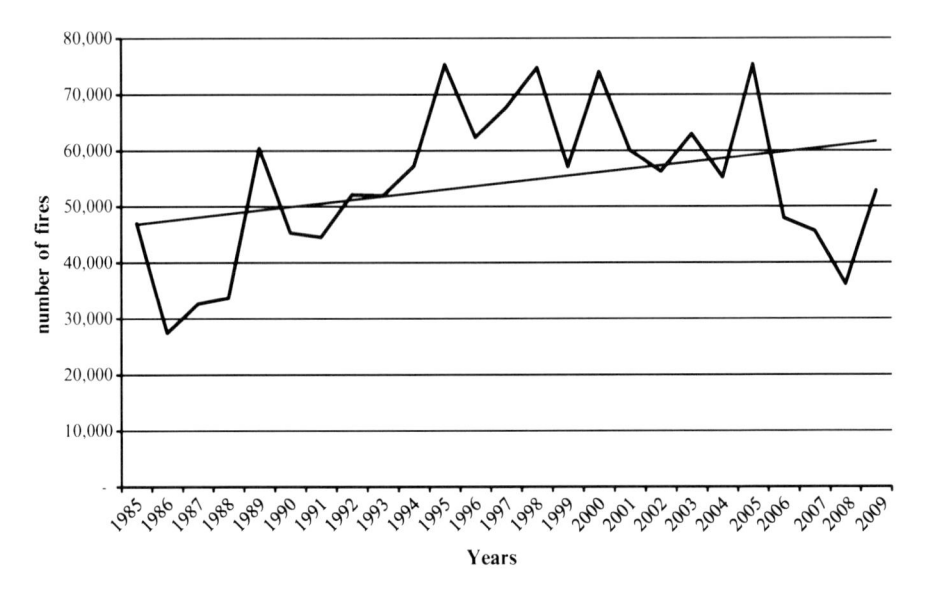

**Fig. 2.1** Total annual number of fires in the EUMed region from 1985 until 2009, and resulting trend line

during this period may be associated with fuel accumulation related to land cover changes such as the expansion of shrublands and abandonment of agricultural lands (Carmo et al. 2011; Lloret et al. 2002; Romero-Calcerrada et al. 2008). The results of the Mann–Kendall test showed that, for the entire study period, the general trend is an increase, but not significant ($S=64$, $P=0.14$). For the last 10 years, on the contrary, a significant decreasing trend was observed ($S=-25$, $P=0.032$).

The burned area, on the other hand, showed a decreasing trend since 1980, with strong annual fluctuations (Fig. 2.2). The results of the Mann–Kendall test show that, for both periods, the general trend was a decrease, but significant only when considering the entire time series ($S=-88$, $P=0.042$). Besides the influence of weather conditions in fire spread and burned area annually, this decrease is likely related to the implementation of fire prevention strategies and to the improvement in fire detection and fire-fighting techniques observed during the last years.

### 2.3.2 Overall Trends by Country

The countries of the EUMed region showed different trends concerning the number of fires (Fig. 2.3).

Comparing the entire time series with the last 10 years, different trends can be observed depending on the country (Table 2.2). Portugal, Spain and Greece showed an increasing trend for the whole study period, while France and Italy had a general decrease.

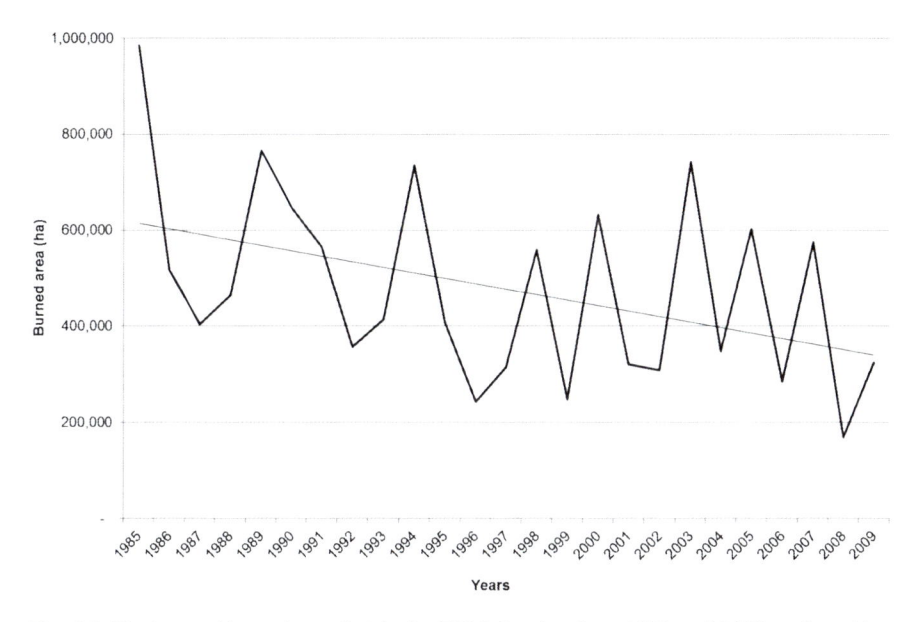

**Fig. 2.2**  Total annual burned area (ha) in the EUMed region from 1985 until 2009, and resulting trend line

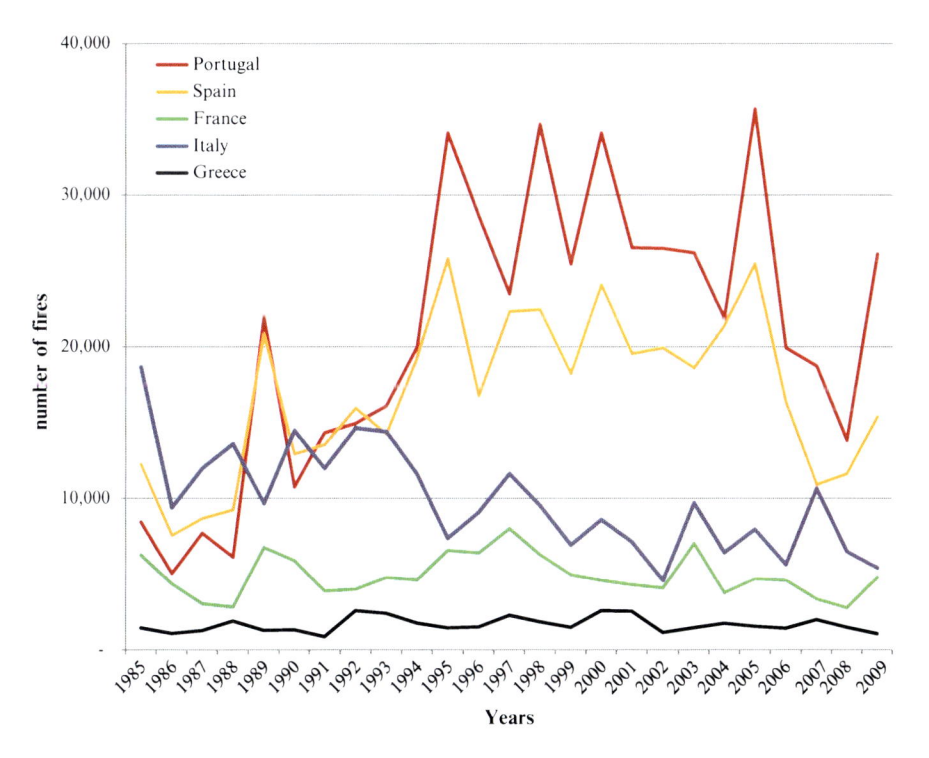

**Fig. 2.3**  Annual number of fires in the countries of the EUMed region from 1985 until 2009

**Table 2.2** Results of the Mann–Kendall test (*S*), associated probabilities (*P*), and Sen slope for the number of fires by country in both periods

| Time period | Portugal | Spain | France | Italy | Greece |
|---|---|---|---|---|---|
| 1985–2009 | | | | | |
| *S* | **110** | 82 | −28 | **−164** | 22 |
| *P* | 0.011 | 0.058 | 0.528 | <0.001 | 0.623 |
| Sen slope | 801.9 | 396.9 | −33.0 | −346.2 | 5.97 |
| 2000–2009 | | | | | |
| *S* | **−27** | −21 | −7 | −7 | −17 |
| *P* | 0.020 | 0.073 | 0.592 | 0.592 | 0.152 |
| Sen slope | −1554.0 | −1134.0 | −157.5 | −201.2 | −118.0 |

*Note*: Negative values mean a decrease and positive values mean an increase. Significant values are signaled in bold

Both the increasing trend observed for Portugal and the decreasing trend of Italy are significant. In the last decade, a decrease was observed for all the countries, significant only for Portugal, which had a median decrease of over 1,500 fires per year (Sen slope).

In relation to the total burned area, the differences among the countries were also evident (Fig. 2.4). Until the end of the 1990s, Spain usually had the highest burned area, but since 2001 Portugal recorded the highest values, particularly in 2003 and 2005, decreasing considerably afterwards. France and Greece showed, in general, the lowest values of area burned for the whole period, but in Greece the years 2000 and 2007 showed a substantial increase in area burned, in the latter case exceeding all the other countries.

Results of the Mann–Kendall test (Sen 1968) suggests a decreasing trend in all countries during both periods (Table 2.3), with the exception of Greece, where an increasing trend was observed for the last decade. However, a significant trend was observed only for Spain and Italy, which showed a median annual decrease in area burned of 5,175 ha for Spain and 3,243 ha for Italy, for the whole period. The test was not significant for Portugal and France. It must be noted that the Mann–Kendall test, as a non-parametric test, does not consider the absolute change in magnitude from year to year, but just the tendency in a rank ordering of the burnt areas for sequential years.

The average fire size showed a dissimilar spatial trend in relation to the number of fires and burned area, with Greece showing the highest values for nearly all the years, with particular incidence in 2007 (Fig. 2.5). For all the other countries, the average fire size decreased continuously since the 1980s, with annual oscillations more evident in Spain in 1994, in Portugal in 2003 and in Italy in 2007.

## 2.3.3   Overall Trends by Province (NUTS3)

The overall trend in the number of fires is very irregular depending on the province, although general patterns can be observed by country (Fig. 2.6). Portugal and Spain

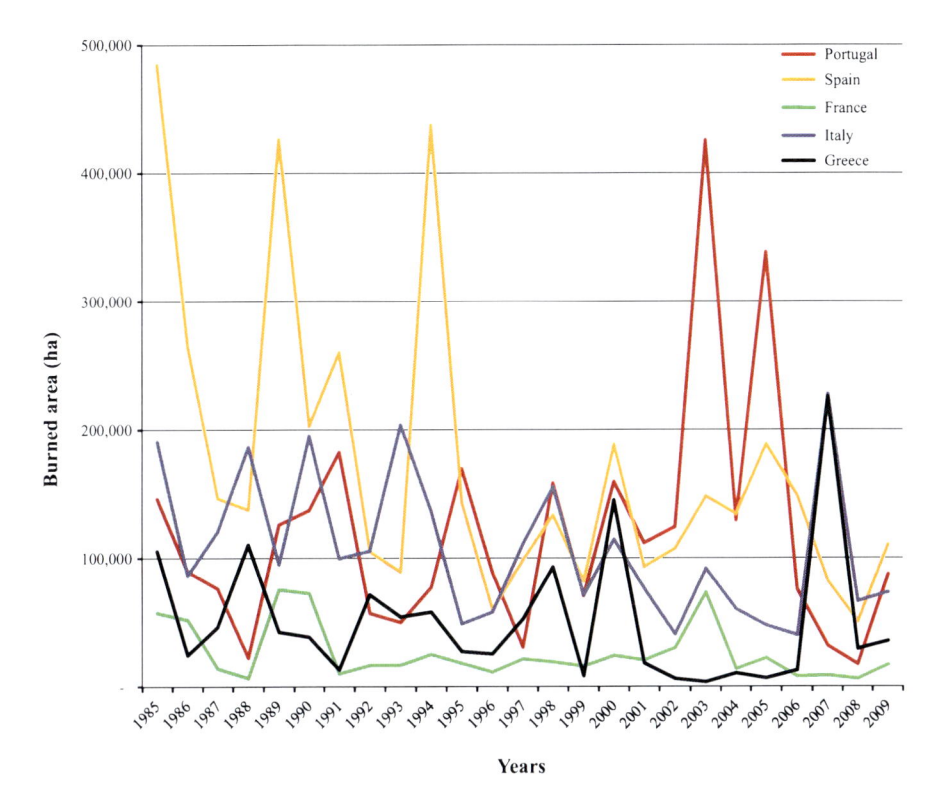

**Fig. 2.4** Annual burned area (ha) in the countries of the EUMed region from 1985 until 2009

**Table 2.3** Results of the Mann–Kendall test ($S$), associated probabilities ($P$), and Sen slope for the burned area by country in both periods

| Time period | Portugal | Spain | France | Italy | Greece |
|---|---|---|---|---|---|
| 1985–2009 | | | | | |
| $S$ | −6 | **−100** | −52 | **−96** | −68 |
| $P$ | 0.907 | 0.020 | 0.233 | 0.026 | 0.117 |
| Sen slope | −101.6 | −5175.0 | −473.5 | −3243 | −1703 |
| 2000–2009 | | | | | |
| $S$ | −19 | −9 | −21 | −5 | 11 |
| $P$ | 0.107 | 0.474 | 0.074 | 0.720 | 0.371 |
| Sen slope | −14016.0 | −6232.0 | −2215.0 | −1443.0 | 2127.0 |

*Note*: Negative values mean a decrease and positive values mean an increase. Significant values are signaled in bold

have the majority of provinces with a significant increasing trend, while Italy and Greece have more provinces with a significant decreasing trend. However it should be noted that the Greek data at NUTS3 level after 1998 are incomplete, because of changes in the reporting system in the country. In the case of Italy, an exception

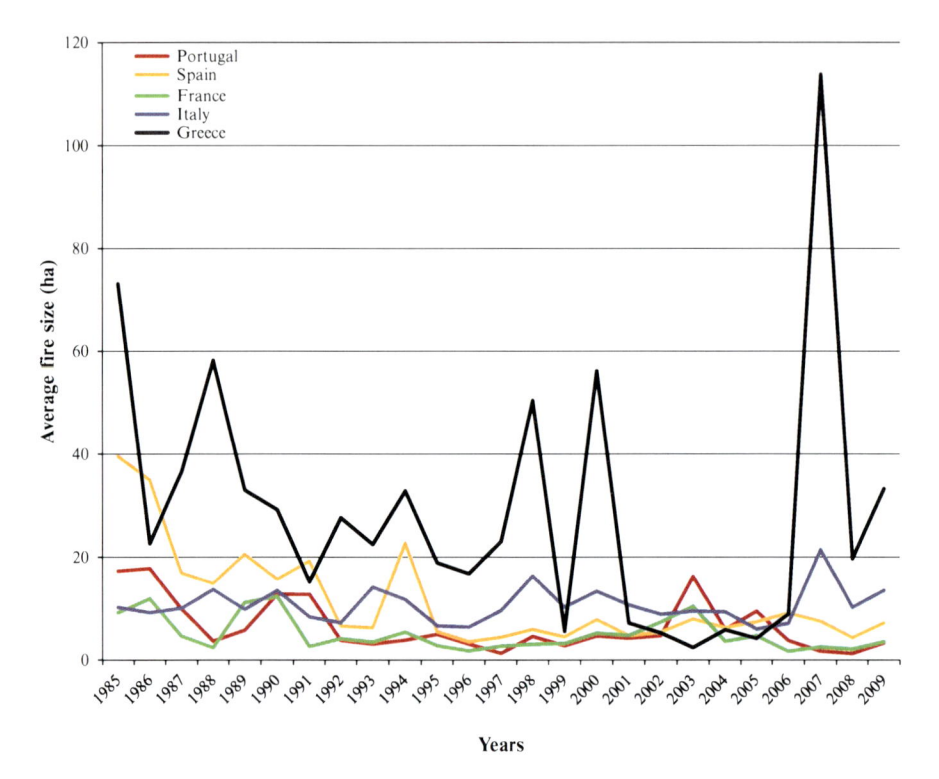

**Fig. 2.5** Annual average fire size (ha) in the countries of the EUMed region from 1985 until 2009

occurs in Sicily, where all provinces showed increasing trend or no trend, while in Sardinia almost all the provinces had a decreasing trend. In France, most of the provinces with available data indicated no trend or a decreasing trend. The situation changed when considering only the data between 2000 and 2009. There are just few provinces in the whole study area with a significant trend, either increasing or decreasing, possibly because the time series is too short at this scale of analysis.

The burned area, on the other hand, evidenced a general significant decreasing trend for most provinces both between 1985 and 2009 and in recent years (Fig. 2.7).

### 2.3.4  Seasonal Trends

Seasonal trends were analyzed at country and regional levels. The average number of fires and average burned area per month between 1985 and 2009 for the EUMed region (Table 2.4) showed that the months with higher number of fires and burned area were August, July and September, respectively. Nearly 73% of the number of

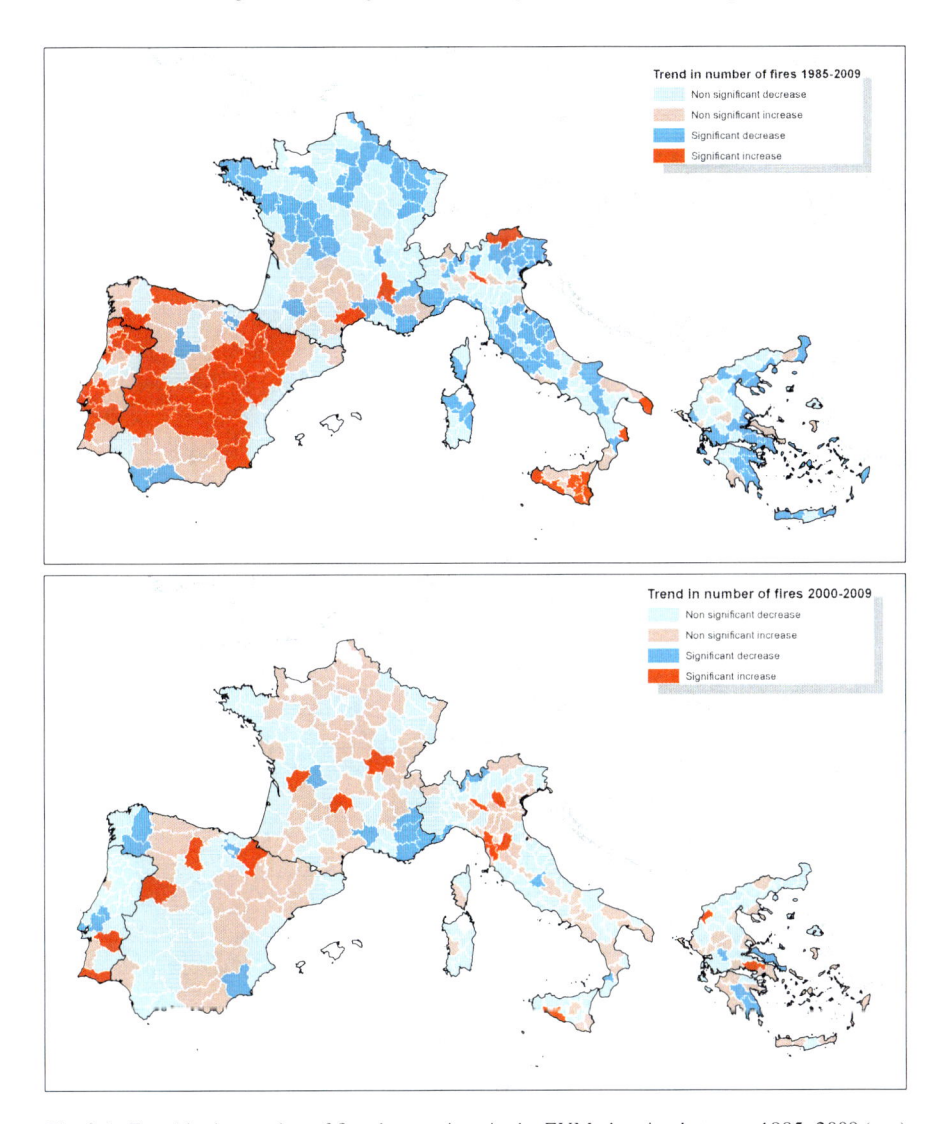

**Fig. 2.6** Trend in the number of fires by province in the EUMed region between 1985–2009 (*top*) and between 2000–2009 (*bottom*) obtained with the Mann–Kendall test

fires and nearly 85% of the burned area occurred between June and October. March showed a higher number of fires and burned area in comparison with the other spring months.

At country level, the average trend across months is similar for all countries, even though the absolute number of fires and burned area is highly variable (Figs. 2.8 and 2.9).

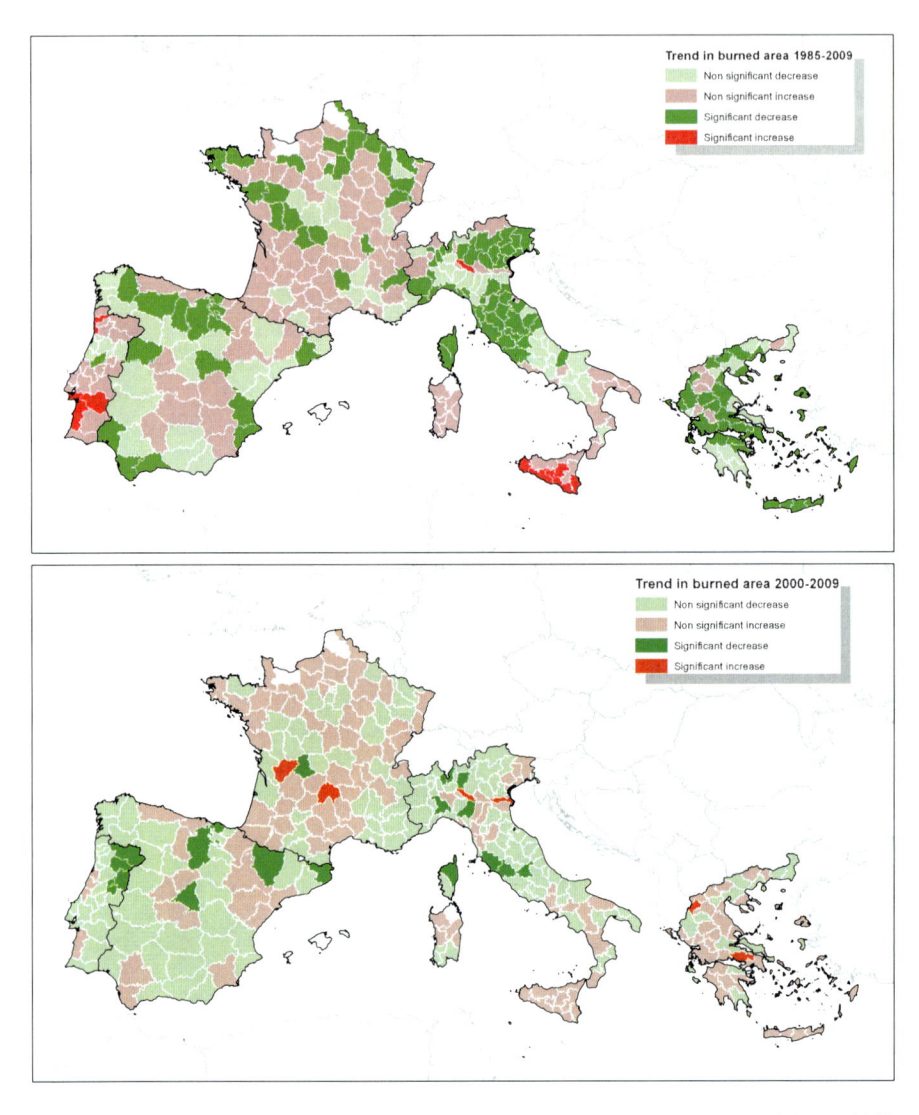

**Fig. 2.7** Trend in the burned area (ha) by province in the EUMed region between 1985 and 2009 (*top*) and between 2000 and 2009 (*bottom*), obtained from the Mann–Kendall test

Portugal showed the highest average number of fires between June and November, while in the other months it is surpassed by Spain and in January and February also by Italy. Greece had the lowest average number of fires recorded in the database in all months; however it must be noted that the detailed data by month for the last 2 years is not yet available for this country.

**Table 2.4**  Annual average (1985–2009) number of fires and burned area per month in the five countries of the EUMed region

| Month | Number of fires | % of total number of fires | Burned area (ha) | % of total burned area |
|---|---|---|---|---|
| January | 196 | 2.0 | 1,396 | 1.6 |
| February | 484 | 4.9 | 2,918 | 3.3 |
| March | 892 | 9.1 | 4,730 | 5.3 |
| April | 509 | 5.2 | 2,270 | 2.6 |
| May | 318 | 3.2 | 1,085 | 1.2 |
| June | 729 | 7.4 | 4,307 | 4.8 |
| July | 1,754 | 17.8 | 23,198 | 26.1 |
| August | 2,548 | 25.9 | 31,451 | 35.4 |
| September | 1,618 | 16.4 | 12,790 | 14.4 |
| October | 4,98 | 5.1 | 3,055 | 3.4 |
| November | 176 | 1.8 | 6,83 | 0.8 |
| December | 134 | 1.4 | 1,028 | 1.2 |

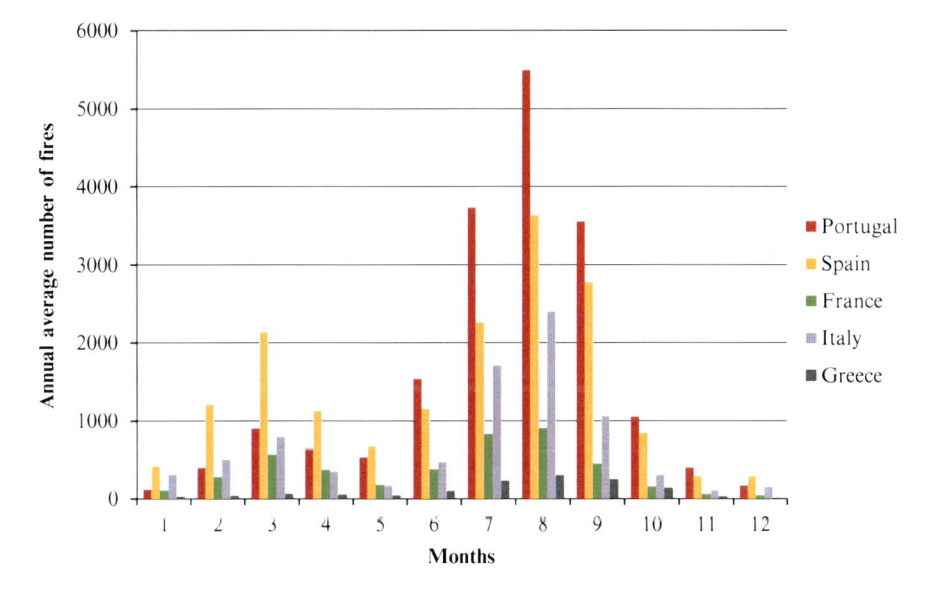

**Fig. 2.8**  Average number of fires per month in the EUMed countries

The burned area had a different trend; Spain had the highest values for all months, followed by Portugal and Italy. France and Greece showed the lowest values (see Fig. 2.9). Between July and September the average burned area increased substantially in all the countries, reflecting the general weather conditions of this period that promote fire occurrence (hot and dry summer).

Based on these results, the data by country were divided in two different seasons: from June to October, corresponding to the season when most of the fires occur, and

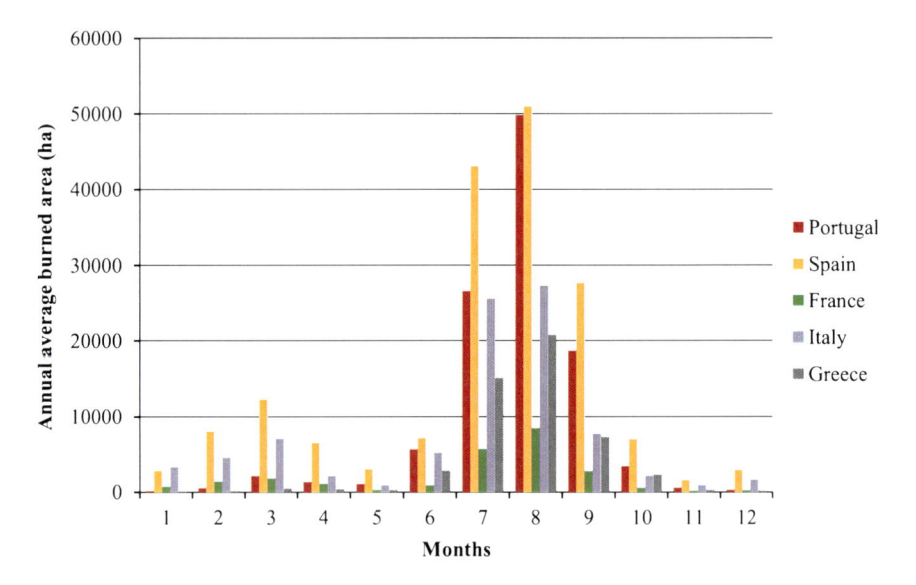

**Fig. 2.9** Average burned area (ha) per month in the EUMed countries

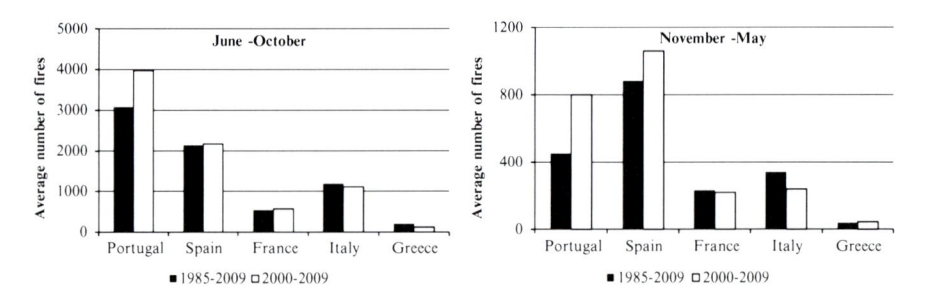

**Fig. 2.10** Comparison of the average number of fires for 1985–2009 and 2000–2009 in both seasons

November to May. Comparing the whole time series with the last decade, the average number of fires has increased in both seasons in Portugal and Spain (Fig. 2.10). In Italy it decreased in both periods. In France it increased in June-October but decreased in November-May, whereas the opposite trend was observed in Greece.

In relation to the burned area, Portugal showed an increase in the last decade in both seasons (although no significant trend was found for the overall season in Table 2.3, so this should be interpreted with caution) and Italy a slight increase in the June-October season (Fig. 2.11). In the season November-May, Spain and Italy have the highest average of burned area in both periods, while Portugal is in third position in spite of the observed increase.

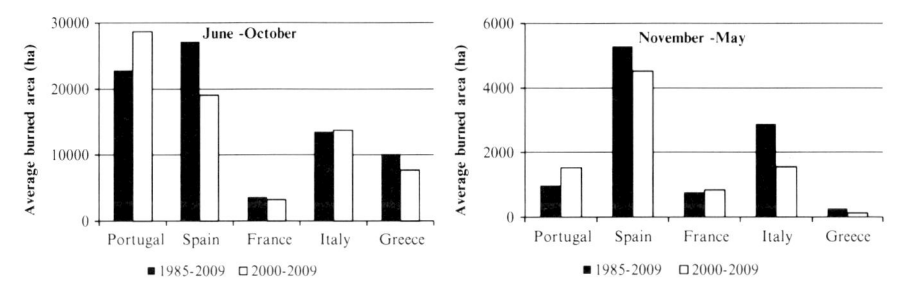

**Fig. 2.11** Comparison of the average burned area for 1985–2009 and 2000–2009 in both seasons

## 2.4   Land Cover Changes in Burned Areas

In this section we aimed to carry out a detailed analysis of land cover changes that occurred in areas affected by fires during the period 2000–2006 in several Southern Europe countries. This was done using the CORINE land cover maps (CLC maps, hereafter) available for 2000 and 2006, and the European Fire Database (EFFIS) containing the annual forest fire information compiled by EU Member States and other European countries (http://effis.jrc.ec.europa.eu). EFFIS database was used for the period 2000–2006. The countries studied included Portugal, Spain, France and Italy. Greece was excluded because the 2006 CLC map was not available for that country.

We worked with the second level of CORINE land cover data and when the results of the analysis indicated the occurrence of a major type of transition at a country level, the third level was used. The areas that were burned in each country and for each year throughout the studied period were obtained after the annual fire maps of the EFFIS database. For each country, a set of seven masks was derived from these fire maps; i.e. one mask for each year from 2000 to 2006. All fires smaller than 50 ha were discarded for the creation of the mask.

Each year the mask was used for extracting two new layers from the two CLC maps of each country. The layer derived from the CLC 2000 map would represent the land cover pre-fire situation in the areas burned that year, whereas the layer derived from the CLC 2006 map would represent the land cover post-fire situation in those same areas. For each year, we combined the two corresponding layers, obtaining a final file in which each polygon would correspond to a given transition of land covers between CLC2000 and CLC2006 and would reflect this information in its attribute table. Based on these data, we generated seven transitional matrices for each country (one per year) and selected, in each case, the major transitional classes to be analyzed at the second CLC data level. An overall transition matrix was also calculated by pooling the data from all countries. Land cover transitions representing less than 50 ha were excluded.

## 2.4.1   Main Land Cover Types Affected by Wildfires

During the study period (2000–2006), the total burned area in the four considered countries was 1,395,119 ha (only considering fires larger than 50 ha). Half of this area (51%) consisted of CLC Level 2 class 32 ("Scrub and/or herbaceous associations"), followed by class 31 ("Forests") (34%). At CLC Level 3 fires affected mainly class 324 ("Transitional woodland-scrub"), corresponding to 23% of the total, class 312 ("Coniferous forest") (15%), followed by classes 311 ("Broad-leaved forest") (12%) and classes 313 ("Mixed forest"), 321 ("Natural grassland"), 322 ("Moors and heathland") and 323 ("Sclerophyllous vegetation"), representing each ca. 9% of the total burned area.

## 2.4.2   Land Cover Changes in Burned Areas

Overall, a total of 1,016,055 ha of burned areas (72.8% of the total) did not change their land cover after fire. The land covers with less persistence in burned areas were Forests, Open spaces with little or no vegetation, and Inland wetlands (Table 2.5). Caution should be taken in interpreting the finding for the latter land cover, as the area with this land cover was very small (150 ha) and thus prone to significant proportional changes even with small variations in polygon boundaries. From the remaining 379,064 ha in which changes occurred, 76.9% became class 32 ("Scrub and/or herbaceous associations") and 19.6% became class 33 ("Open spaces with little or no vegetation").

The transition matrix for the overall burned area (Table 2.5) showed that the main changes driven by fire were the transition from forests to open spaces with little or no vegetation (over 50% of the forests in 2000 suffered this transition). The transition from Open spaces with little or no vegetation to Scrub and/or herbaceous vegetation was also relevant (45%), as well as Inland wetlands to Inland waters (46%) Other important transitions were from Arable land to Artificial, non-agricultural vegetated areas (15%). Of these transitions, only the former can be clearly attributed to fire effects.

In Italy, the total area burned during the study period was 79,118 ha (5.7% of the total burned area in the four countries). Fires affected mainly class 211 "Non-irrigated arable land" (22% of the total area burned in the country), class 321 ("Natural grassland") and class 323 ("Sclerophyllous vegetation") representing each one ca. 19% of the total area burned, and class 311 ("Broad-leaved forest") (15%). 68,621 ha of burned areas (86.7% of the total) did not change land cover after fire. From the 10,497 ha that suffered land cover changes, 31% became class 323 ("Sclerophyllous vegetation"), 19% became class 321 ("Natural grassland"), 11% became class 333 ("Sparsely vegetated areas"), 10% became class 243 ("Land principally occupied by agriculture with significant areas of natural vegetation") and 8% became class 334 ("Burned areas").

**Table 2.5** Transition matrix for the period 2000–2006 in the burned areas in Portugal, Spain, France and Italy

|         |    | CLC2006 |      |      |      |      |      |      |      |      |      |      |      |      |
|---------|----|------|------|------|------|------|------|------|------|------|------|------|------|------|
|         |    | 11   | 12   | 13   | 14   | 21   | 22   | 23   | 24   | 31   | 32   | 33   | 41   | 51   |
| CLC2000 | 11 | *1.00* | 0.00 | 0.00 | 0.00 | 0.00 | 0.00 | 0.00 | 0.00 | 0.00 | 0.00 | 0.00 | 0.00 | 0.00 |
|         | 12 | 0.00 | *1.00* | 0.00 | 0.00 | 0.00 | 0.00 | 0.00 | 0.00 | 0.00 | 0.00 | 0.00 | 0.00 | 0.00 |
|         | 13 | 0.00 | 0.00 | *1.00* | 0.00 | 0.00 | 0.00 | 0.00 | 0.00 | 0.00 | 0.00 | 0.00 | 0.00 | 0.00 |
|         | 14 | 0.00 | 0.00 | 0.00 | *1.00* | 0.00 | 0.00 | 0.00 | 0.00 | 0.00 | 0.00 | 0.00 | 0.00 | 0.00 |
|         | 21 | 0.00 | 0.00 | 0.00 | **0.15** | *0.82* | 0.00 | 0.00 | 0.01 | 0.00 | 0.01 | 0.00 | 0.00 | 0.00 |
|         | 22 | 0.00 | 0.00 | 0.00 | 0.00 | 0.00 | *0.87* | 0.01 | 0.05 | 0.00 | **0.07** | 0.01 | 0.00 | 0.00 |
|         | 23 | 0.00 | 0.00 | 0.00 | 0.00 | 0.00 | 0.00 | *1.00* | 0.00 | 0.00 | 0.00 | 0.00 | 0.00 | 0.00 |
|         | 24 | 0.00 | 0.00 | 0.00 | 0.00 | 0.01 | 0.00 | 0.00 | *0.96* | 0.00 | 0.02 | 0.01 | 0.00 | 0.00 |
|         | 31 | 0.00 | 0.00 | 0.00 | 0.00 | 0.00 | 0.00 | 0.00 | 0.00 | *0.41* | **0.51** | **0.07** | 0.00 | 0.00 |
|         | 32 | 0.00 | 0.00 | 0.00 | 0.00 | 0.00 | 0.00 | 0.00 | 0.00 | 0.00 | *0.94* | 0.05 | 0.00 | 0.00 |
|         | 33 | 0.00 | 0.00 | 0.00 | 0.00 | 0.00 | 0.00 | 0.00 | 0.00 | 0.00 | **0.45** | *0.54* | 0.00 | 0.00 |
|         | 41 | 0.00 | 0.00 | 0.00 | 0.00 | 0.00 | 0.00 | 0.00 | 0.00 | 0.00 | 0.00 | 0.00 | *0.54* | **0.46** |
|         | 51 | 0.00 | 0.00 | 0.00 | 0.00 | 0.00 | 0.00 | 0.00 | 0.00 | 0.00 | 0.00 | 0.00 | 0.00 | *1.00* |

*Note*: Each row provides the proportion of the initial land cover (in 2000) that persisted or changed to other land cover in 2006. Italic cells indicate the persistence values (diagonal) and Bold indicate the main transitions (over 5%). Codes for land cover are: Urban fabric (11); Industrial, commercial and transport units (12); Mine, dump and construction sites (13); Artificial, non-agricultural vegetated areas (14); Arable land (21); Permanent crops (22); Pastures (23); Heterogeneous agricultural areas (24); Forest (31); Scrub and/or herbaceous associations (32); Open spaces with little or no vegetation (33); Inland wetlands (41); Inland waters (51)

In France, the total area burned during the study period was 67.727 ha (4.8% of the total area burned in the four countries). Fires affected mainly classes 323 ("Sclerophyllous vegetation") and 321 ("Natural grassland"), corresponding to 31% and 17%, respectively, of the total area burned in the country. In addition, fires affected also classes 313 ("Mixed forest") and 312 ("Coniferous forest") representing each one ca. 10%. 41,740 ha of the burned areas (61.6% of the total) did not change land cover after fire. From the 25,987 ha that changed land cover trype, 41% became class 334 ("Burned areas"), 30% became class 324 ("Transitional woodland scrub") and 20% became class 323 ("Sclerophyllous vegetation").

In Spain, the total area burned during the study period was 492,243 ha (35.3% of the total area burned in the four countries). Fires affected mainly class 324 ("Transitional woodland scrub"), corresponding to 27% of the total area burned in the country, and class 323 ("Sclerophyllous vegetation") (14%). Classes 321 ("Natural grassland"), 312 ("Coniferous forest") and 313 ("Mixed forest") were also strongly subjected to fires (12%, 11% and 10%, respectively, of the total burned area). 381,982 ha of the burned areas (77.6% of the total) did not change land cover after fire, whereas 110,261 ha did. Among the latter, 47% became class 324 ("Transitional woodland scrub") and 29% became class 334 ("Burned areas").

In Portugal, the total area burned during the study period was 756,031 ha. This country had, therefore, the largest proportion (54.2%) over the total area burned

**Table 2.6** Distribution of the total burned area per country between agradative, degradative and stable land cover transitions

| | Italy | | France | | Spain | | Portugal | |
|---|---|---|---|---|---|---|---|---|
| | ha | % | ha | % | ha | % | ha | % |
| Agradative transitions | 4,187 | 5.3 | 13,331 | 19.7 | 36,736 | 7.5 | 42,717 | 5.7 |
| Degradative transitions | 6,308 | 8.0 | 12,667 | 18.7 | 73,524 | 14.9 | 243,774 | 32.2 |
| Stable transitions | 68,621 | 86.7 | 41,740 | 61.6 | 381,982 | 77.6 | 469,533 | 62.1 |

in the four countries. Fires affected mainly class 324 ("Transitional woodland scrub"), corresponding to 24% of the burned area, class 312 ("Coniferous forest") (19%), class 311 ("Broad-leaved forest") (15%) and class 322 ("Moors and heathland") (11%). 469,533 ha of burned areas (62.1% of the total) did not change land cover after fire, whereas 286,498 ha did, of which 81% became class 324 ("Transitional woodland scrub"), 7% class 334 ("Burned areas") and 3% class 322 ("Moors and heathland").

Classifying all the CLC transition classes into agradative (any transition resulting in an increase of the vegetation cover or leading to a more advanced successional stage), degradative or stable categories, we found clear differences among the considered countries in the distribution of the total burned area among these three types (Table 2.6). In Portugal, Spain and France, the post-fire land cover changes occurred in areas burned between 2000 and 2006 mostly favoured degradative transitions. This degradation trend was particularly strong in Portugal and Spain. In France, agradative transitions represented a slightly larger area than degradative ones.

These results suggest a slow post-fire vegetation dynamics in most of the countries studied. In all of them, except France, degradative transitions accounted for the largest part of the land cover changes that occurred on burned areas. Moreover, a large part of the areas classified as 33 ("Open spaces with little or no vegetation") in CLC 2000 had remained in that same class in CLC 2006, not evolving to classes with increased vegetation cover or towards more mature successional stages. This slow dynamics may be due to various factors. First of all, in Spain and Portugal (the two countries with the smallest proportion of agradative transitions), more adverse climatic conditions (i.e. dryer conditions) in many of the areas affected by fires may have caused lower rates of post-fire vegetation recovery. Secondly, in those two countries, a large part of the fires occurred in the last 2 years of the studied period (2005 and 2006). In Spain and Portugal, these fires accounted for 38% and 33%, respectively, of the total burned area in each case, whereas in Italy and France, these values were much lower (26% and 11%, respectively). In the two former countries, thus, a larger extent of burned areas had a very short time to recover, which, obviously, influenced the results.

In general, the length of the study period was short, as the maximum post-fire period that could be monitored was 6 years. We have to highlight, therefore, that in most cases our results are documenting post-fire land cover dynamics on the short (or sometimes medium) term.

## 2.5 Fire Effects on Land Cover Change Dynamics in the Period 2000–2006

In this section our aim was to evaluate the role of fire in the observed landscape dynamics at European level. The land cover dynamics analysis consisted in the comparison of the observed land cover change and transitions in burned and unburned areas. This analysis was carried out using CORINE Land Cover (CLC) data for 2000 and 2006 and the fire perimeters for fires larger than 500 ha obtained from the EFFIS database.

The results obtained for land cover dynamics analysis are presented in this section as transition matrices. For each of a total of 702 fire perimeters, we considered a paired unburned area with a similar shape, and surrounding the burned patch. To characterize land cover change, the thematic legend (third level) of the CLC layer has been aggregated into a new one composed by six main categories (urban, artificial, agricultural, forest, shrubland and no vegetation areas) in order to simplify the land cover dynamics analysis. This has been done both for CLC 2000 and 2006, and as result two new land use layers were obtained. A transition matrix was computed separately for each burned – unburned patch pair. The differences between these two matrices were then summarized in a new matrix called change-intensity matrix which represented the rates of land cover change in burned versus unburned areas.

In unburned areas (Table 2.7), persistence of the land covers (diagonal values of the matrix) were always larger than 90% with the exception of forests and areas with no vegetation, where it decreased to ca. 70%. The main transitions were from forests to shrublands (26%) and no vegetation to shrublands (30%). The latter transition seems to reflect the process of secondary succession and scrub encroachment, probably in former burned areas, sparsely vegetation areas or even bare ground. The former is probably a consequence of forest logging. In burned areas, the persistence pattern of the different land cover types was similar to the one of unburned areas: also always larger than 90% with the exception of forests and areas with no vegetation, but is this case it was even lower, ca. 35–50% (Table 2.7). Here the main transitions were also from forests to shrublands (57%) but also to areas with no vegetation (6%), from areas with no vegetation to shrublands (49%), and, to a lesser extent, from shrublands to areas with no vegetation (6%). The transition of forests to shrublands and areas with no vegetation could be explained mainly by wildfires. After fires, in a period of 6 years (from 2000 to 2006) areas may not be able to have a significant vegetation development, or only shrublands are able to grow in the early stages of succession. Even if there is forest recovery it will be in an earlier stage of development and would have a shrubland-like physiognomy, or would consist of a transition category between forest and shrub which in this work is categorized as shrubland (see proposed legend). The same driver (fire) can explain the transition from shrublands to areas with no vegetation. In contrast, the significant transition from areas with no vegetation to shrublands may be an evidence of post-fire vegetation recovery, mainly in situations where the areas were burned in the beginning of the study period (2000). It must be taken into account that this land cover class

**Table 2.7** Transition matrices (expressed as percentages) for burned and unburned areas in the 2000–2006 period

| | Difference | | | | | | |
|---|---|---|---|---|---|---|---|
| | To 2006 | Urban | Artificial | Agricultural | Forest | Shrubland | No vegetation |
| From 2000 | Urban | −0.5 | 0.0 | −0.1 | −0.2 | 0.7 | 0.0 |
| | Artificial | −0.7 | −1.2 | 0.7 | 0.9 | 0.0 | 0.3 |
| | Agricultural | −0.8 | 0.0 | −0.9 | −0.3 | 1.4 | 0.6 |
| | Forest | −0.1 | −0.1 | −0.2 | −35.3 | 30.7 | 4.9 |
| | Shrubland | 0.0 | −0.1 | −0.3 | −3.6 | 0.9 | 3.2 |
| | No vegetation | 0.0 | −0.8 | −0.8 | −0.1 | 18.7 | −17.1 |
| | Burned | | | | | | |
| | To 2006 | Urban | Artificial | Agricultural | Forest | Shrubland | No vegetation |
| From 2000 | Urban | 98.5 | 0.0 | 0.6 | 0.0 | 0.9 | 0.0 |
| | Artificial | 0.1 | 96.2 | 1.1 | 1.3 | 0.8 | 0.5 |
| | Agricultural | 0.2 | 0.1 | 95.2 | 0.3 | 3.5 | 0.7 |
| | Forest | 0.0 | 0.1 | 0.4 | 36.3 | 57.1 | 6.2 |
| | Shrubland | 0.0 | 0.1 | 0.9 | 1.1 | 92.3 | 5.6 |
| | No vegetation | 0.0 | 0.3 | 0.3 | 0.2 | 48.6 | 50.6 |
| | Unburned | | | | | | |
| | To 2006 | Urban | Artificial | Agricultural | Forest | Shrubland | No vegetation |
| From 2000 | Urban | 99.0 | 0.0 | 0.6 | 0.2 | 0.2 | 0.0 |
| | Artificial | 0.8 | 97.4 | 0.4 | 0.4 | 0.7 | 0.2 |
| | Agricultural | 1.0 | 0.1 | 96.1 | 0.6 | 2.0 | 0.2 |
| | Forest | 0.1 | 0.2 | 0.6 | 71.6 | 26.4 | 1.2 |
| | Shrubland | 0.0 | 0.2 | 1.2 | 4.7 | 91.4 | 2.4 |
| | No vegetation | 0.0 | 1.1 | 1.1 | 0.3 | 30.0 | 67.6 |

*Note*: Each value in the matrices represents the average value for a sample of 702 wildfires and respective unburned pairs, and each row represents the proportion of the land cover in 2000 that was kept or changed in 2006. The change intensity matrix (Difference matrix, on the top) represents the difference between burned (middle) and unburned (bottom) areas matrices includes also burned areas and sparsely vegetated areas, thus the succession to shrubland-type vegetation is a possible explanation. Alternatively, misclassification of the land cover types in the two different time periods could explain this result, if many areas with no vegetation in 2000 had been classified as shrublands in 2006 (but if that is the case this mistake must have been made also in unburned areas).

The intensity change matrix shows that wildfires have caused changes in land cover dynamics (Table 2.7). In terms of persistence, fire decreased the persistence for all land cover types except shrublands. So, fire promoted faster land cover changes.

The more notorious decrease in persistence was for forests and areas with no vegetation. It is logical that forests are the land cover more easily changed by fire, and thus less persistent. The trend for areas with no vegetation might again be explained by different criteria in classifying the same land cover in the two time periods. The major land cover transitions promoted by fire were the forest to both shrubland cover (+30%) and to areas with no vegetation (+5%). So, as expected, fire causes a much faster change from forests to areas with shrublands or no vegetation in the short term, compared to unburned areas. A similar trend was observed by for specific regions of Portugal and Spain (Lloret et al. 2002; Viedma et al. 2006; Silva et al. 2011). The other significant transition was from shrublands to areas with no vegetation (+19%), although this could be interpreted as a simple maintenance of the same land cover in case the hypothesis of misclassification is confirmed. Other land cover transitions favored by fire included shrublands to no vegetation (+3%) and agricultural areas to shrublands (+1.5%), the latter either reflecting a trend for the abandonment of agriculture in burned areas, as hypothesized by Silva et al. (2011) for three regions in Portugal, or the assignment of different categories to the same land cover (e.g. pastures versus natural grasslands).

Land cover transitions promoted by the absence of fire were less notorious. Larger differences were registered for the transition from shrublands to forest (−4%), reflecting secondary succession in the vegetation, from areas with no vegetation to artificial and agricultural areas (−0.8%), and from agricultural to urban areas (−0.8%). The latter transitions suggest that urbanization processes are more common in unburned areas, compared to the burned ones.

## 2.6  Key Messages

- Land cover changes in Southern Europe in the period 1990–2006 suggest a decrease in fire hazard in this region, as landscape changes corresponding to increased fire hazard occur in a smaller geographic area (4.9 million hectares) than transitions corresponding to decreased fire hazard (5.4 million hectares). This might be explained by disturbances such as logging, drought, wildfires, as well as urbanization;
- Compared with the overall period 1985–2009, changes in the fire regime have been observed in the last 10 years (2000–2009) in Southern Europe. The long-term

trend for the number of fires was an increase, but in the last 10 years the trend was the opposite, a high decrease. In relation to the total burned area, the general trend is for a decrease, lower when considering the entire time series and more pronounced in the last 10 years. For the period 1980–2009, the provinces with a high increase in both number of fires and burned area, were located in Portugal, Central Spain, Southern Sicily and Southeast France. The decreasing trends were found mostly in the Northern provinces of Spain and in Central Greece. The majority of the provinces of Italy and Greece showed no trend. For the period 2000–2008, the majority of provinces in all the countries show a decreasing trend, with a few exceptions in France and Italy;

- The average number of fires has substantially increased in Portugal and Spain in both the "fire season" (June to October) and the rest of the year, while for the other countries the trend is more constant;
- In the period 2000–2006, fires burned mainly areas of forest and shrublands. The main CORINE land cover categories affected were "Transitional woodland-scrub" (23% of the total burned areas, "Coniferous forest" (15%), followed by "Broad-leaved forest", "Mixed forest", "Natural grassland", "Moors and heath-land" and "Sclerophyllous vegetation" (ca. 10% each). Almost 97% of the areas burned during 2000–2006 changed their land cover to "Scrub and/or herbaceous associations" or "Open spaces with little or no vegetation";
- Wildfires affected landscape change dynamics. Fire decreased the persistence for all land cover types except shrublands. The major land cover transitions promoted by fire were the forest to both shrublands (+30%) and to areas with no vegetation (+5%). The other significant transition was from shrublands to areas with no vegetation (+19%), although this could be interpreted as a simple maintenance of the same land cover that was classified differently. Land cover transitions promoted by the absence of fire were less obvious. Larger differences were registered for the transition from shrublands to forest (−4%), reflecting secondary succession in the vegetation.

# References

Camia A, Durrant Houston T, San-Miguel-Ayanz J (2010) The European fire database: development, structure and implementation. In: Viegas DX (Ed.) Proceedings of the VI International conference on forest fire research, Coimbra, Portugal

Carmo M, Moreira F, Casimiro P, Vaz P (2011) Land use and topography influences on wildfire occurrence in northern Portugal. Landsc Urban Plan 100:169–176

EC, European Commission (2010) Forest fires in Europe 2009. Report nr. 10. EUR 24502 EN – Joint Research Centre, Institute for Environment and Sustainability, Office for Official Publications of the European Communities, Luxembourg. ISSN 1018-5593, pp 84

Falcucci A, Maiorano L, Boitani L (2007) Changes in land-use/land-cover patterns in Italy and their implications for biodiversity conservation. Landsc Ecol 22:617–631

Feranec J, Jaffrain G, Soukup T, Hazeu G (2010) Determining changes and flows in European landscapes 1990–2000 using CORINE land cover data. Appl Geogr 30:19–35

Fernandes P (2009) Combining forest structure data and fuel modelling to classify fire hazard in Portugal. Ann For Sci 66(4):415

Kendall MG (1975) Rank correlation methods, 4th edn. Charles Griffin, London

Lloret F, Calvo E, Pons X, Diaz-Delgado R (2002) Wildfires and landscape patterns in the Eastern Iberian Peninsula. Landsc Ecol 17:745–759

Moreira F, Vaz P, Catry F, Silva JS (2009) Regional variations in wildfire susceptibility of land-cover types in Portugal: implications for landscape management to minimize fire hazard. Int J Wildland Fire 18:563–574

Romero-Calcerrada R, Novillo CJ, Millington JDA, Gomez-Jimenez I (2008) GIS analysis of spatial patterns of human-caused wildfire ignition risk in the SW of Madrid (Central Spain). Landsc Ecol 23:341–354

Rothermel R (1983) How to predict the spread and intensity of forest and range fires. USDA, Forest Service, Intermountain Forest and Range Experiment Station, General Technical Report INT-143, Ogden, UT

San-Miguel Ayanz J, Camia A (2009) Forest fires at a glance: facts, figures and trends in the EU. In: Birot Y (ed) Living with wildfires: what science can tell us. A contribution to the Science-Policy dialogue. European Forest Institute, Discussion paper 15, pp 11–18

Sen PK (1968) Estimates of the regression coefficient based on Kendall's tau. J Am Stat Assoc 63:1379–1389

Silva JS, Vaz P, Moreira F, Catry F, Rego F (2011) Wildfires as a major driver of landscape dynamics in three fire-prone areas of Portugal. Landscape and Urban Planning 101:349–358

Van Doorn A, Bakker M (2007) The destination of arable land in a marginal agricultural landscape in South Portugal: an exploration of land use change determinants. Landsc Ecol 22:1073–1087

Viedma O, Moreno JM, Rieiro I (2006) Interactions between land use/land cover change, forest fires and landscape structure in Sierra de Gredos (Central Spain). Environ Conserv 33:212–222

# Chapter 3
# Economic, Legal and Social Aspects of Post-Fire Management

**Robert Mavsar, Elsa Varela, Piermaria Corona, Anna Barbati, and Graham Marsh**

## 3.1 Introduction

In the past, wildland fires mainly caused concern amongst resource managers and local communities that were directly affected by these disastrous events. However, in the last decades the increasing attention paid by the media to large scale fires that occurred in different parts of the world and their consequences also triggered increased concern among the general public. These major fire events also clearly showed that they are not only an environmental problem, but are of significant social dimensions, affecting millions of people, having major economic impacts and causing significant human casualties (González Cabán 2007). For example, the wildfires affecting vast forest areas of Portugal in 2005 caused damages worth almost 800 million € and took 13 lives. Even worse, the large fires affecting Greece during the summer of 2007 caused 64 casualties and damages estimated to be worth over five billion €. The social aspect of forest fires is further underlined by the fact that around 90% of forest fires in Europe are caused by people (Velez 2009).

R. Mavsar (✉) • E. Varela
European Forest Institute, Mediterranean Regional Office (EFIMED), Barcelona, Spain
e-mail: robert.mavsar@efi.int

P. Corona (✉) • A. Barbati
Department for Innovation in Biological, Agro-Food and Forest Systems,
University of Tuscia, Viterbo, Italy

G. Marsh
Department of Geography, Environment and Disaster Management, Coventry University,
Coventry, UK

F. Moreira et al. (eds.), *Post-Fire Management and Restoration of Southern European Forests*, Managing Forest Ecosystems 24, DOI 10.1007/978-94-007-2208-8_3, © Springer Science+Business Media B.V. 2012

The increased public interest and concern about fires also prompted the responsible agencies to develop and implement improved policies and management measures at different levels, which should minimize the negative environmental, economic and social impacts of forest fires (EU 2005). However, the implementation of such measures requires substantial investment of financial, human and organizational resources, which must be justifiable and efficient.

Like any other type of resource management, also post-fire management should be based on three main pillars: ecological, economic and social. Trying to simplify the roles of the three pillars, we could say that the ecological characteristics of an area indicate its present state, vulnerability to different impacts, limitations in the development of management objectives and applicability of management practices. The social aspects will facilitate the identification of relevant stakeholder groups, their needs and preferences, potential conflicts, traditions, institutional set-up and acceptability of certain management measures. Finally, the economics helps to evaluate the efficiency and distributional effects (equity) of feasible management alternatives. Therefore, limiting management decisions on only one of the mentioned pillars can considerably jeopardize the success and the effectiveness of implemented management measures. Rather, the effort should be made that post-fire interventions reflect a balance between the mentioned aspects.

While the rest of this book mainly focuses on the influence of environmental conditions on post-fire management and restoration, this chapter will look at the economic and social aspects that should be considered when applying post-fire management measures.

## 3.2 Economic Aspects of Post-Fire Management

From an economic perspective, a forest fire (as any other forest disturbances) can be considered as an event that interrupts or reduces the flow of forest goods and services provided by the forests (Holmes et al. 2008). Namely, forests are providing a wide range of goods and services that contribute significantly to the welfare of societies (Farrell et al. 2000). However, a forest's capacity to provide these goods and services depends on its characteristics (e.g., health status, growing stock, location, management objectives), which in the case of a forest fire are very often significantly changed. How a forest fire can affect the flow of goods and services is illustrated in Fig. 3.1. We assume that initially a forest was providing quantity $Q_0$ of a good. Without a fire the forest would be able to secure this level of the good over the whole period (line A). However, if we assume that at $t_1$ a fire would affect the forest and its capacity to provide the good, the provided quantity of the good would decrease to $Q_1$.

After a fire a forest may recover its potential to provide goods and services in an amount as it was before the fire. For example, in Fig. 3.1 we assume that the forest would recover his full potential to provide the good at $t_2$, following line B. This would mean that as a consequence of the fire during the period $t_1$ to $t_2$ we would suffer a loss of the good equal to the area shaded in light and dark grey.

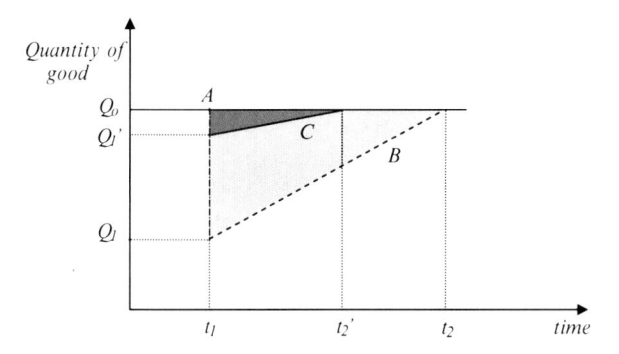

**Fig. 3.1** Illustration of effect of fire on the quantity of a forest good, its recovery process, and the effects of post-fire management practices. For further explanations, see text

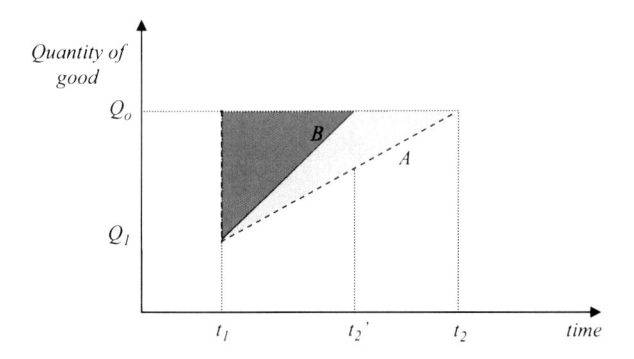

**Fig. 3.2** Illustration of the accelerated recovery process of a forest good as a result of post-fire management measures

On one hand, the effect of the fire on the flow of goods and services depends mainly on the use of the forest before the fire (i.e., which goods and services it provided), fire severity and size, and management measures taken before (prevention), during (suppression) and after (post-fire measures) the fire. Furthermore, it should be also acknowledged that in most cases multiple goods and services are affected simultaneously. For example, the reduced biomass of a forest can at the same time have negative consequences for wood production (net annual increment), carbon sequestration, soil protection, biodiversity, etc.

On the other hand also the recovery process depends on various factors, such as type of good, fire severity and size, and post-fire management measures. The post-fire management measures are mainly aimed at preventing further loss of the flow of certain goods and services, and shortening the recovery period. Figure 3.2 shows an example of how we can represent the effects of post-fire management on the recovery of the provision of a certain good.

We assume that initially the forest was providing $Q_0$ of a good. At $t_1$ as a consequence of fire the flow of the good was reduced to $Q_1$. Next, if no post-fire management measures would have been applied, it would take until $t_2$ that the same quantity of the good would be provided as before the fire. On the contrary, if post-fire management measures would be applied the recovery period would have been reduced to $t_2'$. This means that in this example, without post-fire management we would lose the total amount of the good that equals the sum of the light and dark gray areas; while in the case of post-fire management the loss would be equal to the dark gray area.

To summarize, how the post-fire management influences the recovery process depends on the type and timing of the implemented measures. Managers have in most cases various alternatives as to how to treat a forest after a fire event. Decision on which of them is going to be implemented, depends on the management objectives that should ideally be based upon the ecological, economic and social criteria. From an economic perspective the post-fire management alternatives are evaluated according to the (i) economic efficiency and (ii) distributional effects.

### 3.2.1   Economic Efficiency of Post-Fire Management

In economics, efficiency is the most often used criterion for evaluating whether a situation or change could be improved in social welfare terms. In this respect the so-called Pareto criterion (Arrow and Hahn 1971; Mas-Colell et al. 1995), considers a new situation as preferable, or a change in the allocation of resources efficient, if someone's welfare could be improved without diminishing the welfare of anyone else. A somehow more relaxed criterion is the Hicks-Kaldor criterion (Hicks 1939; Kaldor 1939), which assumes that a new situation would be socially preferable (more "efficient") if the welfare of some people can be improved, and the winners could potentially compensate those who would lose some welfare, and still be better off. This chapter adopts the label efficiency in terms of the latter meaning, which corresponds to the one applied in cost-benefit analysis.

When addressing post-fire management alternatives, the economic efficiency[1] measures their total contribution to society's welfare. As the most efficient, we would consider the alternative that has a bigger positive contribution to the social welfare.

The methodology which is commonly applied to evaluate the contribution of a certain management option is the Cost Benefit Analysis (CBA).[2] CBA is a technique

---

[1] This differs from the term cost efficiency, which measures how well the inputs in a production process are used to produce a fixed set of outputs.

[2] For an overview of alternative and complementary evaluation methods see for example EFTEC (2006).

for the assessment of the relative desirability of competing alternatives (e.g., alternative post-fire management measures). In most cases the assessment involves the comparison of the current (base case) situation to one or more alternatives. For example, if one is interested in evaluating the social welfare impact of regeneration as a post-fire measure for a particular forest, the base case (without regeneration) would be compared to the alternative scenario (with regeneration). The analysis would focus on the differences in costs and benefits, in the situations with and without regeneration. The CBA compares the costs and benefits measured in monetary terms.

Benefits (the positive impacts) refer to the increases in the quantity or quality of goods or services that generate positive utility[3] or a reduction in the price at which they are supplied. The costs (negative impacts) stand for any decreases in the quality or quantity of such goods or services, or increases in their price. The costs also include the usage of resources (e.g., costs of the implementation of post-fire management measures) in the alternative, since they cannot be simultaneously used in any other way.

It is important to acknowledge, that in contrast to other evaluation methods where the costs and benefits can be expressed in different units, in a CBA all the benefits and costs are valued in monetary units. The resulting net benefits from the alternatives will reflect the summation of the changes in the net income of the society as a whole from undertaking an alternative compared with the decision of not undertaking it or undertaking a different one.

An important concept of the CBA is additionality, which refers to the net impacts of the project (Hanley and Spash 1993). This means that the costs of the project that are relevant for the assessment are those that would be incurred if an alternative is undertaken, but that would not be incurred otherwise. Similarly, the benefit of an alternative is the extra amount of a good (e.g. money, time, etc.) that would be gained if the project were undertaken rather than not undertaken (Sugden and Williams 1978).

Another central feature of the CBA is that it can compare benefits and costs that appear at different stages of a project (management alternative), by converting them into a common metric, their present value. This process is called discounting and it is based on the fact that the individuals have time preferences between consumption in different periods. The rate at which an individual is willing to exchange the present consumption for the future consumption is called the discount rate. The higher is the discount rate, the greater preference is given to the present consumption.

The CBA can be conducted from a private or a social perspective. A private CBA considers only the costs and benefits of the analyzed change which are imposed onto or accrue to a private agent (e.g. individual or firm). Thus, it considers only (market) costs and benefits, which are transmitted through prices and would affect agents directly involved (consumers and providers) in the implementation of the considered alternative. This approach is also often called financial appraisal.

---

[3] In economics, utility is a measure of relative satisfaction.

On the contrary, a social CBA attempts to assess the overall impact of a project on the welfare of the society as a whole. Thus, it also considers external (market and non-market) costs and benefits, which are mostly not transmitted through market transactions and are affecting parts of the society not directly involved in the implementation of the alternative (externals). Thus, a social CBA differs from the private in terms of both (i) the extent of the identification and evaluation of inputs and outputs, and (ii) the measure of costs and benefits.

Forest fires in most cases affect the welfare of wide range of stakeholders (private and externals). For example, they can affect the benefits forest owners obtain from a forest (e.g., timber), but also the benefits for the rest of the society, like ecosystem services (e.g., recreation, carbon sequestration, biodiversity) that are mostly not traded in conventional markets, but nevertheless they contribute to the societies welfare. Hence, for the evaluation of efficiency of post-fire management alternatives a social CBA should be applied; where all the cost and benefits of the analyzed alternatives are considered.

For the implementation of a social CBA of post-fire management measures we would need at least the following information: (i) type and quantity of lost goods and services as a consequence of the fire, (ii) the value of all the lost goods and services, (iii) the recovery pattern and period for each of the affected goods and services, (iv) the impact of post-fire management alternatives on the recovery pattern and period for each of the affected goods and services, and (v) the costs for each post-fire management alternative. The subsequent sections focus on those issues. Regarding the general application of a CBA we suggest the reader consults some specialized literature (e.g. Hanley and Spash 1993).

### 3.2.1.1 Fire Impacts on Goods and Services

Forest fires impact a wide range of goods and services. An important concern in estimation of fire related impacts is the proper consideration of environmental (forest) goods and services that might have been damaged (or enhanced) as a consequence of a wildfire. In estimating the economic impacts of fires, only a few of them have traditionally been considered, mainly the decreased quantity/quality of timber.

However, in the last decades it has become obvious that forests are also important as providers of environmental and social goods and services (Farrell et al. 2000). For example, a recent review of economic assessments of fire damages (JRC 2009) found that the major part of the total damage can be ascribed to the loss of different environmental goods and services, like biodiversity, recreation, and soil protection (Fig. 3.3). This emphasises the importance of including a broad range of goods and services when estimating the fire induced damages, and in particular the consideration of both market and non-market goods and services.

Thus, nowadays, estimating the impacts of fires should include the following categories: (i) environmental goods and services that represent the benefits humans derive directly or indirectly from forest ecosystems functions (e.g. wood, biodiversity,

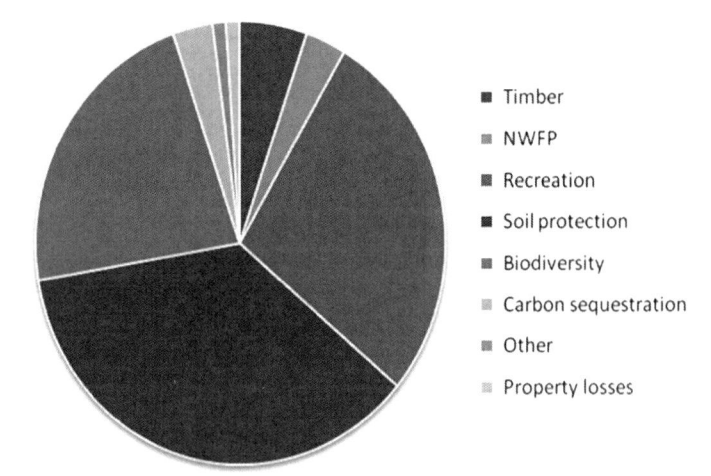

**Fig. 3.3** Average shares of the damages to different goods and services in fire damage studies

recreation); (ii) property, which includes all property directly impacted in a forest fire, such as structures (e.g. houses, cottages, buildings, and infrastructure), personal property (e.g. cars, furnishing, cloths), and production capacities (e.g. agricultural and other machines, transportation means); and (iii) direct health impacts, because forest fires often produce significant health impacts that can result in morbidity or mortality (e.g. fire fighters or residents), and can be suffered locally or on a broader scale (e.g. smoke and air pollution). Although we mostly speak about damages that are provoked by forest fires, it should be noted that the effects of fires can be also beneficial for the provision of certain goods (e.g., fodder production).

In the next section we present an overview of the impacts that forest fires can have on goods and services that were identified to be affected in previous studies.

Timber

The value of standing timber is one of the main categories when estimating the impacts provoked by wildfires (González-Cabán 1998). For example, Abt et al. (2008) reported that in the case of four studied wildfires timber losses were around 20% of all costs and losses. However, the magnitude of timber losses can differ significantly from case to case (JRC 2009). Stand characteristics (e.g. tree species, composition, stand structure) (González et al. 2007; Velez 2000) play a determinant role in determining fire damages and therefore the losses owed to the fire.

Wildfires can have short-and long-run effects on timber and timber markets. Short-run effects (1–2 years) include the immediate destruction of valuable standing timber, and in particular after bigger events, economic disequilibrium associated with the flooding of markets with salvaged timber. The big amount of salvaged timber drives prices downward temporarily, which affects owners of the killed timber, owners of undamaged timber, and timber consumers. Long-run effects on timber markets can arise from the loss of a large portion of standing volume, a loss that tends to drive prices upward for extended periods and produce an extra income for owners of undamaged timber. However, fires also can create conditions favourable for a reduction in timber demand. Therefore, large-scale catastrophes often redistribute wealth among producers and consumers and cause a net economic loss (Prestemon and Holmes 2004).

Hence, when calculating timber damages various aspects must be considered. On the short term, at least quantity and value of the lost timber should be taken into account, where the value is based on market prices. On the long term the quantity and the value of future lost timber production (because of decreased increment) should be considered (Zybach et al. 2009).

## Non-Wood Forest Products

Non-wood forest products (NWFP) consist of goods of biological origin other than wood, derived from forests, other wooded land and trees outside forests (FAO 1999). Forests provide a great diversity of NWFP like cork, edible mushrooms, pine nuts, acorns, resins, medicinal plants, among others. According to the Millennium Environmental Assessment (Hassan et al. 2005) *"At least 150 NWFPs are of major significance in international trade."* Especially in the Mediterranean area non-timber forest goods can be a major income resource (Merlo and Croitoru 2005).

The high diversity of NWFP is also reflected in the different effects that forest fires have on the quantity and/or quality of them. For example, fires can improve the understory forage and grazing (Wade and Lunsford 1988), or it can decrease the availability of fruits, like pine nuts and acorns (Molina et al. 2009). Therefore, in general we cannot assume that all the NWFP production is always damaged.

Unlike the case of timber, the production of particular NWFPs is site specific, and thus, it cannot be generalised to all fire affected areas. To estimate the loss of NWFP, data about the size of the affected area, production capacity, market price and harvesting costs for each NWFP are needed.

## Recreation

Recreation refers to organised or free activities that contribute to human health and well-being. These services include walking, hunting, mountain biking, etc. Because forest fires affect the biotic and abiotic characteristics of forests, subsequently they also affect the potential of these forests to provide non-market benefits, like recreation (González-Cabán 1998). There is still no consensus as to how recreation use changes immediately after a forest fire or during the restoration period (Loomis et al. 1999).

The impact of fires on forest recreation demand has been evaluated from two perspectives (Englin et al. 2008). The first approach estimates the losses in the recreation-tourism related economic sectors of local economies (e.g. food and lodging), that are impacted upon during and after the fire (Barrio et al. 2007; Butry et al. 2001).

The second approach is directed towards the estimation of impacts of wildfires on recreation demand and the value of recreational sites (Loomis et al. 2001).

Similar to NWFPs, forest recreation is limited in its extent. To estimate the economic impacts of forest fires on recreation, the relevant area for recreation activities in the affected forest must be estimated. Furthermore, site-specific data on the number of forest visits and value per visit are needed to accurately estimate recreation losses. To estimate the later, in most cases non-market valuation methods[4] are applied (Loomis et al. 2001).

## Soil Erosion

Soil protection is mainly based on the structural aspects of forests. The vegetation root system and cover play an important role in soil retention and formation. Soil retention is assured by the root system, which stabilises the soil, and foliage, which intercepts rainfall, preventing soil compaction and erosion. On the other hand, soil formation is also supported by the root system that disintegrates the rocky material, while the vegetation cover plays an important role in the fertilisation processes.

The impacts of fire on soils can manifest themselves in significant changes in soil physical, chemical, or biological properties. These include breakdown in soil structure, reduced moisture retention and capacity, development of water repellence, changes in nutrient pools cycling rates, atmospheric losses of elements, offsite erosion losses, combustion of the forest floor, reduction or loss of soil organic matter, alterations or loss of microbial species and population dynamics, reduction or loss of invertebrates, and partial elimination (through decomposition) of plant roots (Beyers et al. 2005). The same authors also claim that impacts of fire on soil are a function of fire severity.

To assess the damages related to soil erosion, we need to estimate the affected area, the level (a function of fire intensity/severity) and quantity of lost soil,[5] and the value of lost soil. To value the damages of soil erosion different approaches can be taken. For example, the study by Görlach et al. (2004) estimates the total costs of soil degradation, considering both private and social costs and also on-site and off-site impacts of soil degradation. A different approach was applied by Mavsar and Riera (2007) estimating the social value of reduced soil erosion in Spain.

## Carbon Emissions

Forests have the capacity to directly influence the composition and the chemical balance of the atmosphere (e.g. $CO_2/O_2$ balance, maintenance of $O_3$ level and $SO_X$ level).

---

[4] See also Sect. 3.2.1.2.

[5] For an example see http://eusoils.jrc.ec.europa.eu/library/themes/erosion/ClimChalp/Rusle.html

This service includes clean air provision and prevention of diseases. One of the specific services, related to gas regulation, is *carbon sequestration*. By capturing and storing the excessive carbon dioxide from the atmosphere, forests contribute to the mitigation of global warming. In the last decade, this service attracted significant attention of policy makers.

In this respect fires have two main impacts. On one hand, air pollution effects of fires include emissions of particulate, noxious gases and $CO_2$, visibility impacts to road and air transportation, public health effects, property damage, and compromised recreation opportunities (Zybach et al. 2009). In light of climate change and its consequences, one of the fire effects drawing most attention is $CO_2$ emission and their economic impacts. Fires account for approximately one-fifth of the total global emissions of carbon dioxide (Sandberg et al. 2002). However, in the European context, annual emissions are only about 0.22% of the total greenhouse gas emissions in the European Union (EEA 2009). On the other hand, the decreased biomass production also influences forests present and future capacity to sequester carbon.

Therefore, to assess the damages from increased carbon emissions from fires, data on the affected area, growing stock and carbon value are needed. For the estimation of the losses related to decreased carbon sequestration capacity, the lost biomass production (e.g., net annual increment) and the value of carbon should be included. In both cases, the value of carbon can be estimated by using the voluntary market prices for carbon credits (e.g. www.endscarbonoffsets.com, http://www. ecosystemmarketplace.com).

## Biodiversity

Biodiversity is an essential factor in sustaining the functioning of the ecosystem and hence underpinning for many other forest goods and services. It generally encompasses different levels, like genetic diversity or differences of genes among populations/individuals of the same species (e.g. varieties of crops), species diversity refers to the variety of plants, animals and micro-organisms in an ecosystem and ecosystem diversity refers to the variety of different ecosystems. According to Mayer (1995) forests are the most important terrestrial ecosystems for conservation and protection of biodiversity.

The impacts of fires on biodiversity can vary significantly and are in general dependent on fire type, severity, intensity, size and frequency, and species characteristics (Brown and Smith 2000; Smith 2000). To estimate the economic impact of losses associated with biodiversity we have to consider the extent of the biodiversity loss (e.g., number of damaged specimens, species or ecosystem types) and the economic value of the respective biodiversity element (e.g., value of specimen). The later is in most cases estimated by applying non-market valuation methods.[6]

---

[6] For more details about non-market valuation methods see Sect. 3.2.1.2.

## Health

Forest fires can have significant impacts on public health (Zybach et al. 2009). These impacts can be the result of direct exposure to the fire or due to increased air pollution (Rittmaster et al. 2006; Zybach et al. 2009) and can occur on- and off-site.

Fires produce a number of air pollutants; some like concentrations of particulate matters (PM) can increase substantially during wildfire episodes and can have significant mortality and morbidity impacts (Butry et al. 2001; US.EPA 2004). However, it is difficult to estimate the extent of these impacts. Not all wildfire health impact studies find statistically significant impacts of wildfire induced PM contamination (Kochi et al. 2008).

The health outcomes that the fire impacts can cause are mortality, restricted activity days (including work days lost), hospital admissions, respiratory symptoms and self-treatment (Kochi et al. 2008). To estimate the health related economic impacts of fires, we need to estimate the changes in number of mortality or morbidity cases that were caused by the fire and the monetary value of these changes. The assessment of the total change of mortality/morbidity is less complicated when considering the direct exposure to fire, but is challenging in the context of increased air pollution (Kochi et al. 2008).

Concentration response models provide information on changes in health risks associated with changes in pollutant concentrations. Economic methods are used to translate these changes in risks into economic (monetary) values. These economic values are calculated in various ways. The value of health effects include the costs of illness as reflected in hospital costs and lost wages as well as the individual's willingness-to-pay to avoid the change in risk, over and above the costs of illness (Rittmaster et al. 2006). The monetary value of preventing mortality is measured as aggregated society's willingness-to-pay to save one anonymous person's life (e.g. Navrud 2001; Dickie and Messman 2004; Kochi et al. 2008).

## Property

In addition to the loss of environmental goods and services, forest fires also cause property losses (Riera et al. 2008). These might comprise private as well as public assets of different types, including homes, rural buildings and structures, businesses, infrastructure, and other goods (Butry et al. 2001; Ciancio et al. 2007; Riera et al. 2008). The value of such losses is estimated mainly through market prices. Nevertheless, the estimation of these losses is often difficult to determine, because of missing data or restricted data access (Butry et al. 2001; Kent et al. 2003).

Forest fires may also have lagged effects, like reducing property values in adjacent residential areas or other areas with similar characteristics (Loomis 2004). To our knowledge, there are only a few studies on this subject (Loomis 2004; Napoleone and Jappiot 2008; Mueller et al. 2007) and no general conclusions can be made. The paucity of studies precludes making general conclusions, however the scant evidence points to a statistical significant reduction in housing price following a wildfire.

### 3.2.1.2    Valuation of Goods and Services

Once all the relevant goods and services that have been damaged are identified and quantified (in physical terms), their social economic value and the total impact on the social welfare need to be estimated.

One of the challenges related to fire damage assessment is the existence of non-market goods and services. When considering forest goods and services, we can observe that only a part them, such as timber, are traded in markets and their value can be directly observed (market prices). On the contrary, other goods and services (e.g., biodiversity, recreation, soil protection) are provided to the community or to individual consumers either free of charge or at a symbolic fee which is well below production costs (OECD 2000). Nevertheless, these goods and services positively contribute to people's welfare and often represent a significant part of the fire induced losses. Thus, they cannot be omitted from the assessment of the economic impact of forest fires.

To clarify the difference between market and non-market goods and services, it has to be understood that in a free market economy goods and services are sold for prices that reflect a balance between the costs of production and what people are willing to pay, and can be considered as a proxy for the value of these goods and services. In turn, non-market good and service are neither bought nor sold directly, and do not have an observable monetary value. Hence, to estimate their value we need to apply other methods, which have been developed in the frame of economic valuation.

Economic valuation of non-market goods and services relies on the notion of *willingness to pay (WTP)*. Willingness to pay for a particular good is defined as the maximum amount of other goods (e.g. money) an individual is willing to give up in order to having that good. WTP is determined by motivations which can vary considerably, ranging from personal interest, altruism, concern for future generations, environmental stewardship, etc. The economic value of the good to an individual is reflected in the willingness to pay of the individual for that good.

Hence, economic valuation methods provide us with approaches that attempt to elicit the monetary value that a certain change in the quantity and/or quality of non-market goods and services has for a society. These methods can be divided into two main groups: Revealed Preference[7] (RP) and Stated Preference[8] (SP) methods. These are based on the fundamental principles of welfare economics; whereby the changes in the well-being of individuals are reflected in their willingness to pay or willingness to accept compensation for changes in the provision or use of those goods and services (Hanley et al. 2001).

The RP methods (e.g., Market Prices, Hedonic Pricing, Travel Cost Method) are based on actual observed behaviour data, including some techniques that deduce

---

[7] More on revealed preference methods in Bockstael and McConnell (2007).

[8] More on stated preference methods in Kanninen, B. J. (Ed.) (2007).

values indirectly from behaviour in surrogate markets, which are assumed to have a direct relationship with the ecosystem service of interest. The SP methods (e.g., Contingent Valuation Method, Choice Modelling) are based on stated choices rather than people's behaviour on real markets; for the former the value is inferred from people's responses to questions describing hypothetical markets where they state their willingness to pay for an environmental good or service.

Related to fire damage assessment, the reconstruction cost approach, and the almost equivalent replacement and reproduction costs approaches, are the most common and robust methodologies. The methodology is based on the idea that the cost of replacing the goods and services provided by an environmental resource can offer an estimate of the value for that resource. This is based on the supposition that, if people incur costs to restore the services of ecosystems, then those services must be worth at least what people paid to replace them. The main underlying assumptions for this approach refer to the predictability of the extent and nature of the physical expected damage (e.g., there is an accurate damage function available). Besides, the costs to replace or restore damaged assets can be estimated within a reasonable degree of accuracy. It is assumed that the replacement or restoration costs do not exceed the economic value of the asset. The latter assumption, however, may not be valid in all cases. The value of the service may fall short of the replacement of restoration costs; either because there are few users or because their use of the service is in low-value activities. On the other hand, this approach may lead to an underestimation of the total economic value of a natural resource, because it considers only financial costs, but it does not account for non-market and non-use values of the valued asset.

The choice of a valuation method depends on the nature and the type of change in the goods or service we want to value. In general, measures based on observed behaviour are preferred to those based on hypothetical behaviour. However, the choice of a valuation technique in any given instance is dictated by the objectives and characteristics of the case and data availability

Although in theory it is recognized that non-market goods and services represent an important part of the fire induced losses, in practice they are very seldom included in the damage assessment approaches, as it will be shown in the next section.

## Current Status of Fire Damage Valuation in Europe: Differences Among Countries and Regions

This section is based on the information collected in a survey conducted in twelve European countries in the frame of the COST Action FP0701. The aim of the survey was to collect information about the methodologies applied for the estimation of economic damages caused by forest fires. Hence, the questionnaire contained questions about the existence of a methodology for the estimation of economic damages and its scope (e.g. national, regional).

The methodology was described in terms of goods and services that are considered when estimating fire induced damages and the methods (e.g., market price, economic valuation, replacement costs, benefit transfer, and expert estimation) that are used for estimating the economic values of the damages. Furthermore, questions about estimation time frame (i.e. short or long) and application situations (e.g. compulsory to all fires, optional) were included.

Table 3.1 summarizes the results of the questionnaire, and shows that only a small number of the potentially affected goods and services are valued.

Wood and soil protection are the most frequently valued goods in the surveyed countries. Wood losses are mainly valued through market prices, while soil protection losses are estimated through expert consultation and replacement costs. Infrastructures, homes and other properties are also frequently valued. Probably this is related with the need for replacing those damaged assets and obtaining funds from insurance companies. Furthermore, in the damage assessment many countries also include suppression costs related to fire events. On the contrary, biodiversity and carbon sequestration are included into the damage assessment only in Italy. While biodiversity is valued through replacement cost calculations, carbon sequestration related losses are estimated by market prices (e.g., carbon credit prices). Also, fodder and forage losses are considered only in one country (Spain), where the value of the losses is based on expert estimations.

According to the results of the survey, Italy appears to be the country performing the most complete assessment (e.g., see the methodology proposed by Ciancio et al. 2007). Most surveyed countries have developed damage assessment protocols that are compulsory for all fires. Those protocols are usually applied by forest management departments. However, there are also exceptions, like in the French case, where insurance companies are responsible for the application.

On the other hand, some countries (e.g., Switzerland and Greece) lack standard assessment procedures. In Switzerland, estimations are conducted only where the cause for fire ignition is known; while in Greece after important fire events ad-hoc expert committees are established to estimate the damages, but no systematic protocol is implemented despite the high frequency of fires.

Most countries conduct short term damage evaluations, generally immediately after the fire event, while no assessment of fire impact is done in the long/medium term.

To summarize, from the inputs obtained from the surveyed countries we can conclude that there is room for significant improvement in the assessment of fire damage across Europe. A more complete valuation would consider items that nowadays are rarely tackled (e.g. water, recreation, health impacts) and would take into account medium term effects of fires. These would be desirable goals towards a more complete valuation of fire damages.

### 3.2.1.3 Recovery Patterns and Periods of Goods and Services

The reduced flow of goods and services after a fire can be permanent or temporary. In most cases the recovery pattern and time needed to restore the provision (flow) of

**Table 3.1** European countries surveyed and assets valued by them in forest fire damage assessment

| Countries/assets valued | Bulgaria | Cyprus | Greece | France | Italy | Lithuania | Portland | Portugal | Romania | Spain | Switzerland | Turkey |
|---|---|---|---|---|---|---|---|---|---|---|---|---|
| *Forest goods and services* | | | | | | | | | | | | |
| Industrial wood | x | x | | x | x | x | | | x | x | x | x |
| Fuel wood | x | x | | x | x | | | | x | x | x | x |
| Cork | | | | x | | | | | | x | | |
| Food | | | | x | | | | | | x | | |
| Fodder and forage | | | | | | | | | | x | | |
| Decorative material | | | | | | | | | | | | |
| Hunting and game products | | | | x | | | | | | | | |
| Pharmaceuticals | | | | | | | | | | | | |
| Biodiversity protection | | | | x | | | | | | | | |
| Climate regulation | | | | | | | | | | | | |
| Air quality regulation | | | | | | | | | | | | |
| Carbon sequestration | | | | x | | | | | | | | |
| Health protection | | | | | | | | | | | | |
| Water regulation | | | | x | | | | | | | | |
| Water purification | | | | | | | | | | | | |
| Soil protection | | | | x | x | | | | x | x | x | |
| Recreation | | | | | x | x | | | | | | |
| Sports | | | | | x | | | | | | | |
| Tourism | | | | | x | | | | | | | |
| Spiritual and cultural services | | | | | | | | | | | | |
| Historical and educational services | | | | | | | | | | | | |
| Aesthetic services | | | | | | | | | | x | | |
| *Other goods and services* | | | | | | | | | | | | |
| Infrastructure | x | | | x | x | x | x | | x | | | x |
| Homes | | | | x | x | x | x | | x | | | x |

(continued)

**Table 3.1** (continued)

| Countries/assets valued | Bulgaria | Cyprus | Greece | France | Italy | Lithuania | Portland | Portugal | Romania | Spain | Switzerland | Turkey |
|---|---|---|---|---|---|---|---|---|---|---|---|---|
| Other property | x | | | x | x | x | x | | x | | | x |
| Agriculture and similar products | | | | | x | x | x | | x | | | x |
| Losses to business | | | | | | x | | | | | | x |
| Health impacts | x | | | | | | | | | | | x |
| *Suppression costs* | | | | | | | | | | | | |
| Material costs | x | x | | | x | x | | | | x | x | x |
| Personnel costs | x | x | | | x | x | | | | x | x | x |

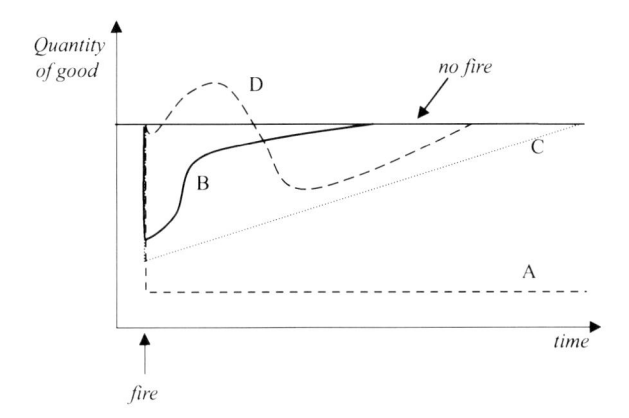

**Fig. 3.4** Valuation of damage costs with an approach based on the missed flow of products and services

goods and services will be case specific, and dependent on different factors, like fire size and severity (Mercer et al. 2007), type of damaged goods, and management measures taken.

Very little is known about the post-fire recovery patterns, because of their heterogeneity. In fact, recovery patterns may have different forms for different goods and services (Fig. 3.4):

- some goods and services may never be recovered (e.g. species going extinct or a monumental tree destroyed by the forest fire – Curve A);
- some goods and services may be recovered very quickly (e.g. water absorption and erosion control in a forest area with no problems of revegetation – Curve B);
- other may recover only in the long term, like recreation in a formally old-growth forest (Curve C);
- some goods and services may even increase after the fire events (Curve D) with respect to the "without scenario" (e.g. herbs production).

Moreover, the function representing the recovery path may be linear (C) or may, more probably, take some other non-linear form (B and D). We should also consider that in most cases various goods will be damaged simultaneously and that their recovery processes could be interrelated.

The estimation of the recovery period and the impact of post-fire management measures on the recovery of the flow of forest goods and services represent another challenge in the process of economic evaluation of post-fire management measures. Although theoretical approaches recognize the role of time in calculating the loss of different forest goods and services, in practice such aspect is not always considered. This is because in most cases, specific studies about post-fire forest recovery dynamics do not exist for the site or forest ecosystem affected by the fire. Furthermore, forest functions are so interconnected that it is difficult to define specific time span for each of them, and some fire damages (soil erosion, debris flow, water pollution, local tax revenues, etc.) do not reach the maximum intensity immediately after the fire, but a certain time later.

Nevertheless, in the literature some suggestions can be found about potential recovery periods for different forest goods and services, even though they should not be generalized and will most likely differ significantly from case to case. For example, Molina et al. (2009) report that in their cases NWFP yields were affected on average over a 4 year period, although the production decrease was not constant. There are a number of studies dealing with the impacts of fire on recreation; however, they differ considerably in their conclusions on how recreation use changes immediately after a forest fire or during the restoration period (Loomis et al. 1999). For example, Flowers et al. (1985) reported that the effects of fire on recreation (number of visitors) can persist from 6 months to 7 years. On the other hand, Loomis et al. (2001) estimated that the effect of forest fires (depending on the type of fire, crown or surface fire) on hiking trips could persist over a very long period (up to 50 years). Another complex issue is the effect of fire on ecosystem biodiversity and the number of years needed to recover to pre-fire levels can vary significantly. Based on experience from two fires in Andalusia, Molina et al. (2009) reported that after a fire, the habitat conditions are re-established somewhere between 2 and 12 years, depending on the fire severity and vegetation composition.

### 3.2.1.4 Impacts of Post-Fire Management on the Flow of Goods and Services

Little is known about the recovery processes of the flow of goods and services, but there is even less information about the impacts of post-fire management measures on this process.

In general we assume that post-fire management measures should accelerate the recovery process and/or reduce potential further losses of goods and services (Fig. 3.1). For example, initially the reduced vegetation cover after a fire does not provoke increased soil erosion, which normally appears after the first heavier precipitation. Thus post-fire measures (e.g., soil stabilization measures) could prevent the loss of soils. This example is also illustrated in Fig. 3.1, where we see that by applying post-fire management measures (line C) the loss of the good is considerably smaller ($Q_1'$) and the recovery process is shorter ($t_2'$), compared to the situation where no such measures were applied (line B).

Thus, to assess the impact on the social welfare of post-fire management we have to know its impact on the flow of goods and services (market and non-market), and the economic value of these changes.

However, as Vega et al. (2008) points out, there is still a lack of knowledge, regarding post-fire management and restoration measures. For example, whether and when burned trees should be removed or not, which logging techniques to be used, the type of management of logging residues and the effects of interaction between fire severity and harvesting. Nevertheless, there are some studies that deal with effects of different post-fire management measures. A few examples are given below.

González-Ochoa et al. (2004) examined the effects of different post-fire silvicultural treatments on tree growth in Aleppo pine (*Pinus halepensis*) forests. They found that tree growth is improved by applying thinning on a good quality site, and thinning and scrubbing on worse quality sites. Furthermore, thinning and scrubbing improved the probability of cone production, which will guarantee the regeneration capacity of the forest. Izhaki and Adar (1997) conducted a study on Mt. Carmel, in Israel and found that different post-fire management measures increased the number of bird species. Furthermore, Spanos et al. (2007) examined the effects of fire and different logging (animals and mechanized) procedures on soil characteristics and loss. They found that fire and logging did not affect the soil pH and caused only a short-term reduction of organic matter content. Logging and particularly the use of skidders for log removal caused an initial increase in the amount of exposed bare ground but later when vegetation cover increased differences were minimized. Two years after the fire, the highest rates of soil loss were observed in the logged area where mules were used for log removal. Soil moisture showed some differences between treatments during the first year after fire but then values were similar. The main woody species showed a species specific response to the treatments and while seeder species were favored in the unlogged sites the same was not true for the re-sprouters. In general, the growth and survival of pine seedlings was not affected by treatments.

González-Ochoa and de las Heras 2002 studied pest outbreaks in restored Aleppo pine (*Pinus halepensis*) stands and the influence of post-fire practices. Due to high densities reached by this species after fire, thinning is necessary. However, they found silvicultural practices could be a trigger for an outbreak of defoliator species (e.g., *Polyporus squamosus*).

When addressing post-fire practices, erosion and fire risk should be considered simultaneously. Marques and Mora (1998) compare soil erosion of clear-cutting and immediate removal of burned trees versus the non-interventions. Although they did not find evidence that clear-cutting increases post-fire erosion, they observed that roads constructed for logging activities accounted for an important part of the erosion in the area. On the other hand, they highlight that left trees could increment the fire risk.

Finally, Vega et al. (2008) evaluated regeneration of Maritime pine (*Pinus pinaster*) stands in Galicia under different post-fire management options. Although harvesting and slash logging resulted in significantly higher seedling mortality, they did not influence the restoration potential of the stands. At the same time they remarked that excessive seedling density may result in a significant increase in the cost of pre-commercial thinning in these stands.

### 3.2.1.5 Costs of Post-Fire Management

Finally, to complete the information needed for the evaluation of efficiency of post-fire management alternatives, we have to know the costs of the implementation for each of them. Here we have to consider all the resources used (e.g., labor, raw materials, equipment) and their costs. It should be acknowledged that we should consider all the costs that appear over the whole restoration period and not only the costs immediately after the fire.

The post-fire management costs should in most cases be relatively easy to predict, because in most countries standardized costs and norms for restoration measures are available. A detailed example of post-fire management measures and their costs for Catalonia is given in Mavsar et al. (2009).

### 3.2.1.6 Efficiency Evaluation of Post-Fire Management

Once all the necessary information is collected we estimate for each post-fire management alternative the net impact on social welfare. For each alternative (including the no management option) we estimate the social net present value. This is done by discounting all costs and benefits to the initial year (e.g., immediately after the fire), and subtracting costs from benefits. After estimating this for all alternatives we evaluate which of them would have the highest positive impact on the social welfare[9] and would thus be the most efficient from the social perspective.

## 3.2.2 Distributional Effects of Post-Fire Management

The economic efficiency criteria presented in the previous section does not consider distributional effects (equity) of post-fire management investments. From a social perspective it does make a difference who profits from the generated benefits or the avoided costs of a post-fire investment (Riera et al. 2005). This is in particular important, because there is evidence that natural disasters, such as forest fires, impact to a greater extent on the poorer parts of the population than the wealthier (Holmes et al. 2007).

Therefore, when evaluating post-fire management measures also the distributional effects should be considered. In this line different aspects of post-fire management could be evaluated. For example, whether new jobs are created and who benefits from this, how the costs and benefits of the investments are distributed in the society, whether the investment helps to alleviate poverty or decrease income distribution in the affected area, whether it affects established land use patterns or land tenure agreements, etc.

There is a wide range of economic effects of a post-fire management investment that can be evaluated. However, which of them will actually be considered in the evaluation process mainly depends on the characteristics of the investment (e.g., size, type, economic and social effects) and the socio-economic characteristics

---

[9] Actually a number of indicators exist (e.g., benefit-cost ratio, internal rate of return) that are complementary to the net present value and that would be considered when conducting a CBA.

of the affected area. This part of the economic evaluation of post-fire management is closely related to the social aspects that will be described in more detail in Sect. 3.3.2. This only underlines the importance of a balanced consideration of different aspects (ecological, economic and social) when taking decisions about post-fire management.

## 3.3   Legal and Social Aspects of Post-Fire Management

The intensity and scale of fire-related problems is highly variable in Europe; thus, it is not surprising that social and legal frameworks on wildfires are influenced by the perceived level of threat (Montiel and Herrero 2010). In the light of this, the main goal of this section is to highlight and examine such frameworks with distinctive reference to post-fire management and related decisions.

### 3.3.1   Legal Issues

A review is presented based on a sample of 18 countries from different regions of Europe and of the Mediterranean basin, that include (one or more) ad hoc post-fire management provisions in national legal frameworks. Information has been collected through specific questionnaires in the framework of the FP0701 Cost Action (see the list of responding countries on Table 3.2) and by reviewing the sparse literature on the subject (FAO, 2007; Rosenbaum, 2007).

Key legislation, programs and regulations encompassing post-fire management are compared among selected countries, with the main goal to get a sense of technical issues currently addressed by national legislation.

Post-fire management issues are considered within laws dealing specifically with forest fires, or by more comprehensive texts addressing forest fire management together with other forest management aspects. In some countries the legal framework on forest fires can be very fragmented and scattered across laws and regulations regarding forests, protected areas, hunting, local governance, civil defense, environmental protection, agricultural land protection, land use and internal affairs (Morgera and Cirelli 2009). Legal provisions regulating post-fire management in fire-affected areas are generally not different between public and private land. Obligations and responsibilities for the implementation of legal provisions are usually allocated to land owners, most frequently private (Fig. 3.5).

Key post-fire management issues considered by the national legal frameworks are (Table 3.2):

1. the interdiction of land use conversion in forest burned areas;
2. the post-fire harvesting and measures to prevent soil erosion;
3. the post-fire forest regeneration: both active forest restoration (planting and seeding) and passive restoration (management of natural regeneration) approaches are foreseen by the law.

**Table 3.2** Key legal provisions applied to forest burned areas in selected European countries

| Country | Interdiction of land use conversion | Post-fire harvesting (salvage logging or management of burned trees/wood) | Soil erosion mitigation and flood prevention | Forest regeneration by planting or seeding | Management of natural regeneration | Grazing/browsing control |
|---|---|---|---|---|---|---|
| Bulgaria | x | | | x | | x |
| Cyprus | | | | x | | |
| Estonia | | | | x | | |
| France | x | x | x | x | | |
| Greece | x | x | x | x | | x |
| Israel | | | | | | x |
| Italy | x | x | | x | | x |
| Latvia | | | | x | x | x |
| Lithuania | x | x | | x | x | x |
| Morocco | x | | x | | | x |
| Poland | x | | | x | x | x |
| Portugal | x[a] | x | x | x | | x |
| Romania | x | x | x | x | x | |
| Spain | x | x | x | | x | x |
| Slovenia | | | | x | | |
| Switzerland (Ticino) | x | | | | | |
| Tunisia | | | | | | x |
| Turkey | x | | | | | |

[a]For selected forest types (e.g. cork oak forest)

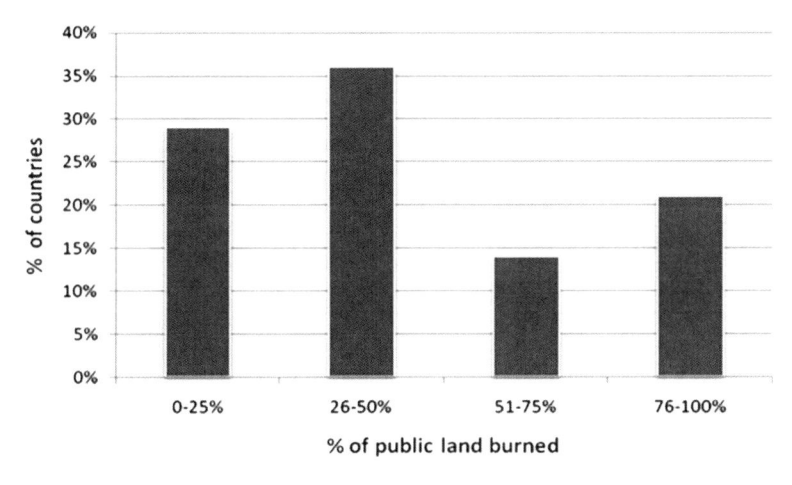

**Fig. 3.5** Split of countries considered in the review by rate of public land burned

### 3.3.1.1 Land Use Conversion

Land use conversion in forest burned areas is explicitly prohibited by law in most countries. There is a general consensus on the illegality of post-fire land use conversion, although with significant differences between countries e.g. in the length of time for non-conversion. The law generally prohibits: (i) the conversion of forest burned areas in urban land uses (e.g. housing, industrial or tourism); (ii) changes to terrain morphology or plant cover of the site for a certain period of time after fire occurrence.

Some examples of the legal provisions are:

- Bulgaria: the national Forest Law forbids the converting of burned forest area to any different land use for a period of 20 years (Bulgaria's Forest Law, Prom. SG. 125/29 Dec 1997, amend. SG 16/03, Art. 14);
- Lithuania: in special cases burned forest areas might be turned into open landscapes, e.g. in protected areas where aesthetical values are of high priority and further used for recreational purposes;
- Poland: the land use conversion of forest areas, whether burned or not, requires a special administrative decision;
- Italy: the national framework law on forest fires (no. 353/2000) forbids land use change in burned areas; in particular it is strictly forbidden: for 15 years after fire to change land-use, for 10 years to establish building or infrastructures, to hunt and graze, for 5 years to reforest using public funds. Land use changes are monitored by the Cadastre of burned areas. Despite this, illegal transformations occur. A similar situation is observed in Greece, despite the law which declares burned areas as afforested areas subject to strict protection measures (Table 3.2).

**Table 3.3** Example of measures to prevent grazing/browsing on forest burned areas

| Country | Fences | Repellent for wildlife control | Tree shelter | Hunting |
|---|---|---|---|---|
| Bulgaria | x | x | | |
| Italy | x | x | x | |
| Latvia | | | x | |
| Lithuania | | x | | |
| Poland | x | x | | |
| Portugal | x | | x | x |
| Romania | x | | | |
| Spain | | | x | |

### 3.3.1.2 Post-Fire Harvesting and Other Protection Measures

Other restrictions applied to forest burned areas mainly concern harvesting of damaged trees (salvage logging) and measures to protect soil and natural regeneration. Italian legislation foresees, as a general measure, early post harvesting, in general in the first months after fire. Post-fire harvesting criteria are defined by laws also in Bulgaria, Romania and Portugal. Provisions to mitigate post-fire soil erosion in forest burned areas and prevent flood risk are also found in some countries. Under undisturbed conditions forest natural regeneration is common after fire, especially in Mediterranean forests types where many plant species are adapted to fire disturbance. Grazing represents often the main disturbance for regeneration establishment: therefore, interdiction of grazing/browsing is a common measure in the examined legislation. Many national legislations apply grazing restrictions for 5 years after fire; several instruments can be implemented to prevent wild/domestic animals disturbance (Table 3.3). Legislation can also prevent the illegal use of fire to clear grazing areas in forests, by prohibiting grazing of livestock on any land that has been subject to wildfires for a period of 10–20 years (Bourrinet 1992).

### 3.3.1.3 Forest Restoration Measures

Another key issue addressed by the legal instruments is to ensure the forest restoration in fire-affected areas when natural processes are not expected to provide adequate regeneration. On this issue the law generally states rights and obligations and the basic objectives and principles to be implemented: e.g. the use of native species for forest restoration is required in most countries (France, Italy, Portugal, Romania, Spain, Switzerland, Turkey, Greece, Poland, Albania and Tunisia).

Forest restoration funding is generally supplied by specific funds allocated at national or regional level e.g. by the Rural Development Plans under the European common agricultural policy framework. Some countries report the lack of funding as one of the main constraints to effective implementation of forest restoration action.

In some cases, the legislation spells out specific obligations, e.g. the preparation of a rehabilitation plan after extensive fires or to ensure consistency between various rehabilitation operations including restoration of the uses of the area and prevention of new fires (Morgera and Cirelli 2009). Examples of legal provisions to ensure the forest restoration process are:

- Switzerland (Canton Ticino): in case of dangerous events, the Municipality, after the authorization of the State Council, can promptly decide the most appropriate measures for protection and rehabilitation; all actions and measures should be taken in the highest respect of the environment; in case of a lack in the involvement of the Municipality, the State Council officially intervenes;
- Slovenia: the Slovenian Forestry Institute, established to perform public forestry services in all forests, monitors the extent and the level of degradation and damage of forests and reports on this to the competent Ministry, which shall order measures for removing the causes of the degradation or damage to forests.

Legislation may also allocate responsibilities for rehabilitation when fire affected lands are included in protected areas. This is the case of Portugal's Decree-Law 139/88: if burned private forests are located in protected areas, the owner could be called upon to produce a replanting plan to be approved by the protected area authority and, for areas larger than 100 ha, an environmental impact study can be also required.

Legislation may also create specific tree-planting obligations to ensure rehabilitation of forest areas after a fire. In Portugal and Bulgaria, the owner of burned forests is obliged to replant trees within 2 years following a fire (Portugal's Decree-Law No. 139/88; Bulgaria's Forest Law, Prom. SG. 125/29 Dec 1997, amend. SG 16/03, Art.42) and within 5 years in Poland. In Spain regional laws foresee the private landowner reforesting forest burned areas when natural regeneration is not expected in the mid-term.

In Cyprus for the State forest there are no legal constraints regarding restoration; however, all State forest burned areas are usually reforested after fire. In the same way, also for the private forests there are no legal provisions regulating this matter. The final and determinative decision for the future of private forest burned areas is left to land owners of fire affected areas.

A different approach is taken in Italy by the national framework law on forest fires (Law no. 353/2000). Post-fire afforestation and any public-funded action of environmental engineering are forbidden in burned areas for 5 years after fire, unless specific authorization is released from the Regional Administrations or the Ministry of Environment. This is meant to prevent speculation, as investing too much in restoration activities might stimulate further fires in areas having problems of unemployment. Furthermore, the national legislation forces the Municipalities to establish the so-called "Cadastre of burned areas" with the support of the State Forest Service. The Cadastre of burned areas considers a burned area always a consequence of an illegal act and any intervention is forbidden for the following period of 5–15 years.

Technical specifications for the implementation of forest restoration measures are generally left to subsidiary legislation; possibly, subsidiary legislation could expressly prohibit the substitution of certain species with others for certain environmental purposes (Bourrinet 1992). Relevant examples are:

- Lithuania: burned areas have to be reforested by sowing or planting tree species suitable for the site. Administrations of forest enterprises and national parks provide information on available seedlings to help local forest owners to find suitable planting material; generally any change in species composition is not allowed: according to Forest Law, damaged pine stands should be restored by new pine stands; mixture of pine with deciduous tree species (e.g. birch) is allowed only to create fire barriers (Forest Law, 2001-04-10, Nr. IX-240, Regulations on protection of forest from fires, Document Nr. 500);
- Estonia: a forest may be reforested only with tree species suitable for the forest site type and the Minister of Environment shall establish the list of the types by the rules of forest management. The forest owner is required to apply reforestation methods in damaged parts of forest with an area of at least 0.5 ha within 2 years after the damage.

### 3.3.1.4 Legal Outline Remarks

As countries from different European and Mediterranean regions are covered in the sample, legal approaches here presented can be regarded as being sufficiently comprehensive to outline general key findings on national legislation concerning post-fire management.

(i) Legislation appears to be useful as a framework enabling forest owners and responsible authorities to identify necessary action to be taken in fire affected areas. Interdiction of land use conversion, forest restoration and grazing control appear to be common cross-cutting issues across regions.

(ii) Countries of the Northern Mediterranean basin (Spain, Portugal, France, Italy, and Greece) have the most well-constructed legal provisions, touching several issues of post-fire management (Table 3.2); in this regard, other Mediterranean countries facing similar fire conditions might benefit from these examples, when formulating or revising national legislation on fire management.

(iii) Despite the existence of a legal framework on post-fire management, the implementation of the law is acknowledged as a key problem. Especially in some Mediterranean countries legal provisions are not effectively enforced at the field level and the reality is sometime a "deviation" from the legislation. Current (and foreseeable) law enforcement capacities should be therefore taken into account when preparing relevant provisions. Important issues related to law enforcement include also:

- financial instruments supporting the implementation of legal provisions; taking into account that most of the fire affected forest areas in Europe are privately owned, obligations like tree planting provisions for forest restoration are unlikely to be implemented by small-scale owners unless funding is available;

- ensuring compensation for damage caused by forest fires. This calls for sound methodologies to assess the economic impacts of forest fires (see Sect. 3.2).

## 3.3.2   Social Issues

Four vectors interact to determine the social experience of the post-fire management: (i) the extent and characteristics of the fire; (ii) the history of relations between the fire management agency/forest services and the local community; (iii) economic implications of the fire; (iv) the level and consistency of the agency/service communication (Ryan and Hamin 2006). After wildfire, fire management agencies and forest services often undertake forest restoration efforts taking, generally, only a technical perspective. Limited efforts are usually made to take explicitly into account the public's perceptions, both aesthetically and ecologically, to fire recovery treatments, as well as the involvement of the community in post-fire planning. On the other hand, appropriate forest treatments can only be determined in light of the management objectives for that particular area of the forest, and those objectives are determined in part by the local community's needs, desires, and aesthetic preferences. The difficulty lies in what can be thought of as the distributional effects of restoration: the community as a whole may benefit but the direct and indirect costs of carrying out restoration may be paid by a much smaller number of individuals such as those owning or using the land that is treated (Lamb 2009).

The ability to intervene and undertake restoration in the various land units will depend on who owns or uses the land and how these people fit into the various land-owner or land user classes; of course, implementing restoration actions will be easier in the case of public (e.g. Turkey) or prevalently public burned forest land.

But, in addition to these on-site land managers there may also be other stakeholders (e.g. tourists) with legitimate interests in the way these lands are managed (Lamb 2009). Some of the factors that may influence the attractiveness of restoration to landowners are shown in Table 3.4 (see also Sect. 3.2.2).

Rather complex arrangements might be needed where restoration is undertaken as a community activity rather than by individuals. Unlike the situation involving individual farmers, the opportunity costs in this case are mostly low. However, as highlighted by a survey carried out within the framework of the FP0701 COST Action, the direct involvement of local communities in post-fire recovery activities is generally quite unusual in Southern Europe, and it mostly concerns replanting. Where communities are heavily involved in the forest rehabilitation, the efforts have a very positive effect on perceptions of the post-fire recovery, as well as helping to re-build the community spirit eventually hampered by the wildfire event. Ryan and Hamin (2006) describe how different restoration efforts compare to natural revegetation from the public's perspective, and how to effectively communicate with and engage the public in the rehabilitation process. Overall, resident perceptions are reported to be better after the fire than before, and acceptance of hazard mitigation

**Table 3.4** Socio-economic factors that may influence the attractiveness of restoration to farmers and landowners (from Lamb 2009)

| Factor | Significance |
| --- | --- |
| Land tenure | Farmers without secure land tenure or usufruct rights are unlikely to undertake a land use activity such as restoration where the benefits take time to achieve |
| Availability of agricultural land | The commercial viability of a farm will often depend on its size; it may be easier to undertake ecological restoration on a large farm than on a small farm because the initial impact of the change is proportionally smaller |
| Productivity of land | Restoration will be more attractive on land that is regarded as unproductive (because of lost soil fertility, weeds, pests etc. ) than on land that is still highly productive; this is because the opportunity costs incurred in converting productive land would be too high |
| The likelihood of financial or other direct benefits arising from restoration | Landowners are more likely to be interested in a land use activity that benefits them immediately and directly; benefits may come from goods such as timber or services such as improved water supplies or new wildlife habitats |
| Availability of subsidies, incentive payments or tax concessions | Such payments may be especially significant for small, risk-averse landowners or those with low incomes |
| Legal obligations to overcome degradation | There may be legal requirements on landowners to prevent fires or eradicate weeds or pests |
| Availability of alternative sources of off-farm income | Landowners able to obtain income from off-farm employment may be more able to convert part of their land holdings to new uses such as restoration |
| Attitude of neighbours | Neighbors can have positive and negative influences; innovative neighbors can provide examples to be copied but conservative neighbors can also argue against change and diminish the propensity of innovators to take on risky new land uses |

measures increases significantly. A second important aspect might be supporting volunteers in rehabilitation efforts, which both aids the forest and helps the community heal from the trauma of the fire. However, only in a few European countries are volunteers directly engaged in restoration activities, whilst they are much more involved in surveillance and fire fighting.

### 3.3.2.1 Post-Fire Management Practices and Their Social Success

Most people view the question of best restoration and rehabilitation treatments as a question for science and forest service expertise. However, on some items the public have usually clear opinions. A wide majority of stakeholders support salvage logging after wildfire, even in communities where salvage logging did not occur. One point of near consensus is the need to remove hazard trees from trails and to re-open

**Table 3.5** Socio-economic indicators of the success of restoration activities (from Lamb 2009)

| Indicator | Reason |
| --- | --- |
| Incomes of resident households improving | Farmers with declining incomes are unlikely to be able to afford to implement or maintain restoration activities |
| Payments being made for ecological services | External stakeholders willing to pay land managers for on-site restoration activities that yield ecological services such as clean water, biodiversity or wildlife habitats |
| Individual landowners continuing to protect and maintain restored sites. spontaneous restoration at other sites by individual landowners without the need for external subsidies or support | Landowners continue to view restoration on their land as being a valid and beneficial land use; the benefits of restoration are self-evident to individual landowners resulting in increased areas of degraded land being treated; the more of such new sites the greater the "success"; a corollary of this is that no new areas of degradation are evident |
| Development of private enterprises able to carry out or benefit from restoration activities (e.g. seedling nurseries, reforestation companies, weed control companies, ecotourism groups etc.) | Restoration and the land uses it fosters have become commercially profitable and created employment and new economic opportunities |
| Development of institutions and learning networks amongst landowners aimed at fostering rehabilitation | These institutions and networks generate, accumulate and transfer knowledge about restoration and the ways it can be incorporated in local land use systems |
| Validation and community support for policies, regulations and institutions designed to protect restored areas and prevent future degradation | These regulations may be formal government legal regulations, traditional community regulations or new rules established by communities to facilitate the management of newly acquired common property resources; the institutions may be traditional community organizations or government regulatory bodies |

trails and other popular recreation areas as quickly as possible. Another point is that forest stand thinning usually gains much popularity after the fire. Overall, there is general support for forest restoration near the urban interface but much less so in the backcountry. Distinctively, people usually strongly support rehabilitation techniques that stabilize soils and minimize flood damage near developed areas (Ryan and Hamin 2007).

It is a rather difficult task to assess the socioeconomic success of restoration. There may be a large number of stakeholders who may react quite differently to a restoration or rehabilitation project with some judging it a success and others counting it a failure. Perhaps the key indicator is that institutions and learning networks have evolved enabling different experiences to be integrated, synthesized and shared amongst practitioners (Table 3.5).

Such learning networks will assist these communities to withstand future ecological or economic shocks and prevent further degradation (Lamb 2009).

#### 3.3.2.2  Social Outline Remarks

Few Southern European countries have mechanisms in place to educate in the area of post-fire management. On the other hand, the survey carried out in the framework of the FP0701 Cost Action indicates that where there are understanding and awareness of the role of fire and its impacts, the average size of the burned area is diminishing and there is more reporting of fires along with more public interest in rehabilitation of fire burned areas.

It is widely acknowledged that the local communities should be involved in the recovery processes. It is on how to do this where thoughts differ and the lack of clear strategies in many cases which hinder greater participation. There is an obvious need for more awareness raising and planning for integration of communities into the long term planning by fire management agencies and forest services so that there is some ownership of the plans and a feeling of ownership of what happens to and in the forests: the end result would then be that the public shares responsibility for the ongoing preservation of the forests. This argues for fire management agencies and forest services to consider explicitly the social communities when planning post-fire restoration projects. Distinctively, understanding the public's likely response to rehabilitation efforts has several benefits for improving management, and the crucial post-fire period provides a window for significantly improving community/agency relations. The key to successful post-fire rehabilitation from the local community's perspective is for managers to quickly communicate the rehabilitation planning and continue to communicate longer-term restoration efforts to the community.

## 3.4  Key Messages

- To select the optimal post-fire management and restoration measures, the corresponding decisions should be based on environmental, economic and social aspects.
- The economic aspects should contribute to estimate the economic efficiency and distributional (equity) effects of the post-fire management measures. The economic efficiency measures the impacts of the post-fire management measures on societies welfare, while the distributional effects indicate who wins and who loses if a certain type of post-fire management measures are implemented.
- Currently only a fraction of economic effects of fires are considered in European countries. There is no commonly agreed approach how to estimate the economic effects of fires and what aspects should be included. This situation considerably limits the possibilities of decision makers to select optimal post-fire and restoration measures.
- It is recommended that efforts are made at the pan-European level to implement procedures that would enable a sound estimation of economic effects of fires. These procedures would enable on one hand the accurate estimation of economic

losses, and on the other hand more objective selection of post-fire management and restoration measures.

- Legislation appears to be useful as a framework enabling forest owners and responsible authorities to identify necessary action for post-fire management. Interdiction of land use conversion, forest restoration and grazing control appear to be common cross-cutting issues across countries of Southern Europe.
- Despite the existence of a legal framework on post-fire management, the implementation of the law is acknowledged as a key problem. Current (and foreseeable) law enforcement capacities should be therefore taken into account when preparing relevant provisions. Important issues related to law enforcement include financial instruments supporting the implementation of legal provisions and ensuring compensation for damage caused by forest fires.
- There is a need for more awareness raising and planning for integration of communities into the long term post-fire planning by fire management agencies and forest services: the crucial post-fire period provides a window for significantly improving community/agency relations.
- The key to successful post-fire rehabilitation from the local community's perspective is for managers to quickly communicate the rehabilitation planning and continue to communicate longer-term restoration efforts to the community.

# References

Abt KL, Huggett RJ, Holmes TP (2008) Designing economic impact assessments for USFS wildfire programmes. In: Holmes TP, Prestemon JP, Abt KL (eds) The Economics of Forest Disturbances: wildfires, storms, and invasive species. Springer, New York, pp 151–166

Arrow KJ, Hahn FH (1971) General competitive analysis. Holden-Day, San Francisco

Barrio M, Loureiro M, Chas ML (2007) Aproximación a las pérdidas económicas ocasionadas a corto plazo por los incendios forestales en Galicia en 2006. Econ Agraria y Recursos Nat 7:45–64

Beyers JL, Neary DG, Ryan KC, DeBano LF (2005) Wildland fire in ecosystems. Effects of fire on soil and water. United States Dept. of Agriculture, Forest Service, Rocky Mountain Research Station, Fort Collins, Co

Bockstael NE, McConnell KE (2007) Environmental and Resource Valuation with Revealed Preferences. In: Bateman IJ (ed.) Springer. Dordrecht, The Netherlands

Bourrinet J (1992) Wildland fires and the law: legal aspects of forest fires worldwide. Nijhoff Publishers, Dordrecht

Brown JK, Smith JK, (2000) Wildland fire in ecosystems: effects of fire on flora, Gen. Tech. Rep. RMRS-GTR. U.S. Department of Agriculture, Forest Service, Rocky Mountain Research Station, Ogden, UT (USA), p 257

Bulgaria's Forest Law, Prom. SG. 125/29 Dec 1997, amend. SG 16/03, Art. 14

Butry DT, Mercer DE, Prestemon JP, Pye JM, Holmes TP (2001) What is the price of catastrophic wildfire? J Forest 99:9–17

Ciancio O, Corona P, Marinelli M, Pettenella D (2007) Evaluation of forest fire damages in Italy. Accademia Italiana di Scienze Forestali, Firenze, p 60

Dickie M, Messman VL (2004) Parental altruism and the value of avoiding acute illness: Are kids worth more than parents? J Environ Econ Manag 48(3):1146–1174

EEA (2009) Greenhouse gas emission trends and projections in Europe 2009: tracking progress towards Kyoto targets. European Environmental Agency, Copenhagen, p 180

EFTEC i.c.w. Environmental Futures Limited (2006) Valuing our natural environment. Final Report. London

Englin J, Holmes TP, Lutz J (2008) Wildfire and the Economic Value of Wilderness Recreation. In: Holmes TP, Prestemon JP, Abt KL (eds.) The Economics of Forest Disturbances – Wildfires, Storms, and Invasive Species. Forestry Sciences, vol. 79, p. 191–208

EU (2005) Council Regulation (EC) No 1698/2005 of 20 September 2005 on support for rural development by the European Agricultural Fund for Rural Development (EAFRD). In: C.o.t.E. (ed.) Union, Offical Journal L 277, Brussels, pp 1–40

FAO (1999) Towards a harmonized definition of non-wood forest products. Unasylva 50 (1999/3)

FAO (2007) Fire management. Global assessment 2006. FAO, Rome

Farrell EP, Fuhrer E, Ryan D, Andersson F, Huttl R, Piussi P (2000) European forest ecosystems: building the future on the legacy of the past. For Ecol Manag 132:5–20

Flowers P, Vaux H, Gardner P, Mills TJ, (1985) Changes in recreation values after fire in the RockyMountains, Res. Note PSW. USDA Forest Service, Pacific Southwest Forest and Range ExperimentStation, Albany, CA, p 15

González JR, Kolehmainen O, Pukkala T (2007) Using expert knowledge to model forest stand vulnerability to fire. Comput Electron Agric 55:107–114

González-Cabán A (1998) Aspectos Económicos de la Evaluación del Danō de Incendios. USDA Forest Service, Pacific Southwest Research Station

González-Cabán A (2007) Wildland fire management policy and fire management economic efficiency in the USDA Forest Service, Wildfire 2007-4th international wildland fire conference, Seville, Spain

González-Ochoa A, de las Heras J (2002) Effects of post-fire silviculture practices on Pachyrhinus squamosus defoliation levels and growth of Pinus halepensis Mill. For Ecol Manag 167:185–194

González-Ochoa AI, López-Serrano FR, de las Heras J (2004) Does post-fire forest management increase tree growth and cone production in Pinus halepensis? For Ecol Manag 188:235–247

Görlach B, Landgrebe-Trinkunaite R, Interwies E, Bouzit M, Darmendrail D, Rinaudo JD (2004) Assessing the economic impacts of soil degradation. Ecologic, Berlin

Hanley N, Spash CL (1993) Cost-benefit analysis and the environment. Edward Elgar Publishing Ltd., Cheltenham

Hanley N, Shogren JF, White B (2001) Introduction to environmental economics. Oxford University Press, Oxford

Hassan RM, Scholes RJ, Ash N, Millennium Ecosystem Assessment (Program). Condition and Trends Working Group (2005) Ecosystems and human well-being: current state and trends: findings of the Condition and Trends Working Group of the Millennium Ecosystem Assessment. Island Press, Washington, DC

Hicks JR (1939) The foundation of welfare economics. Econ J 49:696–712

Holmes TP, Abt KL, Huggett RJ, Prestemon JP (2007) Efficient and equitable design of wildfire mitigation programs. In: Daniel TC, Carroll MS, Moseley C, Raish C (eds) People, fire and forests. Oregon State University, Corvallis, pp 143–156

Holmes TP, Prestemon JP, Abt KL (2008) An Introduction to the Economics of Forest Disturbance. In: Holmes TP, Prestemon JP, Abt KL (eds.) The Economics of Forest Disturbances – Wildfires, Storms, and Invasive Species. Forestry Sciences, vol. 79, p. 3–14

Izhaki I, Adar M (1997) The effects of post-fire management on bird community succession. Int J Wildland Fire 7:335–342

JRC (2009) Review of the methodologies for the assessment of economic impacts of forest fires and economic efficiency of fire management. In: Pettenella D, Florian D, Mavsar R, Goio I, González-Cabán A (eds.) JRC – ISE, Ispra, p 108

Kaldor N (1939) Welfare propositions of economics and interpersonal comparisons of utility. Econ J 49:549–552

Kanninen BJ (ed) (2007) Valuing Environmental Amenities Using Stated Choice Studies. In: Bateman IJ (ed.) Springer. Dordrecht, The Netherlands

Kent B, Gebert K, McCaffrey S, Martin W, Calkin D, Schuster E, Martin I, Wise Bender H, Alward G, Kumagai Y, Cohn PJ, Carroll M, Williams D, Ekarius C (2003) Social and economic issues of the Hayman Fire. In: Service UF (ed.) General technical report RMRS-GTR. USDA Forest Service, pp 315–396

Kochi I, Loomis JB, Champ P, Donovan GH (2008) Health and economic impact of wildfires: literature review, III international symposium on fire economics, planning and policy: common problems and approaches, Carolina, Puerto Rico

Lamb D (2009) Economic, social and cultural factors affecting landscape restoration. In: Bautista S, Aronson J, Vallejo VR (eds) Land restoration to combat desertification. CEAM, Valencia, pp 35–46

Loomis J (2004) Do nearby forest fires cause a reduction in residential property values? J For Econ 10:149–157

Loomis JB, Englin J, Gonález-Cabán A (1999) Effects of fire on the economic value of forest recreation in the Intermountain West: preliminary results. In: Gonález-Cabán A, Omi PN (eds.) Symposium on fire economics, planning, and policy: bottom lines. USDA Forest Service, Pacific Southwest

Loomis JB, Englin J, Gonález-Cabán A (2001) Testing for differential effects of forest fires on hiking and mountain biking demand and benefits. J Agr Res Econ 26:508–522

Marques MA, Mora E (1998) Effects on erosion of two post-fire management practices: clear-cutting versus non-intervention. Soil Till Res 45:433–439

Mas-Colell A, Whinston M, Green J (1995) Microeconomic theory. Oxford University Press, Oxford

Mavsar R, Riera P (2007) Valoración Económica de las Principales Externalidades de los Bosques Mediterráneos Españoles: Informe final. Ministerio de Medio Ambiente, Barcelona, p 93

Mavsar R, Farreras V, Kovac M, Japelj A (2009) Written description of the private and social evaluation models. Deliverable D4.5-2 of the Integrated project "Fire Paradox", Project no. FP6-018505, European Commission, p 76

Mayer A (1995) Forest valuation for decision making: lessons of experience and proposals for improvement. Food and Agriculture Organization of the United Nations, Rome, February 1997

Mercer DE, Jeffrey PP, David TB, John MP (2007) Evaluating alternative prescribed burning policies to reduce net economic damages from wildfire. Am J Agric Econ 89:63–77

Merlo M, Croitoru L (2005) Valuing Mediterranean forests: towards total economic value. CABI Publishing, Wallingford

Molina JR, Rodriguez Y, Silva F, Herrera MA, Zamora R (2009) A simulation tool for socio-economic planning on forest fire suppression management. In: Gomez E, Alvarez K (eds) Forest fires: detection, suppression and prevention. Nova Science Publishers Inc, New York

Monteil C, Herrero G, (2010) An overview of policies and practices related to fire ignitions at the European Union level. In: Sande Silva J, Rego F, Fernandes P, Rigolot E (eds.) Towards integrated fire management – outcomes of the European Project Fire Paradox. European Forest Institute Research Report 23, 2010

Morgera E, Cirelli MT (2009) Forest fires and the law. A guide for national drafters based on the fire management voluntary guidelines. FAO, Rome

Mueller J, Loomis J, González-Cabán A (2007) Do repeated wildfires change homebuyer's demand for homes in high risk areas? A hedonic analysis of the short- and long-term effects of repeated wildfires on home prices in southern California. J Real Estate Finan Econ 38:155–172

Napoleone C, Jappiot M (2008) Et si l'efficacité de la lutte contre les incendies jouait comme une incitation aux comprtements les plis risqués? Forêt méditerranéenne 29 (1):3–11

Navrud S (2001) Valuing health impacts from air pollution in Europe. Env Res Econ 20(4):305–329

OECD (2000) Services: statistics on value added and employment, OECD 2000 edition, Introduction, p 11

Portugal's Decree-Law No. 139/88; Bulgaria's Forest Law, Prom. SG. 125/29 Dec 1997, amend. SG 16/03, Art.42

Prestemon JP, Holmes TP (2004) Market dynamics and optimal timber salvage after a natural catastrophe. For Sci 50(4):495–511

Riera P, García D, Kriström B, Brännlund R (2005) Manual de Economía Ambiental y de los Recursos Naturales. International Thomson Editores Spain. 355 pp

Riera P, Mavsar R, Mogas J (2008) Forest fire valuation and evaluation: a survey. In: Cesaro L, Gatto P, Pettenella D (eds) The multifunctional role of forests – policies, methods and case studies. European Forest Institute, Padova, pp 367–380

Rittmaster R, Adamowicz WL, Amiro B, Pelletier RT (2006) Economic analysis of health effects from forest fires. Can J For Res 36:868–877

Rosenbaum KL (2007) Legislative drafting guide: a practitioner's view. FAO Legal Paper (on-line) #64 (available at www.fao.org)

Ryan RL, Hamin E (2006) Engaging communities in post-fire restoration: Forest treatments and community-agency relations after the Cerro Grande Fire. In McCaffrey S (ed.) The public and wildland fire management: social science findings for managers. General technical report, NRS-1. USDA Forest Service, Northern Research Station, Newtown Square, pp 87–96

Ryan RL, Hamin E (2007) After the fire: local residents' perceptions of post-fire forest restoration. In: Extended abstracts from the human dimensions of wildland fire conference, 10/23-25, Colorado S, McCaffrey P, Woodward M, Robinson, compilers. International Association of Wildland Fire, Fort Collins pp 54–56

Sandberg DV, Ottmar RD, Peterson JL, Core J (2002) Wildland fire on ecosystems: effects of fire on air, Gen. Tech. Rep. RMRS-GTR. U.S. Department of Agriculture, Forest Service, Rocky Mountain Research Station, Ogden, UT, p 79

Smith JK (2000) Wildland fire in ecosystems: effects of fire on fauna, Gen. Tech. Rep. RMRS-GTR. U.S. Department of Agriculture, Forest Service, Rocky Mountain Research Station, Ogden, UT (USA), p 83

Spanos I, Raftoyannis Y, Goudelis G, Xanthopoulou E, Samara T, Tsiontsis A (2007) Effects of post-fire logging on soil and vegetation recovery in a Pinus halepensis Mill. forest of Greece. Eco- and Ground Bio-Engineering: the use of vegetation to improve slope stability developments in plant and soil sciences 103:345–352

Sugden R, Williams A (1978) The principles of practical cost-benefit analysis. Oxford University Press, Oxford

US.EPA (2004) Air quality criteria for particulate matter (final report, Oct 2004). In: Agency UEP (ed.) US Environmental Protection Agency, Washington D.C

Vega JA, Fernández C, Pérez-Gorostiaga P, Fonturbel T (2008) The influence of fire severity, serotiny, and post-fire management on Pinus pinaster Ait. recruitment in three burnt areas in Galicia (NW Spain). For Ecol Manag 256:1596–1603

Velez R (2000) La defensa contra los incendios forestales: Fundamentos y experiencias. McGraw-Hill Inc., Madrid

Velez R (2009) The causing factors: a focus on economic and social driving forces. In: Birot Y (ed) Living with wildfires: what can science tell us. EFI Discussion Papers, 15, p 21–26

Wade DD, Lunsford JD (1988) A guide for prescribed fire in southern forests, Technical Publication R8-TP. USDA Forest Service, Southern Region, Atlanta, p 56

Zybach B, Dubrasich M, Brenner G, Marker J (2009) U.S. Wildfire Cost-Plus-Loss Economics Project: the "One-Pager" checklist. John Wildland Fire Lessons Learned Center, p 20

# Chapter 4
# Fire Hazard and Flammability of European Forest Types

Gavriil Xanthopoulos, Carlo Calfapietra, and Paulo Fernandes

## 4.1 Fire Hazard, Flammability and Their Influencing Factors

Fire danger assesses both fixed and variable environmental factors that determine the inception and behavior of forest fires, as well as their difficulty of suppression and impact (FAO 2006). In this classic approach, *fire hazard* is defined as "a measure of that part of the *fire danger* contributed by the fuels available for burning" (FAO 2006). *Fuel type, volume, condition, arrangement and location, determine the degree both of ease of ignition and of fire suppression difficulty and characterize the fire hazard that a fuel complex represents. Fire hazard is directly related to fuel flammability.* Distinguishing between the two terms is largely a matter of definitions as seen below.

*Flammability* of plant biomass is the main parameter considered in order to assess fire hazard and fire risk, the latter having been defined as "the probability of fire initiation due to the presence and activity of a causative agent". Flammability is a property of plants readily appreciated in a general sense but difficult to define scientifically (Gill and Moore 1996). According to Anderson (1970) and Trabaud

G. Xanthopoulos (✉)
National Agricultural Research Foundation, Institute of Mediterranean Forest Ecosystems and Forest Products Technology, Athens, Greece
e-mail: gxnrtc@fria.gr

C. Calfapietra
Institute of Agro-Environmental & Forest Biology, National Research Council, Porano, Italy

P. Fernandes
Centre for the Research and Technology of Agro-Environmental and Biological Sciences, Universidade de Trás-os-Montes e Alto Douro, Vila Real, Portugal

F. Moreira et al. (eds.), *Post-Fire Management and Restoration of Southern European Forests*, Managing Forest Ecosystems 24, DOI 10.1007/978-94-007-2208-8_4, © Springer Science+Business Media B.V. 2012

(1976) flammability includes three components: *ignitability*, *combustibility* and *sustainability*. Ignitability refers to the capacity of a fuel to be ignited and is often measured by the time until ignition of the fuel once it is exposed to a heat source. Combustibility indicates how plants burn after they have been ignited. Sustainability indicates the capacity of a fuel to burn over time (flame duration) (Saura-Mas et al. 2010). Martin et al. (1993) added one more component, *consumability*, defining it as "the degree to which a fuel is consumed by fire".

In light of the above, flammability relates to the intrinsic properties of fuels and fuel types in regard to the four components above whereas fire hazard additionally includes fuel quantity, condition, arrangement and location. Thus data of flammability of vegetation are very important to calculate fire hazard indices combining those data with the meteorological ones of the area. For instance, the combination of the data of flammability of the fine dead fuel with wind speed data is routinely used by local agencies to forecast fire risk in several areas of the Mediterranean basin.

All four flammability components above are influenced in different ways by a multitude of factors, depending on if flammability is assessed at the leaf, plant or forest stand level. *These factors include the structural characteristics of the fuels, their moisture content, their chemical composition and the overall fuel arrangement.* Furthermore, they are not equally important to each of the four components of flammability, and they are not all independent of one another.

In regard to the structural characteristics of fuel particles, *ease-of-ignition* is mainly related to their thickness, often expressed through their *surface-area-to-volume ratio*. It is well established that thinner fuel particles ignite more easily (Gill and Moore 1996; Montgomery and Cheo 1971; Rundel 1981). *Fuel density*, measured as the weight per unit volume of fuel, is also important as it affects the total amount of heat a particle can absorb before it ignites (Papio and Trabaud 1990; Rothermel 1972). Rotten wood is a well known case of easy-to-ignite fuel due to its low density (Hyde et al. 2011). Moving up from the fuel particle, *fuel loading* per unit of ground, and finally the *porosity* of the fuel bed (Rundel 1981) are also important. Latham and Schlieter (1989) found the probability of ignition of duff from short needle conifer species subjected to an electric arc discharge that simulated the continuing current of a lightning, was almost entirely a function of the depth of the fuel bed. Obviously, more depth usually also means more quantity of fuel in such a naturally compacted fuel layer.

Fernandes (2009) studied 19 forest types in Portugal through fire behaviour simulations and concluded that potential fire behaviour is primarily *driven by stand structure, rather than by cover type*. Schwilk (2003) experimentally manipulated canopies of *Adenostoma fasciculatum*, a California shrub that naturally retains dead branches, to mimic degrees of self-pruning and then conducted experimental burns, measuring temperatures at various points in the burn plots. These measurements showed that the temperatures reached were a result of canopy architecture and not merely of total fuel load. Further refinements to this knowledge have been recently proposed by an Australian study on fuel flammability which has examined and modelled, among other parameters, fuel positioning in space as this affects the potential for vertical and horizontal flame spread (Zylstra 2011).

*Fuel moisture content* is another important factor influencing flammability (Massari and Leopoldi 1998; Trabaud 1976; Xanthopoulos and Wakimoto 1993). Measuring ignitability of leaves of Australian plants Gill and Moore (1996) found that their results could be largely explained in a statistically-significant way using only two variables – moisture content and surface-area-to-volume ratio. Fires, as a rule, start in fine dead fuels which require considerably lower amount of heat, often called "heat of pre-ignition", to reach ignition (Rothermel 1972) than live fuels. Dead fuel can have fuel moistures that range from 2% to over 40% on a dry weight basis, whereas, in live fuels, the range of fuel moisture varies from 30% to over 300%. In addition to ignition, fuel moisture also plays a critical role in fire rate of spread and flame length as evidenced by its inclusion as an input parameter in all fire behavior prediction systems (Rothermel 1972; Taylor et al. 1997).

Chemical characteristics which might affect flammability usually refer to the content in plant tissues of lignin (Mackinnon 1987), carbohydrates (Nimour Nour 1997) and minerals (Mutch and Philpot 1970). As described by Dimitrakopoulos and Panov (2001), the parameters which are considered for flammability assessment are the *fuel heat content* (calorific value) which refers to the thermal energy potentially released during the fire, and the *total ash content* which is practically reducing the amount of fuel generally represented only by the organic part.

Recently, attention has been focused on the relationship between *isoprenoids content and flammability* (Owens et al. 1998), for different reasons. These compounds are highly flammable (Nuñez-Regueira et al. 2005) and are usually quite abundant in Mediterranean species because they protect plants against heat shocks and oxidative stress (Llusià and Peñuelas 2000).

Moreover, it is likely that isoprenoids are the first compounds to be ignited and thus could be responsible for the spread of a forest fire (De Lillis et al. 2009). This is mainly related to the emission burst of a considerable amount of these compounds during the first stage of a fire. Isoprenoids are then ignited during the "glowing combustion" which occurs when both an optimal ratio in the air between oxygen and isoprenoids and an ignition temperature are reached (Ciccioli, personal communication). As a result, volatile isoprenoids can create an expanding fireball that ignites the trees downwind (Button 1987; Brown et al. 2003).

A relationship between isoprenoid and flammability has been found in different experiments and in different plant species, especially from those of Mediterranean regions, although not in all cases (Alessio et al. 2008; De Lillis et al. 2009; Ormeño et al. 2009).

Flammability can be studied in both laboratory and field experiments. One parameter which has been adopted quite recently especially in laboratory experiments is the oxygen consumption to measure the heat released due to combustion of the vegetation. This methodology represents an improvement in characterizing plant flammability, however translation of laboratory results to field conditions can be problematic and moreover experiments are usually expensive (White and Zipperer 2010).

## 4.2 Forest Composition Versus Forest Structure as Related to Flammability

Any type of forest stand is composed of a large variety of plants, either herbaceous or woody. The flammability of these plants is determined from the physical and chemical properties of their parts (leaves, needles, stalks, thin woody parts) and their structural arrangement. Due to differences between plants in total biomass and structure, the variation of leaf flammability does not directly translate into variation in plant flammability. The same concept can be extended from plant flammability to forest flammability, only at this scale things are even more complex as forest structure also affects other parameters such as wind profile, dead and live fuel moisture and soil moisture (Zylstra 2011). This is an important realization when trying to characterize the flammability of a natural plant association or, in case of forest, a forest type.

As the plant composition of a natural forest type is to a large extent defined by the environmental conditions it can be said that a "baseline" flammability level is determined by the species that make up this type of forest. On the other hand, the structure of a particular forest stand which is the result of factors such as age, fire history, grazing history, silvicultural treatments, fuel management treatments etc. can have a very pronounced effect on its fire potential modifying the baseline flammability that would be expected on the average, and of course the fire hazard that this stand represents (Blasi et al. 2005). For example, Mediterranean pine forests have always been described as flammable. However, as use of these forests for production of various forest products, such as resin, fuelwood, and forage, has declined in the last few decades, fuel accumulation and increased horizontal and vertical continuity result in much more intense fire behavior than in the past. Through fire behaviour simulation, Fernandes (2009) showed that fire hazard is widely variable in stands composed of flammable forest species, depending on stand structure and surface fuel accumulation. This has important management implications, as it demonstrates the worth of fuel and stand management to modify fire behaviour and severity.

Obviously, a "baseline" flammability assessment of the various forest types is an important requirement for understanding the fire potential over large forested areas. A more specific fire hazard assessment at the stand level, taking into consideration the factors described above, is one of the main tools that can be used in support of local fire management planning.

## 4.3 Flammability of European Forest Types

As mentioned earlier, flammability is a property of plants readily appreciated in a general sense but difficult to define scientifically (Gill and Moore 1996). This true statement and the need to develop a general flammability assessment of the various European forest types led to the preparation of a questionnaire that was distributed to a good number of forest fire experts in the European countries and North African countries. The questionnaire aimed to contribute towards an assessment of the forest, shrubland and grassland types in these countries in respect to their fire hazard and flammability. Such a ranking can be useful for promoting and planning less fire prone forests and landscapes, and for quantifying the relative fire potential in a country or region.

The main focus of the questionnaire was overall fire hazard assessment in five classes (1–5) as follows:

- Very low: Few fire starts. Low flammability even under extreme conditions (e.g. temperature >38°C, relative humidity < 15%, windspeed > 30 km/h).
- Low: Flammable under extreme conditions when the number of fire starts can be considerable and the fires quite active.
- Medium: Fires can be numerous under hot summer conditions and they are likely to present significant control problems requiring effort and resources.
- High: Fires are very likely under hot summer conditions. They can escape initial attack if not attacked effectively and then will defy direct frontal attack.
- Very high: Fires can be plentiful even under relatively mild fire season conditions. They are likely to escape all but the best initial attack efforts under high fire danger conditions in which case they will defy direct control efforts at their front until conditions change. Under extreme fire weather conditions they can result in mega-fires.

Twenty different experts or teams of experts responded to the questionnaire. They were from Cyprus, France, Greece, Israel, Italy, Lithuania, Morocco, Poland, Portugal, Romania, Spain, Switzerland, Tunisia and Turkey. A database of 150 responses with ratings for 60 different vegetation types was created. The number of fire hazard assessments for each vegetation type varied from one, for the least common types up to 13 for the most common ones. Based on these results a general expert assessment of the typical fire hazard that the various European forest types represent is presented in Table 4.1. Correspondence between the vegetation types and the European forest types classification was made mostly by the experts following the description and keys provided in European Environment Agency (2007).

Although there are clearly many variations on the same vegetation between countries, as shown by the range of fire hazard values of certain types listed in the table, and the table is not exhaustive as not all European countries were represented among the pool of experts who responded to the questionnaire, it can be a useful tool for post-fire management decisions. The table provides also a framework to assess the need and priorities for fuel and silvicultural treatments to mitigate fire hazard.

**Table 4.1** Expert assessment of typical fire hazard (minimum, maximum and mean) for 60 vegetation types of Europe and North Africa and their corresponding European Forest type classification

| No. | Vegetation type | European forest type classification | Min | Max | Mean |
|---|---|---|---|---|---|
| 1 | Hemiboreal Mountain pine (Pinus mugo) forest | 2.1 Hemiboreal forest | 3 | 3 | 3.0 |
| 2 | Hemiboreal Scots pine (Pinus sylvestris) forest | 2.1 Hemiboreal forest | 2 | 2 | 2.0 |
| 3 | Nemoral Scots pine (Pinus sylvestris) forest with mix of Picea excelsa, Betula sp., Fagus sp., Oak sp. etc. depending on site quality – Varying fire hazard | 2.2 Nemoral Scots pine forest | 2 | 4 | 3.0 |
| 4 | Tall Scots pine forest with mix of Birch and other deciduous trees | 2.5 Mixed Scots pine-birch forest | 1 | 1 | 1.0 |
| 5 | Tall Scots pine forest with mix of oak and other deciduous trees | 2.6 Mixed Scots pine-pedunculate oak forest | 1 | 1 | 1.0 |
| 6 | Subalpine larch forest | 3.1 Subalpine larch-arolla pine and dwarf pine forest | 1 | 2 | 1.5 |
| 7 | Spruce and mixed spruce-fir forests | 3.2 Subalpine and mountainous spruce and mountainous mixed spruce-Silver fir -forest | 2 | 2 | 2.0 |
| 8 | Alpine Scots pine and Black pine forest | 3.3 Alpine Scots pine and Black pine forest | 3 | 3 | 3.0 |
| 9 | Tall deciduous oak (Quercus pedunculata/robur) forest | 5.1 Pedunculate oak–hornbeam forest | 2 | 2 | 2.0 |
| 10 | Tall deciduous oak (Quercus sp.) forest | 5.2 Sessile oak-hornbeam forest | 1 | 2 | 1.5 |
| 11 | Beech (Fagus sp.) forest (Lowland) | 6.1 Lowland beech forest of southern Scandinavia and north central Europe | 1 | 1 | 1.0 |
| 12 | Beech (Fagus sp.) forest | 6.4 Central European submountainous beech forest | 1 | 1 | 1.0 |
| 13 | Beech (Fagus sp.) forest with Quercus pedunculata | 6.5 Carpathian submountainous beech forest | 2 | 2 | 2.0 |
| 14 | Beech (Fagus sylvatica & moesiaca) forest with oaks and other Mediterranean broadleaved trees | 6.7 Moesian submountainous beech forest | 2 | 2 | 2.0 |
| 15 | Beech forests | 7.1 SW-European montane beech forest | 1 | 1 | 1.0 |
| 16 | Beech forests | 7.2 Central European montane beech forest | 1 | 1 | 1.0 |
| 17 | Beech (Fagus sylvatica) forests with Taxus baccata, Ilex aquifolium,Tilia cordata etc. | 7.3 Apennine-Corsican mountainous beech forest | 2 | 2 | 2.0 |

(continued)

**Table 4.1**   (continued)

| No. | Vegetation type | European forest type classification | Min | Max | Mean |
|-----|-----------------|-------------------------------------|-----|-----|------|
| 18 | Beech (Fagus sylvatica) forest | 7.5 Carpathian mountainous beech forest | 2 | 2 | 2.0 |
| 19 | Oak (Quercus pubescens) forest | 8.1 Downy oak forest | 2 | 2 | 2.0 |
| 20 | Oak (Quercus cerris) forest | 8.2 Turkey oak, Hungarian oak and Sessile oak forest | 2 | 3 | 2.5 |
| 21 | Oak (Quercus pyrenaica) forest | 8.3 Pyrenean oak forest | 1 | 2 | 1.5 |
| 22 | Oak (Quercus faginea) forest | 8.4 Portuguese oak and Mirbeck's oak Iberian forest | 2 | 2 | 2.0 |
| 23 | Oak (Quercus trojana) forest | 8.5 Macedonian oak forest | 2 | 3 | 2.3 |
| 24 | Oak (Quercus ithaburensis) forest | 8.6 Valonia oak forest | 3 | 4 | 3.5 |
| 25 | Chestnut (Castanea sativa) forest | 8.7 Chestnut forest | 2 | 4 | 2.7 |
| 26 | Forest of other thermophilous deciduous trees (Carpinus orientalis, Ostrya carpinifolia, etc.) | 8.8 Other thermophilous deciduous forests | 2 | 3 | 2.5 |
| 27 | Broadleaved evergreen forest (oak dominated maquis) | 9.1 Mediterranean evergreen oak forest | 3 | 5 | 4.0 |
| 28 | Broadleaved evergreen forest (oak coppice) | 9.1 Mediterranean evergreen oak forest | 3 | 5 | 3.3 |
| 29 | Broadleaved evergreen forest (cork oak) | 9.1 Mediterranean evergreen oak forest | 3 | 5 | 3.3 |
| 30 | Broadleaved evergreen forest (holm oak) | 9.1 Mediterranean evergreen oak forest | 3 | 4 | 3.5 |
| 31 | Broadleaved evergreen forest (Olive-Carob) | 9.2 Olive –Carob forest | 3 | 5 | 4.3 |
| 32 | Laurisylva forest | 9.4 Macaronesian laurisilva | 1 | 1 | 1.0 |
| 33 | Broadleaved evergreen forest (maquis) not dominated by oak (tall) | 9.5 Other sclerophyllous forest | 3 | 4 | 3.7 |
| 34 | Broadleaved evergreen forest (maquis) not dominated by oak (short to medium) | 9.5 Other sclerophyllous forest | 2 | 4 | 2.7 |
| 35 | Mixed heathland (Erica, Ulex, Pterospartium) | 9.5 Other schlerophyllous forests | 5 | 5 | 5.0 |
| 36 | Garrigue | 9.5 Other sclerophyllous forest | 2 | 5 | 3.0 |
| 37 | Broadleaved evergreen - semideciduous forest | 9.6 Moroccan macaronesian arganian forest | 1 | 1 | 1.0 |
| 38 | Aleppo pine (Pinus halepensis) forest | 10.1 Mediterranean Pine Forest | 3 | 5 | 4.1 |
| 39 | Brutia pine (Pinus brutia) forest | 10.1 Mediterranean Pine Forest | 4 | 5 | 4.8 |
| 40 | Mixed Aleppo pine-oak forest | 10.1 Mediterranean Pine Forest | 4 | 5 | 4.3 |
| 41 | Stone pine (Pinus pinea) forest | 10.1 Mediterranean Pine Forest | 3 | 3 | 3.0 |
| 42 | Maritime pine (Pinus pinaster) forest | 10.1 Mediterranean Pine Forest | 5 | 5 | 5.0 |

(continued)

**Table 4.1** (continued)

| No. | Vegetation type | European forest type classification | Min | Max | Mean |
|---|---|---|---|---|---|
| 43 | Black pine (Pinus nigra) forest | 10.2 Mediterranean and Anatolian black pine forest | 3 | 5 | 3.8 |
| 44 | Scots pine (Pinus sylvestris) forest | 10.4 Mediterranean and Anatolian Scots pine forest | 3 | 3 | 3.0 |
| 45 | Bosnian pine (Pinus leucodermis) forest | 10.5 Alti-Mediterranean pine forest | 2 | 2 | 2.0 |
| 46 | Fir (Abies sp.) forest | 10.6 Mediterranean and Anatolian fir forest | 1 | 3 | 2.0 |
| 47 | Juniper (Juniperus sp.) forest | 10.7 Juniper forest | 1 | 4 | 2.8 |
| 48 | Cypress (Cupressus sp.) forest | 10.8 Cypress forest | 2 | 5 | 3.5 |
| 49 | Cedar Forest (Cedrus libani var brevifolia) | 10.9 Cedar Forest | 1 | 5 | 3.0 |
| 50 | Tetraclinis articulata forest | 10.10 Tetraclinis articulata stands | 2 | 2 | 2.0 |
| 51 | Alder Forest | 13.1 Alder Forest | 3 | 3 | 3.0 |
| 52 | Plantations of Pinus halepensis and Pinus brutia | 14.1 Plantations of site-native species | 4 | 5 | 4.5 |
| 53 | Plantations of fir (Abies alba) | 14.1 Plantations of site-native species | 3 | 3 | 3.0 |
| 54 | Plantations of Pinus pinea | 14.2 Plantation of not-site-native species | 5 | 5 | 5.0 |
| 55 | Eucalyptus (Eucalyptus sp.) plantations | 14.2 Plantation of not-site-native species | 4 | 5 | 4.5 |
| 56 | Gorse (Ulex parviflorus) | | 4 | 4 | 4.0 |
| 57 | N. African Desertic Acacia Woodland | | 3 | 3 | 3.0 |
| 58 | Grassland | | 3 | 3 | 3.0 |
| 59 | Managed grassland | | 2 | 3 | 2.3 |
| 60 | Steppic grassland (Stipa tenacissima, Artemisia inculta, Noaea mucronata …) | | 4 | 4 | 4.0 |

## 4.4 Fire Hazard and Forest Flammability in a Global Warming Scenario

Climate and global changes involve numerous processes such as the increase of $CO_2$ concentration in the atmosphere, of air temperature and tropospheric $O_3$ levels, the variation in the UV-B radiation, in precipitation regimes and in the atmospheric events such as depositions of pollutants, etc. Simulating all these processes together to study the response of plants especially in field experiments is not feasible also because some processes vary their extent in the different parts of the planet or during the day. Even more difficult is to predict what effect these changes will have on fire hazard both because it is impossible to extrapolate each factor in real conditions and particularly relate the variation of each factor to the change in fire hazard.

It is evident that particularly increase in air temperature and decreased precipitation will directly influence flammability and fire regimes, thus in turn influencing plant biomass and species composition (Mouillot et al. 2002). This means altering both fuel quantity and quality. However, changes may be more complex as changes in species composition are quite likely and other indirect effects can occur, such as on fire recurrence and fire severity (Cary 2002). *With climate change and changed fire regimes there are likely to be changes to fuels in ways that are presently unknown* (Gill and Zylstra 2005) which can in turn affect also the parameters related to flammability.

Warmer climates connected with decreased precipitation will certainly affect fire frequency like expected for Mediterranean Basin regions (IPCC 2007). Double $CO_2$ climate scenarios are expected to cause in regions like California increase in wildfire events by 40–50% due mainly to structural change in vegetation such as conifers to broadleaves and tree to grasses (Fried et al. 2004).

The relationship between fire occurrence and climate is often analysed using meteorological fire hazard indices. However these indices should be applied to very short time period and moreover need very detailed atmospheric data which are available only in few places in southern Europe.

The effects of warmer and drier climates on fire hazard are definitely the most investigated especially through the use of fire behavior models and General Circulation Model predictions of future climate (Torn and Fried 1992; Davis and Michaelsen 1995).

Several estimations on the effects of climate change on fire hazards have been carried out through modeling related to climate scenarios. Westerling and Bryant (2008) modeled wildfire risks for California under four climatic change scenarios comparing different scenarios for 2005–2034, 2035–2064, and 2070–2099 against a modelled 1961–1990 reference period in California and neighboring states and showing how wildfires generally increase but this is more evident under one of the four scenarios considered.

Giannakopoulos et al. (2009) modelled climatic changes over the Mediterranean basin for the years 2031–2060, considering also fire frequency using the Canadian Fire Weather Index (FWI) which takes into account the effects of fuel moisture and wind on fire behavior. According to this analysis and considering a 2°C mean increase in temperature, fire risk will increase by 2 6 additional weeks over all land areas and a large part of this increase in fire risk is related to extreme fire risk (FWI > 30). The study showed also that South of France, and coastal areas of the rest of Mediterranean Region will experience a significant increase in the number of days with fire risk (1–4 weeks), but not in the number of days under extreme fire risk.

However it shouldn't be neglected that also the increase in $CO_2$ concentration can have important implications both on vegetation structure, species composition and amount of fuel load. Several manipulative studies have shown that productivity (and thus potentially fuel load) will increase in forests ecosystems by 23% on average of the different forest ecosystems regardless of productivities and ecosystems as a consequence of the increase in $CO_2$ concentration (Norby et al. 2005).

Few studies have focused on the effects of climate change on the contents of terpenoids compounds which, as described above, are strongly affecting flammability in several plant species. As these compounds are considered protectors against thermal shocks and more generally against oxidative stress, it is expected that the production of these compounds will increase under warmer climate, despite increase in $CO_2$ concentrations as well as $O_3$ levels at leaf level will probably inhibit terpenoid production and emission from tree species (Calfapietra et al. 2008).

Drought, often combined with warmer climates, often exacerbates the tree mortality which further increases forest flammability inducing a positive fire feedback loop that often compromises the survivability of forests especially in regions characterized by high stress conditions (Nepstad et al. 1999; Cochrane et al. 1999).

It has been also noticed that those forests characterized by drought-induced reductions in leaf area index are also the most susceptible to fire and are likely to suffer logging and further increases in flammability (Uhl and Kauffman 1990), as well as are likely to be ignited through escaped agricultural fires (Nepstad et al. 2001).

## 4.5 Incorporating Fuel Flammability and Fire Hazard Estimates in Post-Fire Management Decisions

Table 4.1 can be used for broad area estimation of fire hazard, such as for assessing the overall fire problem in a country, or for providing general guidance on forest type selection when planning post-fire re-vegetation of a burned site.

In regard to the former, estimating fire hazard over broad areas may support decision making about resources and funds distribution within countries, and it may help in comparing the fire problem between countries based on their vegetation without the influence of fire suppression effectiveness which affects the size of yearly burned area. The information about expected changes due to global warming should also be taken into consideration in planning for the future.

In regard to post-fire re-vegetation, although changing forest type is not an easy decision and should be done very carefully, the opportunity often arises when enrichment or complete change of vegetation during reforestation works may be clearly justified. One such example is the reforestation of the site of ancient Olympia in Peloponnese Greece which was burned in August 2007, during the worst fires ever experienced in Greece. When planning urgent site reforestation in preparation for the lighting of the Olympic torch for the Olympic games of 2008 in Beijing, the scientists of the Institute of Mediterranean Forest Ecosystems and Forest Products Technology in Athens studied the ancient texts describing the vegetation on the site in antiquity which proved that the modern day *Pinus halepensis* forest was not the prevailing vegetation at that time. Accordingly, when replanting the site they re-introduced the broadleaved species that occupied the site in ancient times resulting in reduced fire hazard for the regenerating forest (Lyrintzis et al. 2010).

On the other hand, assessments of fire hazard for fire management purposes should ideally be quantitative and based on fire behavior modeling. This implies usage of a fuel-complex description typology, which is specific of the fire model being employed. The fuel model concept, as per Rothermel (1972), is well suited for objective, quantitative fire hazard classification. A fuel model is a set of fuel descriptors that 'feeds' Rothermel's fire spread model. The original set of stylized fuel models (Anderson 1982) covers the variety of wildland fuel in coarse terms, according to the main fire spread vector (litter, herbs, shrubs, slash), and can accordingly be used to broadly classify the baseline fire potential of a given vegetation type. Custom fuel models (Brown et al. 1982; Burgan and Rothermel 1984) portray the specificities of a fuel-complex, and should as much as possible be fine tuned with real-world fire behavior data (Cruz and Fernandes 2008). Fuel models for local, regional or national application have been developed in Europe (e.g. Allgöwer et al. 1998; Dimitrakapoulos 2002; Cruz and Fernandes 2008), but a unifying approach has not been attempted. Fuel models depict surface fire characteristics and have to be combined with canopy fuel descriptors in order to describe stand-level flammability, at least in crown-fire prone forest types. Furthermore, in post-fire management decisions, changes of fire hazard with years after fire should also be considered, incorporating into the planning some type of simple modeling of biomass growth with age (e.g. Kazanis et al. 2011) or looking more carefully into the various theoretical models of fuel accumulation (McCarthy et al. 2001; Zylstra 2011). It is this type of assessment of fire hazard that is appropriate for presuppression planning at local level and for supporting decisions about the need for forest fuel management, the place where it should be applied and the standards that should be followed.

## 4.6 Key Messages

Summarizing, it can be said that the basics of what affects fuel flammability and fire hazard are quite well known for some time now. However:

- Recent research findings in regard to fuel chemical properties and on the influence of fuel distribution in flame propagation within fuel strata appear to offer new insights, allowing fine tuning of fuel management methods for fire hazard reduction;
- Post-fire management can certainly benefit from this knowledge as it may affect planting standards, silvicultural treatments, fuel treatment for hazard reduction, etc.;
- The broad classification of European forest vegetation types can be a valuable tool for policy decisions allowing macroscopic analyses such as damage potential assessment, firefighting strength requirements, etc.
- Both approaches of fire hazard assessment above can help in guiding post-fire management decisions in regard to species selection for post-fire site reforestation.

# References

Alessio G, Peñuelas J, Lusia J, Ogaya R, Estiarte M, De Lillis M (2008) Influence of water and terpenes on flammability in some dominant Mediterranean species. Int J Wildland Fire 17:274–286

Allgöwer B, Harvey S, Rüegsegger M (1998) Fuel Models for Switzerland: Description, Spatial Pattern, Index for Crowning and Torching. In: Viegas DX (ed.) Proceedings of the 3rd International Conference on Forest Fire Research, Nov. 16–20, 1998, Luso, Portugal. ADAI, Univ of Coimbra, Portugal, pp 2605–2620

Anderson HE (1970) Forest fuel ignitibility. Fire Technol 6:312–319

Anderson HE (1982) Aids to determining fuel models for estimating fire behavior. USDA For. Serv. Gen. Tech. Rep. INT-122, Ogden, Utah

Blasi C, Bovio G, Corona P, Marchetti M, Maturani A (eds.) (2005) Fires and ecosystem complexity. From forest assessment to habitat restoration. Ministero dell'Ambiente e della Tutela del Territorio, Società Botanica Italiana, Palombi & Partner, Rome

Brown JK, Oberhew RD, Johnson CM (1982) Handbook for inventorying surface fuels and biomass in the interior west. USDA Forest Service, Gen. Tech. Rep. INT-129, Ogden, UT

Brown AL, Hames BR, Daily JW, Dayton DC (2003) Chemical analysis of solids and pyrolytic vapors from wildland trees. Energ Fuel 17(4):1022–1027

Burgan RE, Rothermel RC (1984) BEHAVE: fire behaviour prediction and fuel modelling system-FUEL subsystem. USDA Forest Service, Gen. Tech. Rep. INT-167, Ogden, UT

Button D (1987) The smell of Christmas. In 'Alaska Science Forum', 21 Dec, Article 852. (Geophysical Institute, University of Alaska: Fairbanks, AK) Available at http://www.gi.alaska.edu/ScienceForum/ ASF8/852.html

Calfapietra C, Scarascia Mugnozza G, Karnosky DF, Loreto F, Sharkey TD (2008) Isoprene emission rates under elevated $CO_2$ and $O_3$ in two field-grown aspen clones differing for their sensitivity to O3. New Phytol 179:55–61

Cary GC (2002) Importance of a changing climate for fire regimes in Australia. In: Bradstock RA, Williams JE, Gill AM (eds) Flammable Australia: the fire regimes and biodiversity of a continent. Cambridge University Press, Cambridge, pp 26–46

Cochrane MA, Alencar A, Schulze MD, Souza CM, Nepstad DC, Lefebvre P, Davidson EC (1999) Positive feedbacks in the fire dynamic of closed canopy tropical forests. Science 284:1832–1835

Cruz MG, Fernandes PM (2008) Development of fuel models for fire behaviour prediction in maritime pine (Pinus pinaster Ait.) stands. Int J Wildland Fire 17:194–204

Davis FW, Michaelsen JC (1995) Sensitivity of fire regime in chaparral ecosystems to global climate change. In: Oechel WC, Moreno JM (eds) Global Change and Mediterranean-type Ecosystems. Springer-Verlag, New York, pp 435–456

De Lillis M, Bianco PM, Loreto F (2009) The influence of leaf water content and isoprenoids on flammability of some Mediterranean woody species. Int J Wildland Fire 18:203–212

Dimitrakopoulos AP (2002) Mediterranean fuel models and potential fire behaviour in Greece. Int J Wildland Fire 11:127–130

Dimitrakopoulos AP, Panov PI (2001) Pyric properties of some dominant Mediterranean vegetation species. Int J Wildland Fire 10:23–27

European Environment Agency (2007) European forest types: categories and types for sustainable forest management reporting and policy. European Environment Agency. Technical report No 9/2006 (2nd edn), Copenhagen, Denmark. p 111

FAO (2006) Fire management: voluntary guidelines. Principles and strategic actions. Fire Management Working Paper 17. Rome. Available at www.fao.org/forestry/site/35853/en

Fernandes PM (2009) Combining forest structure data and fuel modelling to assess fire hazard in Portugal. Ann For Sci 66:415p1–415p9

Fried JS, Torn M, Mills E (2004) The impact of climate change on wildfire severity: a regional forecast for northern California. Clim Chang 64(1–2):169–191

Giannakopoulos C, Le Sager P, Bindi M, Moriondo M, Kostopoulou E, Goodess CM (2009) Climatic changes and associated impacts in the Mediterranean resulting from a 2°C global warming. Global and Planetary Change 68(3):209–224

Gill AM, Moore PHR (1996) Ignitability of leaves of Australian plants. CSIRO Division of Plant Industry, Canberra, 34 p

Gill AM, Zylstra P (2005) Flammability of Australian forests. Aust For 68(2):88–94

Hyde JC, Smith AMS, Ottmar RD, Alvarado EC, Morgan P (2011) The combustion of sound and rotten coarse woody debris: a review. Int J Wildland Fire 20(2):163–174

IPCC (2007) Climate Change 2007: the physical science basis. Cambridge University Press, Cambridge

Kazanis D, Xanthopoulos G, Arianoutsou M (2011) Understorey fuel load estimation along two post-fire chronosequences of Pinus halepensis Mill. forests in Central Greece. J For Res. doi: 10.1007/s10310-011-0250-0

Latham DJ, Schlieter JA (1989) Ignition probabilities of wildland fuels based on simulated lightning discharges. US Forest Service, Intermountain Forest and Range Experiment Station, Research Paper INT-411, Ogden, UT

Llusià J, Peñuelas J (2000) Seasonal patterns of terpene content and emission from seven Mediterranean woody species in field conditions. Am J Bot 87:133–140

Lyrintzis G, Baloutsos G, Karetsos G, Daskalakou EN, Xanthopoulos G, Tsagari C, Mantakas G, Bourletsikas A (2010) Olympic rebirth. Wildfire 19(1):12–20

Mackinnon AJ (1987) The Effect of the composition of wood on its thermal degradation. Strathclyde University, Glasgow

Martin RE, Gordon DA, Gutierrez MA (1993) Assessing the flammability of domestic and wildland vegetation. In: Proceedings of the 12th Conference on Fire and Forest Meteorology, Oct. 26–28, 1993, USA. Soc. of Amer. For., Jekyll Island, Georgia, Bethesda, pp 130–137, 796

Massari G, Leopaldi A (1998) Leaf flammability in Mediterranean species. Plant Biosyst 132(1):29–38

McCarthy MA, Gill AM, Bradstock RA (2001) Theoretical fire interval distributions. Int J Wildland Fire 10:73–77

Montgomery KR, Cheo PC (1971) Effect of leaf thickness on ignitibility. For Sci 17:475–478

Mouillot F, Rambal S, Joffre R (2002) Simulating climate change impacts on fire frequency and vegetation dynamics in a Mediterranean-type ecosystem. Glob Chang Biol 8:423–437

Mutch RW, Philpot CW (1970) Relation of silica content to flammability in grasses. For Sci 16:64–65

Nepstad DC, Verissimo A, Alencar A, Nobres C, Lima E, Lefebvre P, Schlesinger P, Potter C, Moutinho P, Mendoza E, Cochrane M, Brooks V (1999) Large scale impoverishment of Amazonian forests by logging and fire. Nature 398:505–508

Nepstad D, Carvalho G, Barros AC, Alencar A, Capobianco JP, Bishop J, Moutinho P, Lefebvre P, Silva UL, Prins E (2001) Road paving, fire regime feedbacks, and the future of Amazon forests. For Ecol Manag 154:395–407

Nimour Nour E (1997) Inflammabilité de la végétation méditerranéenne. Thesis report. Université de Provence, Marseille, France

Norby RJ, Delucia EH, Gielen B, Calfapietra C, Giardina CP, King JS, Ledford J, Mccarthy HR, Moore DJP, Ceulemans R, De Angelis P, Finzi AC, Karnosky DF, Kubiske ME, Lukac M, Pregitzer KS, Scarascia-Mugnozza GE, Oren RE, Schlesinger WH (2005) Forest response to elevated $CO_2$ is conserved across a broad range of productivity. Proc Natl Acad Sci USA 102:18052–18056

Nuñez-Regueira L, Rodriguez-Anon JA, Proupin J, Mourino B, Artiaga-Diaz R (2005) Energetic study of residual forest biomass using calorimetry and thermal analysis. J Therm Anal Calorim 80:457–464

Ormeño E, Cespedes B, Sanchez IA, Velasco-Garcia A, Moreno JM, Fernandez C, Baldy V (2009) The relationship between terpenes and flammability of leaf litter. For Ecol Manag 257:471–482

Owens MK, Lin CD, Taylor CA, Whisenant SG (1998) Seasonal patterns of plant flammability and monoterpenoid content in Juniperus ashei. J Chem Ecol 24:2115–2129

Papió C, Trabaud L (1990) Structural characteristics of fuel components of five Mediterranean shrubs. For Ecol Manag 35:249–259

Rothermel R (1972) A mathematical model for predicting fire spread in wildland fuels. USDA, Forest Service, Intermountain Forest and Range Experiment Station, Research Paper INT-115, Ogden, UT

Rundel PW (1981) Structural and chemical components of flammability. In: Mooney HA, Bonnicksen TM, Christensen NL, Lotan JE, WA Reiners (eds) Proceedings of the conference on fire regimes and ecosystem properties. USDA Forest Service, Gen. Tech. Rep. WO – 26. pp 183–207

Saura-Mas S, Paula S, Pausas JG, Lloret F (2010) Fuel loading and flammability in the Mediterranean Basin woody species with different post-fire regenerative strategies. Int J Wildland Fire 19:783–794

Schwilk DW (2003) Flammability is a niche construction trait: canopy architecture affects fire intensity. Am Nat 162:725–733

Taylor SW, Pike RG, Alexander ME (1997) Field guide to the Canadian forest fire behavior prediction (FBP) system. Special Report 11, Fire Management Network, Northern Forestry Centre, Canadian Forest Service, Natural Resources Canada. Edmonton, Alberta. viii + 60 p

Torn MS, Fried JS (1992) Predicting the impacts of global warming on wildland fire. Climatic Change 21(3):257–274

Trabaud L (1976) Inflammabilite et combustibilite des principales especes mediterraneennes. Oecol Plant 11(2):117–136

Uhl C, Kauffman JB (1990) Deforestation, fire susceptibility and potential tree responses to fire in the eastern Amazon. Ecology 71:437–449

Westerling AL, Bryant BP (2008) Climate change and wildfire in California. Climatic Change 87(Suppl 1):S231–S249

White RH, Zipperer WC (2010) Testing and classification of individual plants for fire behaviour: plant selection for the wildland–urban interface. Int J Wildland Fire 19:213–227

Xanthopoulos G, Wakimoto RH (1993) A time-to-ignition – temperature – moisture relationship for branches of three western conifers. Can J For Res 23:253–258

Zylstra P (2011) Forest flammability: modelling and managing a complex system. Ph.D. thesis, School of Physical, Environment and Mathematical Sciences, The University of New South Wales, the Australian Defence Force Academy, Canberra, NSW, Australia, 434 p

# Chapter 5
# Fire Ecology and Post-Fire Restoration Approaches in Southern European Forest Types

**V. Ramón Vallejo, Margarita Arianoutsou, and Francisco Moreira**

## 5.1 Plant Adaptations to Fire and Post-Fire Response Mechanisms

Having suffered the repetitive action of fire in the course of their evolution, many plant species of Mediterranean environments have developed special adaptations to cope with it, ensuring their persistence in time. Plants possess two basic mechanisms to regenerate after fire: (i) vegetative regeneration (resprouting) of the same burned individuals, and (ii) establishment of new individuals through seed germination (Arianoutsou 1999; Bond and van Wilgen 1996; Whelan 1995). Knowing which mechanism exists in a given burned forest or shrubland is critical to evaluate the post-fire management alternatives.

Relatively few Mediterranean species do not show any specific regeneration mechanism after fire, and in this case the recovery of burned populations depends upon colonization from nearby unburned areas.

Resprouting after fire is a widespread trait in all fire prone environments and in all dicotyledonous plant lineages (Pausas and Keeley 2009). Resprouting occurs

V.R. Vallejo (✉)
Fundacion CEAM, Parque Tecnologico, Paterna, Spain
e-mail: ramonv@ceam.es

M. Arianoutsou
Department of Ecology and Systematics, Faculty of Biology, School of Sciences,
National and Kapodistrian University of Athens, Athens, Greece
e-mail: marianou@biol.uoa.gr

F. Moreira
Centre of Applied Ecology, Institute of Agronomy, Technical University of Lisbon,
Lisbon, Portugal
e-mail: fmoreira@isa.utl.pt

F. Moreira et al. (eds.), *Post-Fire Management and Restoration of Southern European Forests*, Managing Forest Ecosystems 24, DOI 10.1007/978-94-007-2208-8_5, © Springer Science+Business Media B.V. 2012

**Fig. 5.1** Geophytes resprouting after fire from underground bulbs; *lower image*: *Urginea maritima*; *upper inner image Sternbergia* sp.; *upper image*: *Crocus* sp. (Source: Margarita Arianoutsou, Univ. of Athens)

usually at the root crown of the burned plants from dormant buds that remained intact after fire, being protected by the insulating soil. Resprouting is also the regeneration mechanism of plants that possess lignotubers, such as *Euphorbia acanthothamnos*, *Erica arborea*, *E. australis*, *E. multiflora*, or underground bulbs as many geophytes do, e.g. *Cyclamen* spp., *Muscari commosum*, *Urginea maritima*, *Crocus* spp. (Fig. 5.1).

Post-fire recovery of resprouters is a rather straightforward process as new shoot growth is supported by the almost intact belowground biomass surviving the fire.

**Fig. 5.2** Woody species resprouting after fire from the root crown (*upper left image*). *Upper image on the right*: *Phlomis fruticosa*; *Lower image on the left*: *Quercus fraineto*. Often oaks can regenerate through epicormic growth, that is by developing new branches directly on the burned tree trunk (*lower image on the right*). (Source: Margarita Arianoutsou, Univ. of Athens)

Resprouts usually reach their reproductive maturity rather quickly, producing flowers and fruits a couple or, at most, a few years after resprouting. All oak trees (*Quercus* spp.) and most shrub species of phrygana and maquis all over the Mediterranean are resprouters (Fig. 5.2).

The species that have resprouting as their only regeneration mode after fire are called *obligatory resprouters*. *Facultative resprouters* are species that primarily regenerate through seed germination but they can also regenerate through resprouting, e.g. *Sarcopoterium spinosum* and *Erica* spp. Maquis species regenerate almost immediately after fire, while seasonal dimorphic species (phrygana) may resprout after the first autumn rains. This difference has been attributed to the different penetration depths of their root systems (Arianoutsou 1999). Resprouting is ensured as long as there is adequate storage of carbohydrates in the root crown, the lignotuber or the bulb (Jones and Laude 1960).

The second adaptation strategy for plant species to cope with fire is seedling recruitment. The pines of the thermo-Mediterranean zone (e.g. *Pinus halepensis* and *Pinus brutia*), most of the rockroses (Cistaceae) and many herbaceous leguminous species are *obligatory seeders* (Fig. 5.3). Seedlings emerge after the first autumn rains from seeds that were either dispersed before fire, having remained dormant in the soil as a soil seed bank, or dispersed after fire from a canopy seed bank.

**Fig. 5.3** Obligate seeders. *Upper image*: (*left*) serotinous cone of *Pinus halepensis*; (*right*) *Pinus halepensis* seedling. *Lower image*: *Cistus creticus* seedlings massively appearing on the burned ground. (Source: Margarita Arianoutsou, Univ. of Athens)

Hard coated seeds normally lie dormant in the litter or topsoil layers and are released from dormancy by the heat shock induced by fire (Doussi and Thanos 1994; Ferrandis et al. 2001; Keeley and Fotherigham 2000; Papavassiliou and Arianoutsou 1993). Seeds forming these seed banks may also be stimulated by other factors related to fire cues, like high concentration of nitrates (Pérez-Fernández and Rodríguez-Echeverría 2003; Thanos and Rundel 1995), which is common in post-fire soils (Arianoutsou-Faraggitaki and Margaris 1982), change in the red/far-red ratio induced by canopy removal (Roy and Arianoutsou-Faraggitaki 1985) or smoke (Dixon et al. 1995; Pérez-Fernández and Rodríguez-Echeverría 2003).

Several pine species of the Mediterranean environments store their seeds in closed cones forming canopy seed banks; these pines are called *serotinous* (Fig. 5.3). High temperatures developing during fire on the plant canopy induce the dehiscence of the hard cones, the melting down of the resin keeping the scales of the cones tight and the subsequent seed dispersal (see Leone et al. 1999; Thanos and Daskalakou 2000; Ne'eman et al. 2004). Seed germination requires imbibition of the embryo and this takes place after the first autumn rains. Seedlings often appear in large numbers; however, high mortality usually occurs after the first dry season, which is mainly due to the drought effect (Arianoutsou and Margaris 1981; Papavassiliou and Arianoutsou 1993; Daskalakou and Thanos 2004). *Pinus halepensis* (Aleppo pine) and *Pinus brutia* (Brutia pine) are the most typical examples of Mediterranean serotinous pines.

Several plant species do not posses any specific post-fire adaptation mechanism, so once a fire occurs they can locally disappear. Typical examples are *Coridothymus capitatus* and *Juniperus phoenicea*. Similarly, non-serotinous pines, like *Pinus sylvestris* and *Pinus nigra,* once they are burned in an intense fire they are dependent for their recovery on seed dispersal from adjacent unburned patches. However, in the case of a light or moderate surface fire in which trees are protected by their bark and their canopy is not affected, seed germination may occur (Retana et al. 2002; Ordóñez et al. 2004; Arianoutsou et al. 2008, 2010a).

Ecological effects are shaped by fire regimes, namely the collective effects of fire frequency, intensity/severity, season, and size (Gill et al. 2002; Gill and Bradstock 2003). Fire frequency and severity are the most critical factors directly affecting plant responses. For the long term survival of the plants it is essential to know not only their adaptive traits towards a 'normally' occurring fire event, but also how they are affected by the fire regime. For example, there is an interplay between the capacity of species to survive and regenerate from fire and the interval between fires (fire recurrence). All species require a characteristic time length to replenish their regeneration capacity. Plants that are killed by fire and regenerate through seeds rely on this seed germination in order to persist at the specific location. For these plants, there must be sufficient time between successive fires for the seedlings to mature and produce seeds and hence replenishing soil and canopy seed banks. This time will vary between species that flower within the first post-fire year (such as herbaceous legumes), to those that may take 6–8 or more years to reach maturity (as pines do). If another fire occurs before these plants have matured, dramatic changes in the vegetation composition and physiognomy may occur (Arianoutsou et al. 2002, 2011). This is what is called the *immaturity risk*. The same holds for resprouters, when time between two consecutive fires is not adequate for them to replenish their carbohydrates reserves necessary for their post-fire regeneration. However, some species show extremely high capacity to resprout after frequent fires, beyond actual fire recurrence, as is the case of *Quercus coccifera* (Trabaud 1991a; Delitti et al. 2005).

Fire severity is a function of the amount of heat released by fire and the duration of the heating. It may determine the proportion of individuals that survive a particular fire, and it may also affect regeneration processes such as seed germination or resprouting potential. Fire-severity dependent mortality in resprouting species has been extensively documented for several species in the Iberian Peninsula (López-Soria and Castell 1992; Lloret and López-Soria 1993). Severe fires usually kill the stems of resprouters, but it seems that their regeneration at the population level is not generally affected. In relation to seeders, seeds lying in the soil seed bank seem not to be negatively influenced by intense fires. On the contrary, there are several references in the literature about heat induced seed germination after fire (e.g. Arianoutsou and Margaris 1981; Thanos and Georgiou 1988; Doussi and Thanos 1994; Keeley and Bond 1997).

## 5.2 Implications of Altered Fire Regimes Induced by Climate Change

Climate has a clear influence on fire regime (Pausas 2004). Consequently, if the expected changes in climate become a reality over the next century (IPCC 2007), an altered fire regime could have serious impacts upon Mediterranean ecosystems and their resilience towards fire (see also Chap. 11). Liu et al. (2010) have investigated the trend in global wildfire potential under the climate change due to the greenhouse effect. They measured fire potential by the Keetch-Byram Drought Index (KBDI), which is calculated using the observed maximum temperature and precipitation, and projected changes at the end of this century (2070–2100) by general circulation models (GCMs) for present and future climate conditions, respectively. Their analysis showed that future wildfire potential increases significantly in the United States, South America, central Asia, southern Europe, southern Africa, and Australia. Fire potential seems to move up by one level in these regions, from currently low to future moderate potential or from moderate to high potential. The largest relative changes were predicted for southern Europe. These findings are calling for increased and pro-active management efforts for preventing the potential catastrophic consequences and for ensuring an effective ecosystem recovery.

    If fires recur more or less regularly, in the case of a climate change induced fire regime, selection pressure will favor those organisms that take advantage of the recurrence at a given interval and eliminate those that cannot follow (Flannigan et al. 2000). The vital attributes scheme developed by Noble and Slatyer (1980) has served as the basis of several predictive models to distinguish key groups of plant species that may be sensitive to changes in fire regime (Arianoutsou 1999, 2004; Arianoutsou et al. 2011; Kazanis and Arianoutsou 2004; Lloret and Vilà 2003; Pausas 1999; Pausas et al. 2004b). A key outcome of the vital attributes system is that differing functional types of plants will have differential sensitivity to recurrent disturbances such as fire. Functional types most sensitive to disturbance are those in which established individuals (adults and juveniles) are prone to death by fire (i.e. no capacity for vegetative recovery) and where seed banks may be depleted by fire. Consequently, sensitive functional types of this kind will be characterized by species that exhibit a high probability of mortality of juveniles and adults irrespective of fire intensity, plus seed bank types where germination is negatively affected by fire.

## 5.3 Fire Prone European Forest Types

In an overview of natural disturbances in European forests during the nineteenth and twentieth centuries, Schelhaas et al. (2003) estimated that forest fires were responsible for 16% of the total wood volume lost per year (35 million m³), with larger damages caused only by storms (53%). The same authors have shown that during the period 1960–2000, most of the forest fires occurred in west Mediterranean, namely Spain and Portugal (44.9% of the total area burned), followed by the central Mediterranean (26.1%), mainly Italy and Slovenia, and east Mediterranean (17%), mainly Greece and Turkey. The remaining areas affected by wildfires included the sub-Atlantic (France mainly), with 7.3%, and the Central Pannonic (2.2%), mainly Poland and Romania.

These results highlight the significance of southern Europe as the major geographic area affected by wildfires in Europe. In terms of countries, Portugal, Spain, France, Italy, Greece and Turkey are the most affected ones (Schmuck et al. 2010), and forests in these countries are particularly fire prone. However, some countries in central and eastern Europe, in some years with dry and hot weather, have thousands of hectares burned (e.g. Bulgaria and Poland) (Schmuck et al. 2010).

Following the EEA (2007) forest type classification, there are two major forest categories – (a) broadleaved evergreen forests, and (b) coniferous forests of the Mediterranean, Anatolian and Macaronesian regions – occurring in the regions of Europe more affected by wildfires.

Broadleaved evergreen forests occur mainly in the thermo and meso-Mediterranean vegetation belt, whose climate determines the dominance of broadleaved sclerophyllous trees. The present distribution and physiognomy of Mediterranean evergreen oakwoods is the result of a long history of anthropogenic disturbance by clearance, coppicing, fires and overgrazing, resulting in vast areas covered today by degraded stages of evergreen oakwoods: arborescent matorral, maquis and garrigues (EEA 2007). The more fire prone types are the Mediterranean evergreen oak forests with *Quercus suber, Q. ilex, Q. rotundifolia* and *Q. coccifera* as the main species. The cork (*Q. suber*) and holm (*Q. ilex/ Q. rotundifolia*) oak forests, corresponding to the drier types, are the most widespread evergreen oak forests in the Mediterranean region and the more fire prone. Kermes oak (*Q. coccifera*) garrigues are another fire prone type. Among these *Quercus* species, the cork oak has the unique feature of having a bark with commercial interest (cork) that is exploited, which makes it very peculiar within the post-fire management context. This species is addressed in a separate chapter in this book (Chap. 9).

The coniferous forests of the Mediterranean, Anatolian, and Macaronesian regions is a broad category of conifer (pines, firs, junipers, cypress, cedar), mainly xerophytic, forest communities distributed throughout Europe (EEA 2007). The more fire prone coniferous forest types are the thermophilous pine forests with *Pinus pinea, P. pinaster, P. halepensis* and *P. brutia*, largely widespread in the lowlands of the circumediterranean region. These correspond to the forest types more affected by wildfires in Europe, in particular the three latter types. Mediterranean and Anatolian black pine (*Pinus nigra*) forests (and plantations), as well as Mediterranean and Anatolian Scots pine (*Pinus sylvestris*) forests in the mountain ranges of the Iberian peninsula and northern Greece, are also fire prone types in Southern Europe. However, in the post-fire management context, these pine forests differ from the former by the fact that they do not have serotinous pines (as *P. pinaster*, *P. halepensis* and *P. brutia* do) that enable the natural post-fire regeneration of these forests. These two broad types of pines (serotinous and non-serotinuous) are addressed in separate chapters (Chaps. 7 and 8).

Thermophilous deciduous forests are a third forest category that is also fire prone, although to a lesser extent. Thermophilous deciduous forests occur mainly in the supra-Mediterranean vegetation belt. Anthropogenic exploitation has modified the natural composition of these forests, leading in most cases to the elimination of species without commercial interest or with poor resprouting capacity or, conversely, the introduction of other forest species that would not occur naturally (chestnut) (EEA 2007). Within this category, the more fire prone types are probably the *Quercus pyrenaica*, *Quercus faginea*, and chestnut *Castanea sativa* forests

In addition to these native forests, plantations have increased in the last decades, mainly in some countries such as Portugal and Spain. Although these are considered a different forest type by EEA (2007), in this book they are pooled in the corresponding forest type, except in cases where the species used is exotic (e.g. eucalyptus plantations in Portugal). Independently of being cultivated or having invasive character, exotic species are an increasingly important post-fire management issue in some regions in Europe. This topic is examined in more detail in Chap. 10.

The total area burned per year is not exclusively composed of forests. Some non-forest land covers, and in particular shrublands, are highly susceptible to wildfires and represent a significant proportion of the total area burned (e.g. Pausas et al. 2008). Furthermore, increased fire frequency is turning former forests into shrublands, and promoting homogeneous landscapes covered by different shrubland types (Moreira et al. in press). The post-fire management of shrublands is dealt with in Chap. 12.

## 5.4   Major Questions in Mediterranean Forest Restoration

This section introduces some key issues in post-fire management which are common to all forest types. These include measures to promote soil protection, the management of burned trees, restoration or conversion, the use of active or indirect restoration, the management of herbivory, alien species, and pests and diseases.

### 5.4.1   Soil Protection to Reduce Erosion Risk

Soil erosion is among the most damaging post-fire processes. Soil degradation and erosion risk may be greatly enhanced by fires through the combined effect of direct soil heating and temporal loss of protective soil cover (Vallejo 1999). Water erosion may produce onsite loss of soil productivity and offsite siltation and flooding causing damages to humans and structures. Soil losses may be irreversible at the ecological time scales if they exceed soil formation rates, which are low in Mediterranean regions as in dry regions in general (Wakatsuki and Rasyidin 1992). Therefore, for ecological and safety reasons, reducing soil erosion and runoff risk should be the first priority in post-fire management (Vallejo and Alloza 1998; Vallejo 1999).

The major factors affecting soil erosion risk (Wischmeier and Smith 1978; Scott et al. 2009) are related to topography (slope grade and length), rainfall intensity, soil erodibility (related to soil properties), plant cover (including litter), and artificial erosion control measures like slope terracing. Fire may significantly affect soil erodibility, depending on fire severity, and, especially, plant cover, although both factors are partially interrelated: for low severity fires, plant cover and litter may partly remain thus protecting the soil from wind and water erosion. At the landscape and even at the hillslope scale (Schoennagel et al. 2008; Gimeno-García et al. 2011), fire severity is usually quite heterogeneous in space and not very high at the soil level for surface and crown fires which are the most common in the Mediterranean Basin. In severely burned areas, plant cover recovery rate is controlling how long after fire the soil will be exposed to the direct impact of raindrops, especially for heavy rains, and to excessive water erosion risk. Modelling these factors will allow identifying areas exposed to high soil erosion risk as a basis for planning post-fire soil protection actions (see Chap. 1 and Alloza and Vallejo 2006).

From the post-fire management perspective, the most critical factor affecting soil erosion risk which is susceptible of manipulation is plant/mulch cover. In Southern Europe, high intensity rainfall is most likely to occur by the end of summer and autumn. Therefore, for summer fires, the most common in the region, there is a high erosion risk right after the fire, when low plant cover and heavy rains may coincide in time (Vallejo 1999). The risk will continuously decrease as plant cover regenerates. Hence soil protection measures should be taken as soon as possible after the fire in areas where high erosion and runoff risk have been identified, and they have to be effective in the very short term. These measures are grouped under the concept of *post-fire emergency rehabilitation* (see Robichaud et al. 2000 for a critical review) or *emergency interventions* (see Chap. 1). The Forest Service of the United States Department of Agriculture has published a comprehensive catalogue of treatments for emergency response after forest fires (BAER, Napper 2006). Following the BAER procedures, the main topics to take into account when defining where to apply those emergency measures include fire severity, presence of water-repellent soils, mapping of effective soil cover, flood or debris risky areas, riparian stability assessment, potential erosion or sedimentation and water quality deterioration, and situation of infrastructures. An emergency plan should be set defining the priorities, time frame for implementation, personnel and funding availability, coordination of active agents (authorities, stakeholders and politicians), economical, social and environmental costs.

Land (hillslope) treatments stabilize the burned areas by preventing or mitigating the negative effects of fire. The most used techniques include mulching, erosion barriers (Fig. 5.4), scarification, slash spreading, planting and seeding, control of

**Fig. 5.4** Forms of barriers to control post-fire soil erosion; *upper image*: branches barriers placed across a stream (Arkadia 2007 fire, Greece. Source Margarita Arianoutsou, Univ. of Athens); *middle image*: log barriers (Mt. Parnitha 2007 fire, Greece. Source Margarita Arianoutsou, Univ. of Athens); *lower image*: fine branches barriers, (Useres 2007 fire, Valencia Region, Spain. Source: Teresa Gimeno, CEAM)

invasive species and protection of special sites and habitats. Stream and channel treatments are used to reduce or mitigate water control and quality, trap sediment and debris and maintain stream and channel characteristics.

The application of an organic layer of mulch, either alone or combined with seeding native grasses, is an effective rehabilitation option on steep slopes with low plant cover and high erosion risk, as it is aimed at reducing rain splash, surface flow, soil crusting and compaction, thereby increasing infiltration. To be effective, this technique should be applied soon after the fire and before the heavy autumn rains, which means that areas vulnerable to erosion must be identified as soon as possible. In burned pine forests on steep slopes and soft soils, log dams or contour-felled barriers may be also effective post-fire management practices for reducing physical soil degradation and erosion.

### 5.4.1.1 Identification of the Conditions that Might Require Emergency Actions: Do They Relate to Forest Type?

Emergency treatments should be implemented in burned areas showing high erosion and runoff risk, with slow natural plant recovery rate, and when there are high values at risk downslope. Among these factors, the only that may be specific to a given forest type is the post-fire recovery rate of vegetation, as it depends on plant regeneration strategies (see Sect. 5.1). Hardwood forests usually resprout after fire, therefore if stand density is high enough, plant cover regeneration would quickly protect the soil in front of erosive agents.

For forests dominated by obligate-seeders, e.g. pine forests, plant cover regeneration in the short-term (up to 1–2 years after fire) mostly depends on understory vegetation, and this may vary within a forest type due to differences in land use and forest management history, stand age and density, soil characteristics, etc. For example, in Aleppo pine forests we can find a shrubby understory dominated by resprouters on limestone, with very low post-fire erosion risk, or an understory dominated by obligate seeders on old agricultural fields, with high erosion risk. Therefore, for pine forests, no species specific approach on post-fire emergency actions can be generalized.

## 5.4.2 Salvage Logging

After forest fires, one of the first decisions to take is how to manage the affected timber. Harvesting of commercially valuable dead or damaged trees (salvage logging) is the most common practice, provided burned timber has enough economic value to pay for the logging operations and to yield some benefit to the forest owner. Timber value continuously decreases as time passes after fire because of the wood decay, thus the forest owner is interested in logging as soon as possible to maximize economic benefit. In practice, for large fires it is almost impossible to rapidly harvest

**Table 5.1** Argued pros and cons of salvage logging

| Potential benefits | Potential negative impacts |
|---|---|
| To obtain some economic benefit of charred logs[a] | |
| To avoid boring insect pests[b] (e.g. Scolitydae) | |
| To improve pine germination (if logging is immediate) and avoid damage to regenerated pines[c] | Logging has detrimental effects on seedling growth[d] |
| | Microsites around burned trees favour regeneration and pine seedling germination[e] |
| | Salvage logging reduces forest breeding birds and their seed dispersal activity which is critical for late successional species[f]; also reduces deadwood associated fauna[g] |
| Trees naturally falling down (usually 2–3 years after fire) expose tree crown to erosion[h] | Dragging charred logs may produce soil surface degradation and soil erosion – rill erosion[i] |
| Risk of accidents by falling trees in inhabited areas | |
| To reduce landscape visual impact | |

[a](SAF 1996); [b]Amman and Ryan (1991); [c] Roy (1956); [d]Gayoso and Iroumé (1991); Beghin et al. (2010); [e]Ne'eman (1997); Ne'eman and Izhaki (1998); Bautista and Vallejo (2002); [f]Castro et al. (2009); [g]Cobb et al. (2010); [h]Poff (1989); [i]Terry (1994); Bautista et al. (2004)

all burned forest area, therefore priorities should be established and this gives time for planning and allows for introducing ecosystem conservation criteria, in addition to the immediate economic value.

Salvage logging is controversial (Lindenmayer et al. 2004). Table 5.1 provides a list of the most commonly argued pros and cons of conducting salvage logging as soon as possible after a fire.

Economic benefits of salvage logging depend on the timber quality, distance from roads, market conditions, and forest ownership. Megafires may saturate charred wood market. Usually, private owners have more interest in taking some revenue from charred wood, and in the European Union they often receive subsidies from forest administrations for salvage logging, on the basis of its assumed beneficial ecological impacts to promote forest recovery.

Some of the potential ecological benefits and/or impacts of salvage logging may largely depend on the specific forest site and fire characteristics (Peterson et al. 2009). However, there is a tendency among forest managers to overgeneralize the pros and cons of logging. For example, charred wood potential to trigger a pest outbreak may depend on the specific threatening insect species biology and on the degree of damage of the burned trees (see Sect. 5.4.7).

In relation to soil erosion, shortly after fire the ash layer and the unprotected soil are highly sensitive to trampling and mechanical operations. Thus, short-term salvage logging on vulnerable soils may cause more erosion than the fire itself.

In Valencia, eastern Spain, Bautista et al. (2004) recorded up to 50 Mg ha⁻¹ year⁻¹ soil loss in rills during the first 3 years after logging on *Pinus pinaster* stands developed over sandstones. Post-fire erosion rates in the same region were less than one order of magnitude lower (Mayor et al. 2007; Llovet et al. 2009). However, in the same study, Bautista et al. (2004) found that logging on stands developed over limestone produced low erosion rates.

Obviously, the impact of logging on soil erosion depends on soil erodibility and topography. For vulnerable soils, logging techniques avoiding log dragging could be applied, such as using a cable yarder or an helicopter, though these techniques are expensive and difficult to justify from an economic point of view in the commonly low stocking Mediterranean forests.

In the region of Valencia, after some 20 years of post-fire monitoring, the following generic recommendations were proposed in unpublished technical reports (CMA 1994), further elaborated by Bautista el al. (2004):

1. Avoid short-term salvage logging on vulnerable soils, at least until a protective vegetation cover has developed (usually after the first post-fire spring). Therefore, logging should be planned according to ecosystems vulnerability, in the same line as any other post-fire management activity (see Chap. 1).
2. In any case, retain some snags in order to keep forest nesting birds and other biodiversity components.
3. Selective logging in patches could combine economic and ecological benefits, avoiding sensitive soil erosion areas and generating mosaics that would optimize biodiversity (Izhaki and Adar 1997).
4. Monitor surviving weakened trees for the risk of pest outbreaks.
5. Use branches as barriers or to produce mulching material (chipped wood) to protect spots with high erosion risk (e.g. gullies, road talus, see Fig. 5.4).

### 5.4.3  *Forest Restoration or Forest Conversion*

Stand replacing fires offer the opportunity to consider alternative forest types to be promoted (Moreira and Vallejo 2009) in the context of land planning. Stand replacing fires in the Mediterranean may occur mostly in the case of conifer forests, when fire affects young, immature stands, or species not having specific post-fire regeneration mechanisms (see Sect. 5.1), or when long dry conditions occur during the first vegetative seasons after fire. For hardwoods, changing the dominant tree species is difficult even after a fire, as it may require uprooting the stumps. This has been practiced sometimes to eliminate eucalypt forests and recuperate native oaks in the Iberian Peninsula, but with high economic investment and ecological impact on the soil.

Forest conversion could be considered for breaking the horizontal continuity in homogeneous forests, usually planted pine forests. These are very abundant in the Mediterranean countries owing to the large plantations conducted along the twentieth century (Vallejo 2005). Changing burned pine forest patches into hardwood forest would in many cases re-naturalize the area, increase gamma diversity and fire resilience, and reduce pest outbreaks. In addition, for fire management planning, introducing patches of different forests types (e.g. promoting riparian vegetation along creeks) or even other land uses (e.g. agricultural areas or pastures) located in strategic areas may reduce fire propagation risk and help fire suppression (see also Chap. 1).

One other situation where conversion may be a suitable alternative is when the composition of fire prone shrublands is changed by planting fire-resistant resprouting trees (see Chap. 12).

### 5.4.4 Active Versus Indirect Restoration

One major decision in the post-fire management of burned forests, when the restoration of the former forest type is the main objective, is whether to use natural regeneration (indirect restoration), if it is present or predicted to occur, or active restoration (plantations and seeding).

There is strong political pressure for reforesting or afforesting burned areas in the Mediterranean region as soon as possible after a wildfire, and this has been a common practice since the late nineteenth century, particularly in conifer forests (Pausas et al. 2004a; Vallejo 2005). As an example, following the 2006 wildfires in Galicia (Spain), which burned 150,000 ha of land, reforestation has been considered a restoration priority (Amil 2007). In the case of Portugal, policies for the reforestation of burned forests have been common in the last decades (Carvalho Mendes 2006). However, reforestation/afforestation may not be the best alternative in many cases and the current political and social paradigm of "compensating" areas burned with active reforestations in a similar or higher number of hectares should be changed (see Chap. 1).

Reforestations are usually carried out through active restoration techniques such as plantation or direct seeding (Lamb and Gilmour 2003; Vallejo et al. 2006). Planting is expensive due to the costs associated with the acquisition of plant seedlings from nurseries, their transport to the burned area (particularly in areas with difficult access), site preparation (usually this is the highest cost), and other costs associated with equipment, fertilizers, tree shelters, replacement of dead seedlings, and human labour. Furthermore, activities associated with soil preparation for planting may increase the risk of soil erosion, and the mortality rates of planted seedlings, although quite variable, are often higher than 50% (e.g. Maestre and Cortina 2004; Pausas et al. 2004a; Vallejo 2005). Direct seeding is less costly and can be applied in extensive areas (e.g. aerial seeding), but the success (seed germination and plant establishment) is also usually very low (e.g. Vallejo and Alloza 1998; Espelta et al. 2003a; Pausas et al. 2004a). Therefore, these active techniques have a low cost-effectiveness and should be considered only when other options are not feasible, e.g. in areas where no natural tree regeneration is expected and where there are no mature trees in the vicinity that might provide seeds to naturally colonise the site in the medium term.

As an alternative to active restoration, other applications of financial and human resources may be much more effective, in particular through exploring the potential of natural regeneration characteristic of many Mediterranean species, i.e. taking advantage of regeneration from seeds left in the ground by the burned vegetation (e.g. Pausas et al. 2004b; Holz and Placci 2005), or from resprouting of burned trees and shrubs (e.g. Espelta et al. 2003b; Lloret et al. 2005). In addition to the lower costs of these passive (natural) restoration techniques (e.g. Lamb and Gilmour 2003; Whisenant 2005; Vallejo et al. 2006), plant survival and growth rates are higher when compared to active restoration and consequently a higher and faster-growing vegetation cover is achieved. Sprouts have many potential advantages over seedlings or planted trees because they have an already established root system and high stored energy reserves, which may confer greater chances of plant survival and recovery. For example, Moreira et al. (2009) have shown that oak *Quercus faginea* and ash *Fraxinus angustifolia* resprouts in burned areas in central Portugal could survive 20% more and grew 4–5 times faster when compared to planted seedlings (see Chap. 1). The use of plant resprouting ability is already acknowledged as a powerful tool to restore ecosystems such as tropical dry forests (Vieira and Scariot 2006) or the Atlantic rainforests of Brazil (Simões and Marques 2007) and this should be extended to Mediterranean ecosystems. Thus, we advocate a much more frequent use of assisted natural restoration, based on the management of natural regeneration from seeds or resprouts. Depending on the objectives, this may involve thinning, the selection of shoots, and the control of unwanted vegetation. The costs associated with assisted natural regeneration can be much lower when compared to active restoration, meaning that with a similar amount of funding available a much larger area can be effectively treated. Of course the decision of opting by active or natural restoration will be constrained by the type of pre-fire vegetation, the ecosystem response to fire, and the objectives for the burned area.

## 5.4.5 The Management of Herbivory

Grazing animals (both domestic and wild herbivores) can be either beneficial or detrimental, depending on the post-fire management objectives.

The beneficial effects of grazing and browsing are that they contribute to reduce fuel loads and, thus, fire hazard (e.g. Nader et al. 2007; Tsiouvaras et al. 1989). In fact, fire and grazing are quite similar disturbances in several aspects of their impact on vegetation (Bond and Keeley 2005). Using animals (mainly domestic animals) in fuel management (what is sometimes called *prescribed herbivory*) requires knowing their feeding strategy. Browsers consuming woody species (shrubs and tree branches), e.g. goats or deer, are better suited for controlling shrubby areas than grazers consuming mostly herbaceous vegetation (e.g. cows or sheep). There are a few studies in southern Europe confirming this positive impact of grazing. For example, in a study in Spain, Valderrábano and Torrano (2000) showed that goats were effective in reducing the survival and growth of an invasive shrub with high ignition capacity. So, grazing animals can be used in post-fire management when the objective is to reduce fuel loads (e.g. in fuel breaks, or in areas at the wildland-urban interface). In some countries (e.g. Portugal, USA), there are commercial enterprises selling prescribed herbivory, with animals being transported and confined to specific target areas. Important factors to take into account to increase the effectiveness of grazing include: (i) the selection of the animal species, (ii) the selection of the grazing season and grazing period, and (iii) the establishment of an appropriate stocking rate. Pastoral fire has been practiced in the Mediterranean for thousands of years to suppress the unpalatable woody species to animals in favor of the herbaceous plants palatable to sheep and cattle (Blondel and Aronson 1999). The coordinated use of this old management tool can also greatly contribute to reduce fire hazard (Rego et al. 2010).

Herbivory can, however, be an important limiting factor when the pos-fire objective is the regeneration of burned areas (Vallejo et al. 2006). After a fire, the regenerating plants are particularly susceptible to animal consumption. The first species to emerge typically have high digestibility and are very attractive to herbivores (Hobbs et al. 1991; Hobbs 2006). Resprouting species support well a slight consumption, but can be affected if there is repeated consumption of their terminal buds, essential to their growth in height. For example, Catry et al. (2007) showed that deer browsing can have a strong negative impact on post-fire basal sprout regeneration of several Mediterranean broadleaved species. Seedlings will obviously die if consumed by animals. Grazing damage is not restricted to browsing, but also trampling. So, if an area is being managed for natural regeneration, or has been planted, and if there is a density of grazers compromising the success of this regeneration, managers will probably have to invest in the protection of regenerating/planted plants. This can be done using three main different alternatives: (i) reducing the animal population size; (ii) protecting individual plants; (iii) protecting areas from herbivores. The reduction of animal densities to levels that are compatible with plant regeneration could be the ideal solution from the vegetation recovery point of view. However, this may not be feasible or compatible with the management objectives, e.g. in areas with wild herbivores where hunting is not allowed. The protection of individual trees is adopted in many countries, regardless of fire occurrence, when animals have access to regeneration areas or plantations. Various types of protections of variable prices and efficiency are available. The most common approach is to protect each tree with a protective cylindrical-shaped wire mesh. Another possible protection method involves the application of chemical repellent but in most cases its effectiveness is short-lived or is still unproved. Fencing parts of the area to regenerate may be also a good option, depending on the size of the area to protect and on tree density. Generally for larger areas and higher tree density, this technique is cheaper than the protection of individual plants. This option has the added advantage of allowing a denser regeneration in the burned area, and a better ground cover, contributing to preventing post-fire soil erosion. Possible disadvantages include the limited access to the area and higher fuel accumulation that may increase fire danger. Temporary protection by electric fencing is most suitable for domestic species.

Finally, the use of nurse plants has been shown to provide protection from grazers in a study in Spain (Castro et al. 2002), and this topic deserves further research.

## 5.4.6   Fire and Alien Species

Alien species may be an important post-fire management issue in situations where fire promotes their occurrence, particularly if they become invasive and compromise the regeneration of native vegetation. Besides their impact on native diversity, if these new plant communities have different fuel properties that increase their combustibility then a feedback loop may be created where fire promotes invasive species that turn the affected area more fire prone (Brooks et al. 2004).

Most Mediterranean environments have experienced a high degree of human interference and disturbance, a process that dates back over 10,000 years, and this has resulted in a marked transformation of the vegetation (Heywood 1995; Thompson 2005). In contrast to other Mediterranean regions of the World, e.g. California and South Africa, where large areas of relatively intact vegetation still remain, much of the Mediterranean Basin has been transformed from its native state (Mooney 1988). The result is that Mediterranean plant communities are rather resistant to biological invasions as native species are likely to be good competitors under the strong selection regime imposed by humans on the Mediterranean flora and that the multiple stresses of drought, fire and grazing present a limitation to prospective alien plant species to establish and further to become invasive (Hulme et al. 2007, Arianoutsou et al., in preparation). For example, very few alien annual species have been recorded in early post-fire communities of *Pinus halepensis* forests in Greece. These taxa represent less than 1% of the regenerating flora and their abundance was not substantial (Kazanis 2005). So, despite the increasing pace of research related to biological invasions in the Mediterranean region (e.g. Vilà et al. 2007; Celesti-Grapow et al. 2009, 2010; Arianoutsou et al. 2010b) very few, if any, paper relates biological invasions to fire, in contrast with other regions of the world, and especially California.

The abundance of an exotic species, *Cortaderia selloana* (Pampas grass), has been related to potential fire occurrence in the Mediterranean (Doménech et al. 2005). *C. selloana* is a South American longlived perennial grass native to Argentina, Brazil and Uruguay which is considered invasive worldwide and it was first introduced as ornamental to Europe between 1775 and 1862 (Bossard et al. 2000). It has escaped from cultivation and it is invading abandoned farmlands, roadsides, shrublands and wetlands. *C. selloana* is considered to increase fire hazard because of the accumulation of dry leaves and flowering stalks on the plant.

Although fire usually opens temporarily the plant canopy in many forest and shrubland ecosystems offering empty space for colonizing species (Arianoutsou et al. 2011), it seems that alien species, if they are available in the vicinity, cannot easily compete with natives to become established in the burned Mediterranean habitats as it seems to be the case for other Mediterranean climate regions, e.g. California (Keeley et al. 2005; Zouhar et al. 2008), Chile (Gómez-González et al. 2011), South Africa (van Wilgen and Richardson 1985; van Wilgen et al. 2010) or Australia (Thomson and Leishman 2005; Miller et al. 2010). However, there is some evidence that in areas of Southern Europe with less dry climates, invasive species are becoming a problem in burned areas. In northern and central Portugal, for example, the genera *Acacia*, *Hakea*, *Ailanthus* and *Eucalyptus* are a growing concern for forest managers, as their prevalence in burned areas is notoriously increasing (Catry et al. 2010).

It has been suggested that plant communities under altered fire regimes are more susceptible to invasion than those under a natural (historical) fire regime (Trabaud 1991b; D'Antonio 2000). For example, in most Mediterranean climate ecosystems, fire has been an important selective force shaping adaptive traits in native plant species (e.g. Pausas et al. 2006). Under conditions of natural fire frequency, the ability of native species to cope with fire leads to a relatively high ecosystem resilience to invasion, because alien species that colonize open areas are rapidly excluded from the system (Trabaud 1991b; Keeley 2001). But this dominance of native relative to alien species may be changed under a different fire regime and under different climatic conditions.

In conclusion, fire does not seem to promote the occurrence of exotic plants in most of the Mediterranean ecosystems, as these systems are rather resilient to disturbance and although aliens seem to prefer disturbed places to establish, they cannot cope with native species in such dry environments. It is true, that in most cases where alien plants have been recorded in natural systems, their presence was mostly concentrated in mesic places where nutrient availability is often higher (Stohlgren et al. 2003; Vilà et al. 2007; Arianoutsou et al. 2010c). This may explain why in some specific regions with less xeric environments, such as central and northern Portugal, exotic species are becoming a major problem in post-fire management (see Chap. 10). Clearly, more research is needed on this topic.

### 5.4.7 Pests and Diseases

Fire and insects are natural disturbance agents in many forest ecosystems, affecting succession, nutrient cycling and forest species composition. There are two basic mechanisms through which fire and pests or pathogens interact. One is the mortality or weakening of trees by fire and the subsequent promotion of damaging insect and pathogen populations. The second is through the mortality of trees caused by these agents, which contributes to dead fuel accumulation and increased fire hazard. However, there is not much information available on the relationships between wildfires and tree insect pests and pathogens in Southern Europe. Most research is focused on bark beetles and wood boring insects (e.g. Fernandez et al. 1999; Fernandez 2006).

Pines are the target of a variety of bark beetles that can cause tree death, branch dieback and reduced productivity (FAO 2009). Pine beetles of the subfamily Scolytinae (mainly the genus *Ips*, *Orthotomicus* and *Tomicus*) are considered a major problem in burned pine forests, but there is contradictory information on how to manage burned trees to minimize this hazard. For example, while forest managers often consider burned trees as the more likely to attract bark beetles, and partially affected trees (crowns partially scorched or consumed) are left hoping that they survive, in a region of Spain, Bautista et al. (2004) have shown that the latter are the ones that have the higher degree of infection thus should the first ones to log. In Northern Spain, Fernandéz (2006) also found that all *P. pinaster* trees with 25–50% burnt crowns were the first to be attacked by *Ips sexdentatus* and consequently die. The black trees (100% burnt crown) were mainly used for the establishment of secondary xylophagous beetles, such as Buprestidae and Cerambycidae. She concluded, however, that the most effective way to reduce the risk of mortality in healthy standing trees was to remove nearby dead and dying trees before the *I. sexdentatus* population grew large enough to attack healthy, less injured or recovering trees.

The moths of the genus *Lymantria* are significant defoliators of a wide range of broadleaf and conifer trees. Severe outbreaks can occur resulting in severe defoliation, growth loss, dieback and sometimes tree mortality. The pine processionary caterpillar *Thaumetopoea pityocampa* is considered the most destructive forest insect pest throughout the Mediterranean Basin. It is a tent-making caterpillar that feeds gregariously and defoliates various species of pine and cedar (FAO 2009). These species may cause intense defoliation with consequent tree weakness and even death. This might contribute to fuel accumulation, increasing fire hazard.

Some insect pests cause problems in specific geographical regions or forest types. For example, eucalyptus longhorned borers of the genus *Phoracanta* (Coleoptera: Cerambycidae) are serious borer pests of eucalypts, particularly those planted outside their natural range (FAO 2009), e.g. the eucalyptus plantations in Portugal. *Phoracantha* species tend to attack damaged or stressed trees; vigorous, well-watered trees are rarely attacked though it does occur. *Gonipterus scutellatus* (Coleoptera: Curculinoidae) is an exotic weevil infesting eucalypt plantations. The nematode *Bursaphelenchus xylophilus* (pine wilt nematode) was found in Portugal for the first time in 1999 and is a major threat for the maritime pine *Pinus pinaster* forests. The spread of this nematode is via wood-boring beetles of the genus *Monochamus*. They become infested with the nematode just before emerging as adults from diseased host trees. Adult beetles can act as vectors for thousands of nematodes (FAO 2009). In a recent review, Lindner et al. (2010) mention the likely increase in the virulence of thermophilic pathogen species in the Mediterranean, in response to climate change. Trees weakened by fire will become more susceptible, thus increased problems are expected. On the other hand, trees weakened or killed by these pathogens may increase fire hazard (Dios et al. 2007).

The fungi *Armillaria* spp. are a common worldwide pathogen of trees, woody shrubs and herbaceous plants causing root rot, root-collar rot and butt rot (FAO 2009). They cause wood decay, growth reduction and mortality, particularly in trees stressed by other factors. Climate change may be implicated in the increasing incidence of oak declines due to *Phytophtora* spp. *Phytophtora* require wet soil conditions to proliferate. In the last few decades floods have occurred more frequently creating favourable conditions for pathogen proliferation in forests. These floods have been followed by drought events that have weakened the trees and made them more susceptible to the pathogen, resulting in higher mortality than ever before (FAO 2009). As with other conifer bark beetle species, *Ips sexdentatus* is a vector for blue-stain fungi (*Ophiostoma* spp.) which hastens the death of trees, discolours the wood and can result in loss of timber value. Fire events by favouring the build-up of bark beetle populations may in particular activate this fungi-insect association (Bueno et al. 2010).

## 5.5  Key Messages

- Most Mediterranean plant species have developed regeneration strategies allowing their efficient recovery after fire. According to these strategies, species are grouped in obligatory resprouters, facultative resprouters, and obligatory seeders. Some species do not show any specific post-fire regeneration mechanism, hence they locally disappear after fire and can only re-colonize burned areas from adjacent unburned patches.
- Climate change projections indicate that wildfire potential will increase in Southern Europe, and this could reduce Mediterranean ecosystems resilience to fire.
- The most fire-affected vegetation types in Europe are thermophilous pine forests with *Pinus pinaster*, *P. halepensis* and *P. brutia*, broadleaved evergreen forests with *Quercus suber*, *Q. ilex*, *Q. rotundifolia* and *Q. coccifera*, and shrublands.
- Post-fire emergency rehabilitation actions should be applied to burned forests showing high erosion and runoff risk, with slow natural plant recovery rate. The forests more prone to fire erosion are pine forests with an understory dominated by obligate seeders. Mulching is one of the more effective techniques to decrease erosion risk.
- Salvage logging has economic and ecological benefits but also negative impacts, depending on the local conditions. General recommendations are: avoiding dragging logs in vulnerable soils, retain some snags for biodiversity purposes, monitor surviving weakened trees for the risk of pest outbreaks, and use charred wood for soil protection where there is high erosion risk.
- Forest conversion after fire could be considered for fire prevention and for the re-naturalization of the landscape.
- Although active reforestation/afforestation is the usual action taken by policy makers and forest administrations, in many cases assisted natural restoration is more efficient and cost-effective.
- Grazing animals contribute to reduce fuel load and fire hazard. However, herbivores may also limit the post-fire vegetation recovery. Therefore, protection actions should be taken where domestic and wild herbivore populations threaten regeneration.
- Alien invasive species are not much favored by fire in xeric Mediterranean environments, but in some moister areas they are becoming an increasing problem for post-fire management.
- Pests and diseases may increase dead fuel accumulation, and consequently fire hazard. Fire may facilitate pest outbreak, especially for pine bark beetles.

**Acknowledgments** Many of the ideas expressed here started being developed at the PHOENIX project centre of the European Forest Institute. CEAM is supported by Generalitat Valenciana and Bancaja. CEAM contribution is based in the research conducted in the projects GRACCIE (CONSOLIDER-INGENIO 2010), PROMETEO-FEEDBACKS and FUME (EC FP7 nr. 243888). ISA contribution benefited from activities within the FIREREG project (contract PTDC/AGR-CFL/099420/2008) funded by FCT (Portugal). The authors would like to thank Dr Ioanna Louvrou Curator of University of Athens Botanical Museum for her support in preparing the photographic plates.

# References

Alloza JA, Vallejo VR (2006) Restoration of burned areas in forest management plans. In: Kepner WG, Rubio JL, Mouat DA, Pedrazzini F (eds) Desertification in the Mediterranean region: a security issue. Springer, Dordrecht

Amil ML (2007) Forest fires in Galicia (Spain): threats and challenges for the future. J Forest Econ 13:1–5

Amman GD, Ryan KC (1991) Insect infestation of fire-injured trees in the greater Yellowstone area. Res. Note INT-398, USDA, Forest Service, Intermountain Research Station, Ogden

Arianoutsou M (1999) Effects of fire on vegetation demography. International symposium on forest fires: needs and innovations, (DELFI), Athens

Arianoutsou M (2004) Predicting the post-fire regeneration and resilience of Mediterranean plant communities. In: Arianoutsou M, Papanastasis VP (eds) Ecology, conservation and management of Mediterranean climate ecosystems of the world. Proceedings of the MEDECOS 10th International Conference, Rhodes, Greece, Millpress, Rotterdam, Electronic Edition

Arianoutsou M, Margaris NS (1981) Producers and the fire cycle in a phryganic ecosystem. In: Margaris NS, Mooney HA (eds) Components of productivity in Mediterranean climate regions – basic and applied aspects. Dr W. Junk, The Hague

Arianoutsou M, Kazanis D, Kokkoris Y, Skourou P (2002) Land-use interactions with fire in Mediterranean *Pinus halepensis* landscapes of Greece: patterns of biodiversity. In: Viegas DX (ed) Proceedings of the IV International Forest Fire Research Conference, Millpress, electronic edition, (2002)

Arianoutsou M, Kazanis D, Copanellou I (2008) Report on the research and the study of the post-fire regeneration of the mixed pine forest of Strofylia, at the area of lake Kaiafa (GR 2330005, NATURA 2000). University of Athens, Department of Ecology and Systematics, (in Greek)

Arianoutsou M, Christopoulou A, Tountas Th, Ganou E, Kazanis D, Bazos I, Kokkoris I (2010a) Effects of fire on high altitude coniferous forests of Greece. In: Viegas DX (ed) Book of proceedings of the VIth international conference on forest fire research, Coimbra, Portugal (electronic edition)

Arianoutsou M, Bazos I, Delipetrou P, Kokkoris Y (2010b) The alien flora of Greece: taxonomy, life traits and habitat preferences. Biol Invasions 12:3525–3549. doi:10.1007/s10530-010-9749-0

Arianoutsou M, Delipetrou P, Celesti-Grapow L, Basnou C, Bazos I, Kokkoris Y, Blasi C, Vilà M (2010c) Comparing naturalized alien plants and recipient habitats across an east–west gradient in the Mediterranean Basin. J Biogeogr 37:1811–1823. doi:10.1111/j.1365-2699.2010.02324.x

Arianoutsou M, Koukoulas S, Kazanis D (2011) Evaluating post-fire forest resilience using GIS and multi-criteria analysis: an example from Cape Sounion National Park, Greece. Environ Manag 47:384–397. doi:10.1007/s00267-011-9614-7

Arianoutsou-Faraggitaki M, Margaris NS (1982) Decomposers and the fire cycle in a phryganic (East Mediterranean) ecosystem. Microb Ecol 8:91–98

Bautista S, Vallejo VR (2002) Spatial variation of post-fire plant recovery in Aleppo pine forest. In: Trabaud L, Prodon R (eds) Fire and biological processes. Backhuys Publishers, Leiden

Bautista S, Gimeno T, Mayor A, Gallego D (2004) El tratamiento de la madera quemada tras los incendios forestales. In: Vallejo VR, Alloza JA (eds) Avances en el estudio de la gestión del monte Mediterráneo. Fundación CEAM, Valencia

Beghin R, Lingua E, Garbarino M, Lonati M, Bovio G, Motta R, Marzano R (2010) *Pinus sylvestris* forest regeneration under different post-fire restoration practices in the northwestern Italian Alps. Ecol Eng 36:1365–1372

Blondel J, Aronson J (1999) Biology and wildlife of the Mediterranean region. Oxford University Press, Oxford

Bond WJ, Keeley JE (2005) Fire as a global "herbivore": the ecology and evolution of flammable ecosystems. Trends Ecol Evol 20:387–394

Bond WJ, van Wilgen BW (1996) Fire and plants. Chapman and Hall, London

Bossard CC, Randall JM, Hshousky MC (2000) Invasive plants of California's wildlands. University of California, Berkeley, Los Angeles

Brooks ML, D'Antonio CM, Richardson DM, Grace JB, Keeley JE, Ditomaso JM, Hobbs RJ, Pellant M, Pyke D (2004) Effects of invasive alien plants on fire regimes. BioScience 54:677–688

Bueno A, Diez JJ, Fernández MM (2010) Ophiostomatoid fungi transported by Ips sexdentatus (Coleoptera; Scolytidae) in *Pinus pinaster* in NW Spain. Silva Fenn 44:387–397

Carvalho Mendes A (2006) Implementation analysis of forest programmes: some theoretical notes and an example. Forest Policy Econ 8:512–528

Castro J, Zamora R, Hódar JA, Gómez JA (2002) Use of shrubs as nurse plants: a new technique for reforestation in Mediterranean mountains. Restor Ecol 10:297–305

Castro J, Moreno-Rueda G, Hódar JA (2009) Experimental test of post-fire management in pine forests: Impact of salvage logging versus partial cutting and non-intervention on bird-species assemblages. Conserv Biol 24:810–819

Catry FX, Bugalho M, Lopes T, Rego FC, Moreira F (2007) Post- fire effects of ungulates on the structure, abundance and diversity of the vegetation in a Mediterranean Ecosystem. In: Rokich D, Wardell-Johnson YC, Stevens J, Dixon K, McLellan R, Moss G (eds) Proceedings of the international Mediterranean ecosystems conference – Medecos XI, Kings Park and Botanic Garden, Perth, Australia

Catry FX, Bugalho M, Silva JS, Fernandes P (2010) Gestão da vegetação pós-fogo. In: Moreira F, Catry FX, Silva JS, Rego F (eds) Ecologia do Fogo e Gestão de Áreas Ardidas. ISAPress, Lisbon

Celesti-Grapow L, Alessandrini A, Arrigoni PV et al (2009) The inventory of the non-native flora of Italy. Plant Biosyst 143:386–430

Celesti-Grapow L, Alessandrini A, Arrigoni PV et al (2010) Non-native flora of Italy: species distribution and threats. Plant Biosyst 144:12–28

IPCC – Intergovernmental Panel on Climate Change (2007) Climate change 2007: impacts, adaptation, and vulnerability. Parry ML, Canziani OF, Palutikof JP, van der Linden PJ, Hanson CE (eds), Contribution of working group II to the 3rd assessment report of the intergovernmental panel on climate change Cambridge University Press, Cambridge

CMA, Conselleria de Medio Ambiente (1994) Circular de 21 de enero de 1994 sobre la extracción de madera quemada por un incendio forestal. CMA, Valencia

Cobb TP, Morissette JL, Jacobs JM, Koivula MJ, Spence JR, Largor DW (2010) Effects of post-fire salvage logging on deadwood-associated beetles. Conserv Biol 25:94–104

D'Antonio CM (2000) Fire, plant invasions, and global changes. In: Mooney HA, Hobbs RJ (eds) Invasive species in a changing world. Island Press, Washington, D.C

Daskalakou EN, Thanos CA (2004) Post-fire regeneration of Aleppo pine – the temporal pattern of seedling recruitment. Plant Ecol 171:81–89

Delitti W, Ferran A, Trabaud L, Vallejo VR (2005) Effects of fire recurrence in *Quercus coccifera* L. Shrublands of the Valencia Region (Spain): I. Plant composition and productivity. Plant Ecol 177:57–70

Dios VR, Fischer C, Colinas C (2007) Climate change effects on mediterranean forests and preventive measures. New Forest 33:29–40

Dixon KW, Roche S, Pate JS (1995) The promotive effect of smoke derived from burnt native vegetation on seed germination of Western Australian plants. Oecologia 101:185–192

Doménech R, Vila M, Pino J, Gesti J (2005) Historical land-use legacy and *Cortaderia selloana* invasion in the Mediterranean region. Glob Change Biol 11:1054–1064. doi:10.1111/j.1365-2486.2005.00965.x

Doussi MA, Thanos CA (1994) Post-fire regeneration of hardseeded plants: ecophysiology of seed germination. In: Viegas DX (ed) Proceedings of the 2nd international conference on forest fire research, Coimbra, Viegas DX Publisher, Coimbra

Espelta JM, Retana J, Habrouk A (2003a) An economic and ecological multi-criteria evaluation of reforestation methods to recover burned *Pinus nigra* forests in NE Spain. For Ecol Manag 180:185–198. doi:10.1016/S0378-1127(02)00599-6

Espelta JM, Retana J, Habrouk A (2003b) Resprouting patterns after fire and response to stool cleaning of two coexisting Mediterranean oaks with contrasting leaf habits on two different sites. For Ecol Manag 179:401–414. doi:10.1016/S0378-1127(02)00541-8

European Environmental Agency (EEA) (2007). European forest types. EEA Technical Report nr. 9. Copenhagen

FAO (2009) Global review of forest pests and diseases. FAO Forestry Paper 156, Rome

Fernández MM (2006) Colonization of fire-damaged trees by *Ips sexdentatus* (Boerner) as related to the percentage of burnt crown. Entomol Fenn 17:381–386

Fernandez Fernandez MM, Salgado CJM (1999) Susceptibility of fire-damaged pine trees (*Pinus pinaster* and *Pinus nigra*) to attacks by Ips sexdentatus and Tomicus piniperda (Coleoptera: Scolytidae). Entomol Gener 24:105–114

Ferrandis P, De las Heras J, Martínez Sanchez JJ, Herranz JM (2001) Influence of a low-intensity fire on a *Pinus halepensis* Mill. forest seed bank and its consequences on early stages of plant succession. Israel J Plant Sci 49:105–114

Flannigan MD, Stocks BJ, Wotton MB (2000) Climate change and forest fires. Sci Total Environ 262:221–229

Gayoso J, Iroumé A (1991) Compaction and soil disturbances from logging in Southern Chile. Ann Sci For 48:63–71

Gill AM, Bradstock RA (2003) Fire regimes and biodiversity: a set of postulates. Proceedings of the Australian national university fire forum, CSIRO Publishing, Melbourne, Feb 2002

Gill AM, Bradstock RA, Williams JE (2002) Fire regimes and biodiversity: legacy and vision. In: Bradstock RA, Williams JE, Gill AM (eds) Flammable Australia: the fire regimes and biodiversity of a continent. Cambridge University Press, Cambridge

Gimeno-García E, Pascual JA, Llovet J (2011) Water repellency and moisture content spatial variations under *Rosmarinus officinalis* and *Quercus coccifera* in a Mediterranean burned soil. Catena 85:48 57

Gómez-González S, Torres-Díaz C, Valencia G, Torres-Morales P, Cavieres LA, Pausas JG (2011) Anthropogenic fires increase alien and native annual species in the Chilean coastal matorral. Divers Distrib 17:58–67. doi:10.1111/j.1472-4642.2010.00728.x

Heywood VH (1995) The Mediterranean flora in the context of world biodiversity. Ecol Medit 20:11–18

Hobbs NT (2006) Large herbivores as sources of disturbance in ecosystems. In: Danell K, Bergstrom R, Duncan P, Pastor J (eds) Large herbivore ecology, ecosystem dynamics and conservation. Cambridge University Press, Cambridge

Hobbs NT, Schimel DS, Owensby CE, Ojima DS (1991) Fire and grazing in the Tallgrass Prairie: contingent effects on nitrogen budgets. Ecology 72:1374–1382

Holz S, Placci G (2005) Stimulating natural regeneration. In: Mansourian S, Vallauri D, Dudley N (eds) Forest Restoration in Landscapes. Beyond Planting Trees, Springer, New York

Hulme PE, Brundu G, Camarda I, Dalias P, Lambdon P, Lloret F, Medail F, Moragues E, Suehs C, Traveset A, Andreas Troumbis A, Vilà M (2007) Assessing the risks to Mediterranean islands ecosystems from alien plant introductions. In: Tokarska-Guzik B, Brundu G, Brock JH, Child LE, Pyšek P, Daehler C (eds) Plant invasions. Backhuys Publishers, Leiden

Izhaki I, Adar M (1997) The effects of post-fire management on bird community succession. Int J Wildland Fire 7:335–342

Jones MR, Laude HM (1960) Relations between sprouting in chamise and the physiological condition of the plant. J Range Manag 13:210–214

Kazanis D (2005) Post-fire succession in *Pinus halepensis* forests of Greece: patterns of vegetation dynamics. PhD thesis, University of Athens (in Greek with an English summary)

Kazanis D, Arianoutsou M (2004) Long-term post-fire vegetation dynamics in *Pinus halepensis* forests of central Greece: a functional-group approach. Plant Ecol 171:101–121

Keeley JE (2001) Fire and invasive species in Mediterranean-climate ecosystems of California. In: Galley P, Wilson TP (eds) Proceedings of the invasive plant workshop: the role of fire in the control and spread of invasive species. Tall Timbers Research Station, Tallahassee

Keeley JE, Bond WJ (1997) Convergent seed germination in South African fynbos and Californian chaparral. Plant Ecol 133.153–167

Keeley JE, Fotherigam CJ (2000) The role of fire in regeneration from seed. In: Fenner M (ed) Seeds: the ecology of regeneration in plant communities. CAB International, Wallingford

Keeley JE, Baer-Keeley M, Fotheringham CJ (2005) Alien plant dynamics following fire in Mediterranean-climate California shrublands. Ecol Applic 15:2109–2125. doi:10.1890/04-1222

Lamb D, Gilmour D (2003) Rehabilitation and restoration of degraded forests. IUCN, Gland, Switzerland and Cambridge, UK and WWF, Gland, Switzerland

Leone V, Logiurato A, Saracino A (1999) Serotiny in *Pinus halepensis* Mill., recent issues. In: Ne'eman G, Izhaki I (eds) Abstracts of MEDPINE, international workshop on Mediterranean pines. Beit Oren, Israel

Lindenmayer DB, Foster DR, Franklin JF, Hunter ML, Bnoss RF, Schmiegelow FA, Perry D (2004) Salvage harvesting policies after natural disturbance. Science 303:1303

Lindner M, Maroschek M, Netherer S, Kremer A, Barbati A, Garcia-Gonzalo J, Seidl R et al (2010) Climate change impacts, adaptive capacity, and vulnerability of European forest ecosystems. For Ecol Manag 259:698–709. doi:10.1016/j.foreco.2009.09.023

Liu Y, Stanturf GS (2010) Trends in global wildfire potential in a changing climate. For Ecol Manag 259:685–697. doi:10.1016/j.foreco.2009.09.002

Lloret F, López-Soria L (1993) Resprouting of *Erica multiflora* after experimental fire treatments. J Veg Sci 9:417–430

Lloret F, Vilà M (2003) Diversity patterns of plant functional types in relation to fire regime and previous land use in Mediterranean woodlands. J Veg Sci 14:387–398

Lloret F, Médail F, Brundu G, Camarda I, Moragues E, Rita J, Lambdon P, Hulme PE (2005) Species attributes and invasion success by alien plants on Mediterranean islands. J Ecol 93:512–520

Llovet J, Ruiz-Valera M, Josa R, Vallejo VR (2009) Soil responses to fire in Mediterranean forest landscapes in relation to the previous stage of land abandonment. Int J Wildland Fire 18: 222–232

López-Soria L, Castell C (1992) Comparative genet survival after fire in woody Mediterranean species. Oecologia 91:493–499

Maestre FT, Cortina J (2004) Are *Pinus halepensis* plantations useful as a restoration tool in semi-arid Mediterranean areas? For Ecol Manag 198:303–317. doi:10.1016/j.foreco.2004.05.040

Mayor AG, Bautista S, Llovet J, Bellot J (2007) Post-fire hydrological and erosional responses of a Mediterranean landscape: Seven years of catchment-scale dynamics. Catena 71:68–75

Miller G, Friedel M, Adam P, Chewings V (2010) Ecological impacts of buffel grass (*Cenchrus ciliaris* L.) invasion in central Australia – does field evidence support a fire-invasion feedback? Rangeland J 32:353–365. doi:10.1071/RJ09076

Mooney HA (1988) Lessons from Mediterranean climate regions. In: Wilson EO (ed) Biodiversity. National Academy of Sciences/Smithsonian Institution, Washington DC

Moreira F, Vallejo VR (2009) What to do after fire? post-fire restoration. In: Birot Y (ed), Living with fires, European Forest Institute Discussion Paper 15, EFI, Joensuu, Finland

Moreira F, Catry FX, Lopes T, Bugalho MN, Rego F (2009) Comparing survival and size of resprouts and planted trees for post-fire forest restoration in central Portugal. Ecol Eng 35:870–873. doi:10.1016/j.ecoleng.2008.12.017

Moreira F, Viedma O, Arianoutsou M, Curt T, Koutsias N, Rigolot E, Barbati A, Corona P, Vaz P, Xanthopoulos G, Mouillot F, Bilgili E (2011) Landscape-wildfire interactions in Southern Europe: implications for landscape management. J Env Manag 92:2389–2402

Nader G, Henkin Z, Smith E, Ingram R, Narvaez N (2007) Planned herbivory in the management of wildfire fuels. Rangelands 29:18–24

Napper (2006) BAER – Burned Area Emergency Response Treatments Catalog. USDA Forest Service. Watershed, Soil, Air Management 0625 1801 –STDTDC. San Dimas, California

Ne'eman G (1997) Regeneration of Natural Pine Forests-Review of work done after the 1989 fire in Mount Carmel, Israel. Int J Wildland Fire 7(4):295–306

Ne'eman G, Goubitz S, Nathan R (2004) Reproductive traits of *Pinus halepensis* in the light of fire—a critical review. Plant Ecol 171:69–79

Ne'eman G, Izhaki I (1998) Stability of pre- and post-fire spatial structure of pine trees in Aleppo pine forest. Ecography 21:535–542

Noble IR, Slatyer RO (1980) The use of vital attributes to predict successional changes in plant communities subject to recurrent disturbances. Vegetatio 43:5–21

Ordóñez JL, Franco S, Retana J (2004) Limitation of the recruitment of *Pinus nigra* in a gradient of post-fire environmental conditions. Ecoscience 11:296–304

Papavassiliou S, Arianoutsou M (1993) Regeneration of the leguminous herbaceous vegetation following fire in a *Pinus halepensis* forest of Attica, Greece. In: Trabaud L, Prodon R (eds) Fire in Mediterranean ecosystem, ecosystem research report no 5, Commission of the European Communities

Pausas JG (1999) Mediterranean vegetation dynamics: modelling problems and functional types. Plant Ecol 140:27–39

Pausas JG (2004) Changes in fire and climate in the eastern Iberian Peninsula (Mediterranean basin). Clim Chang 63:337–350

Pausas JG, Keeley JE (2009) A burning story: the role of fire in the history of life. BioScience 59:593–601. doi:10.1525/bio.2009.59.7.10

Pausas JG, Bladé C, Valdecantos A, Seva J, Fuentes D, Alloza J, Milagrosa A, Bautista S, Cortina J, Vallejo R (2004a) Pines and oaks in the restoration of Mediterranean landscapes of Spain: new perspectives for an old practice – a review. Plant Ecol 171:209–220

Pausas JG, Bradstock RA, Keith DA, Keeley JE, GCTE (2004b) Plant functional traits in relation to fire in crown-fire ecosystems. Ecology 85:1085–1100

Pausas JG, Keeley JE, Verdú M (2006) Inferring differential evolutionary processes of plant persistence traits in Northern Hemisphere Mediterranean fire prone ecosystems. J Ecol 94:31–39

Pausas JC, Llovet J, Rodrigo A, Vallejo R (2008) Are wildfires a disaster in the Mediterranean basin? – A review. Int J Wildland Fire 17:713–723

Pérez-Fernández MA, Rodríguez-Echeverría S (2003) Effect of smoke, charred wood, and nitrogenous compounds on seed germination of ten species from woodland in central-western Spain. J Chem Ecol 29:237–251

Peterson DL, Agee JK, Aplet GH, Dykstra DP, Graham RT, Lehmkuhl JF et al (2009) Effects of timber harvest following wildfire in western North America. USDA Forest Service, Gral. Technical Report PNW-GTR-776. Washington DC

Placci G (2005) Stimulating natural regeneration. In: Mansourian S, Vallauri D, Dudley N (eds) Forest restoration in landscapes. Beyond planting trees. Springer, New York

Poff RJ (1989) Compatibility of timber salvage operations with watershed values. In: Berg NH (Tech. coord), Proceedings of the symposium on fire and watershed management, Gen. Technical Report PSW-109. USDA, Forest Service, Pacific Southwest Forest and Range Experiment Station, Berkeley

Rego F, Rigolot E, Fernandes P, Montiel C, Silva JS (2010) Towards integrated fire management. EFI Policy Brief 4

Retana J, Espelta JM, Habrouk A, Ordóñez JL, de Solà-Morales F (2002) Regeneration patterns of three Mediterranean pines and forest changes after a large wildfire in northeastern Spain. Ecoscience 9:89–97

Robichaud PR, Beyers JL, Neary DG (2000) Evaluating the effectiveness of post-fire rehabilitation treatments. General Technical Report RMRS-GTR-63. U.S. Department of Agriculture, Forest Service, Rocky Mountain Research Station, Fort Collins

Roy DF (1956) Salvage logging may destroy Douglas-fir reproduction. For. Res. Note 107. California Forest and Range Experiment Station

Roy J, Arianoutsou-Faraggitaki M (1985) Light quality as the environmental trigger for the germination of the post-fire species *Sarcopoterium spinosum* L. Flora 177:345–349

SAF (1996) The role of "salvage" and "sanitation" harvesting in the restoration and maintenance of of healthy forests. Society of American Foresters. www.safnet.org

Schelhaas M, Nabuurs G, Schuck A (2003) Natural disturbances in the European forests in the 19th and 20th centuries. Glob Chang Biol 9:1620–1633. doi:10.1046/j.1365-2486.2003.00684.x

Schmuck G, San-Miguel J, Camia A, Tracy D, Santos de Oliveira S, Boca R, Whitmore C, Giovando C, Liberta G, Schulte E (2010) Forest fires in Europe 2009. JRC Report 10, European Commission, Ispra, Italy

Schoennagel T, Smithwick EAH, Turner MC (2008) Landscape heterogeneity following large fires: insights from Yellowstone National Park, USA. Int J Wildland Fire 17:742–753

Scott DF, Curran MP, Robichaud PR, Wagenbrenner JW (2009) Soil erosion after forest fire. In: Cerdà A, Robichaud PR (eds) Fire effects on soils and restoration strategies. NH Science Publishers, Enfield

Simões C, Marques M (2007) The role of sprouts in the restoration of Atlantic rainforest in Southern Brazil. Restor Ecol 15:53–59

Stohlgren TJ, Barnett DT, Kartesz JT (2003) The rich get richer: patterns of plant invasions in the United States. Front Ecol Environ 1:11–14

Terry JP (1994) Soil loss from erosion plots of differing post-fire forest cover. In: Sala M, Rubio JL (eds) Soil erosion and degradation as a consequence of forest fires. Geoforma ediciones, Logrono

Thanos CA, Daskalakou EN (2000) Reproduction in *Pinus halepensis* and *P.brutia*. In: Ne'eman G, Trabaud L (eds) Ecology, Biogeography and Management of *Pinus halepensis* and *P. brutia* Forest Ecosystems in the Mediterranean Basin. Backhuys Publishers, Leiden

Thanos CA, Georghiou K (1988) Ecophysiology of fire-stimulated seed germination in *Cistus incanus* ssp. *creticus* (L.) Heywood and *Cistus salvifolius* L. Plant Cell Environ 11:841–849

Thanos CA, Rundel PW (1995) Fire-followers in chaparral: nitrogenous compounds trigger seed germination. J Ecol 83:207–216

Thompson JD (2005) Plant evolution in the Mediterranean. Oxford University Press, Oxford

Thomson VP, Leishman MR (2005) Post-fire vegetation dynamics in nutrient-enriched and non-enriched sclerophyll woodland. Austral Ecol 30:250–260

Trabaud L (1991a) Fire regimes and phytomass growth dynamics in a *Quercus coccifera* garrigue. J Veg Sci 2:307–314

Trabaud L (1991b) Is fire an agent favouring plant invasion? In: Groves RH, Di Castri F (eds) Biogeography of Mediterranean invasions. Cambridge University Press, Cambridge

Tsiouvaras C, Havlik N, Bartolome J (1989) Effects of goats on understory vegetation and fire hazard reduction in a coastal plain in California. For Sci 35:1125–1131

Valderrábano J, Torrano L (2000) The potential for using goats to control *Genista scorpius* shrubs in European black pine stands. For Ecol Manag 126:377–383

Vallejo VR (1999) Post-fire restoration in Mediterranean ecosystems. In: Eftichidis G, Balabanis P, Ghazi A (eds) Wildfire management. European Commission, Algosystems, Athens

Vallejo VR (2005) Restoring Mediterranean forests. In: Mansourian S, Vallauri D, Dudley N (eds) Forest restoration in landscapes. Beyond planting trees. Springer, New York

Vallejo VR, Alloza JA (1998) The restoration of burned lands: the case of eastern Spain. In: Moreno JM (ed) Large forest fires. Backhuys Publishers, Leiden

Vallejo VR, Aronson J, Pausas JG, Cortina J (2006) Restoration of Mediterranean woodlands. In: Van Andel J, Aronson J (eds) Restoration ecology. The new frontier. Chapter 14, Blackwell Publications, Oxford van Wilgen BW, Richardson DM (1985) The effect of alien shrub invasions on vegetation structure and fire behaviour in South African fynbos shrublands: a simulation study. J Appl Ecol 22:955

van Wilgen BW, Forsyth GG, de Klerk H, Das S, Khuluse S, Schmitz P (2010) Fire management in Mediterranean-climate shrublands: a case study from the Cape fynbos, South Africa. J Appl Ecol 47:631–638. doi:10.1111/j.1365-2664.2010.01800.x

Vieira DL, Scariot A (2006) Principles of natural regeneration of tropical dry forests for restoration. Restor Ecol 14:11–20

Vilà M, Pino J, Font X (2007) Regional assessment of plant invasions across different habitat types. J Veg Sci 18:35–42

Wakatsuki T, Rasyidin A (1992) Rates of weathering and soil formation. Geoderma 53:251–263

Whelan RJ (1995) The ecology of fire. Cambridge studies in Ecology. Cambridge, Cambridge University

Whisenant S (2005) Managing and directing natural succession. In: Mansourian S, Vallauri D, Dudley N (eds) Forest restoration in landscapes. Beyond planting trees. Springer, New York

Wischmeier WH, Smith, DD (1978) Predicting rainfall erosion rates. A guide to conservation planning. USDA Handbook number 537. Washington, DC

Zouhar K, Smith JK, Sutherland S, Brooks ML (2008) Wildland fire in ecosystems: fire and nonnative invasive plants. General Technical Report RMRS-GTR-42-vol 6. Department of Agriculture, Forest Service, Rocky Mountain Research Station, Ogden

# Chapter 6
# Post-Fire Management of Serotinous Pine Forests

Jorge de las Heras, Daniel Moya, José Antonio Vega, Evangelia Daskalakou, V. Ramón Vallejo, Nikolaos Grigoriadis, Thekla Tsitsoni, Jaime Baeza, Alejandro Valdecantos, Cristina Fernández, Josep Espelta, and Paulo Fernandes

## 6.1 Ecological Context

### 6.1.1 Serotinous Pines

Fire response of pines was categorized by Keeley and Zedler (1998) according to two different strategies that are generally considered alternatives or complementary: individual survival (*fire resistant species*) and stand resilience (*fire evader species*). The first strategy is basically characterized by thick bark, long needles, thick protected buds, self-pruning, deep rooting, rapid growth and, in a limited number of species, resprouting capability. The second is characterized by the presence of a large canopy seed bank that ensures abundant post-fire seedling recruitment (*serotiny*); seeds are stored in the fruit or cones after maturation and the seed release occurs in response to an environmental trigger, fire in this case.

*Fire resilient pines*, characterized by the production of a high number of serotinous cones, small seeds and early reproduction capability, include a Mediterranean group of species primarily formed by *Pinus halepensis* Mill. (Aleppo pine), *P. brutia* Ten.

J. de las Heras (✉) • D. Moya
ETSI Agronomos, University of Castilla-La Mancha, Albacete, Spain
e-mail: Jorge.heras@uclm.es

J.A. Vega • C. Fernández
Centro de Investigación Forestal (CIF) Lourizán, Pontevedra, Spain

E. Daskalakou
Institute of Mediterranean Forest Ecosystems and Forest Product Technology, Athens, Greece

V.R. Vallejo • J. Baeza • A. Valdecantos
Fundacion CEAM, Parque Tecnologico, Paterna, Spain

F. Moreira et al. (eds.), *Post-Fire Management and Restoration of Southern European Forests*, Managing Forest Ecosystems 24, DOI 10.1007/978-94-007-2208-8_6, © Springer Science+Business Media B.V. 2012

(Brutia pine) and *P. pinaster* Aiton. (Maritime pine). According to the EEA (2006) forest type classification, they are all included in the category of the coniferous forests of the Mediterranean, Anatolian and Macaronesian regions.

*Pinus halepensis* and *P. brutia* are the main pine tree species in the Mediterranean Basin, especially in low altitudinal ranges, and both have high fire resilience (Thanos and Daskalakou 2000). *P. pinaster* has been formerly considered a non-serotinous species (Keeley and Zedler 1998) due to the polymorphism of this trait among and within populations (Tapias et al. 2001, 2004) but there are several populations showing a high degree of serotiny (Vega et al. 2010), therefore this species has been considered in this chapter. However, as this chapter is focused on serotinous pine species distributed in fire-prone habitats, other species such as *P. cembra, P. uncinata* and *P. mugo* are not included. *P. radiata* is an introduced species for production purposes (plantations) and is not included as well.

## 6.1.2 Species Distribution

Serotinous pine trees are important components of many landscapes in the Mediterranean Basin and played a major role in the origin of the flora and vegetation (Barbero et al. 1998). The most common pine species, *Pinus halepensis* and *P. brutia*, are widely distributed around the Mediterranean Basin (Quezel 2000). Both species show taxonomical proximity and could be difficult to distinguish them out of their natural habitats where is frequent their hybridization. Together with *P. pinaster* Ait. which mainly grows in western Mediterranean Basin (Carrion et al. 2000), they are the three major conifers of the Mediterranean zone in terms of ecological and economic value and total area covered (Fig. 6.1).

*P. halepensis* cover is about 2.5 million ha mainly in western Mediterranean Basin (Iberian Peninsula, France and Italy) although it also occurs in northern Africa (mainly Morocco, Algerian and Tunisia) and locally in some locations of the eastern area (Egypt, Greece, Former Yougoslav Republic of Macedonia, Turkey, Syria,

N. Grigoriadis
Forest Research Institute (FRI), National Agricultural Research Foundation (NAGREF), Thessaloniki, Greece

T. Tsitsoni
Faculty of Forestry and Natural Environment, Laboratory of Silviculture, Aristotle University of Thessaloniki, Thessaloniki, Greece

J. Espelta
CREAF (Centre for Ecological Research and Forestry Applications), Universidad Autonoma de Barcelona, Bellaterra, Spain

P. Fernandes
Centre for the Research and Technology of Agro-Environmental and Biological Sciences, Universidade de Trás-os-Montes e Alto Douro, Vila Real, Portugal

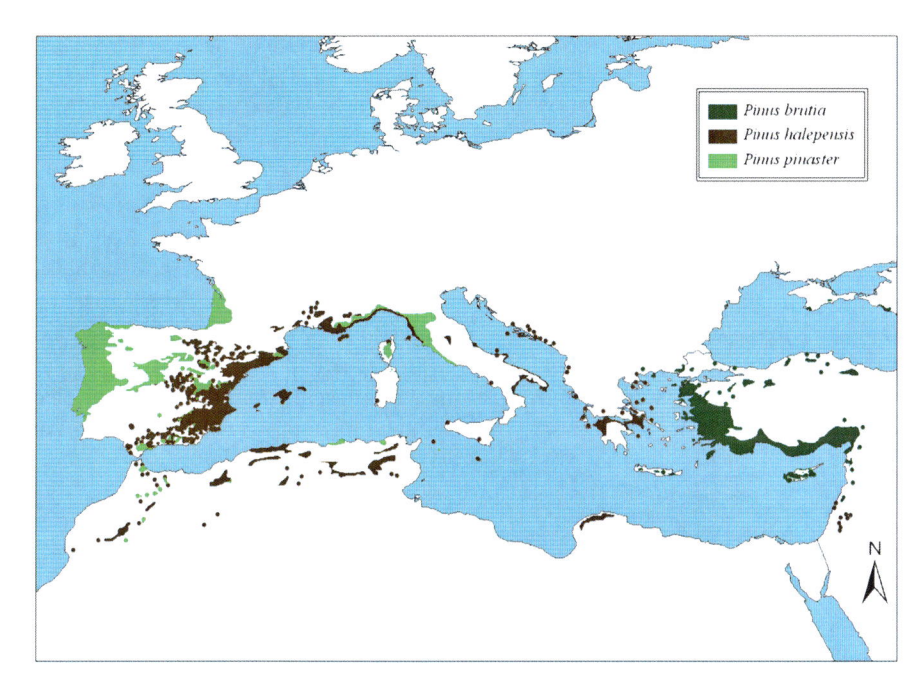

**Fig. 6.1** Distribution of *P halepensis*, *P brutia* and *P pinaster* in the Mediterranean Basin (Source: Euforgen.org)

Jordan and Israel). It usually grows on marly limestones and marls, from the sea level up to high altitudes (more than 2000 m asl), although it prefers lower altitudes occurring mostly in the Thermo and Meso-Mediterranean. Its habitat ranges from the lower arid or semiarid to humid bioclimates.

*P. brutia* covers more than four million ha in the eastern Mediterranean Basin (Greece, Turkey, Kurdistan, Afghanistan, Iran, Syria, Jordan and Israel). Its cover has been increased with extensive afforestations in the nineteenth century, in France, Italy and Morocco. Similarly to *P halepensis*, it grows on marly limestones and fissured soils with high altitudinal range (up to 2000 m amsl) although its optimal is in the Thermo and Meso-Mediterranean. This is stricter species in terms of water requirements and it is not frequent in arid or semiarid bioclimates.

*P. pinaster* covers more than four million ha in western Mediterranean Basin, being found in the Iberian Peninsula, France, Italy and northern Africa (mainly in Morocco and Tunisia). Although it prefers sandy and acid soils, it also grows in basic soils and even in sandy and poor soils. The altitudinal range includes from the sea level to high altitudes (about 1500 m amsl) up to the Supra-Mediterranean belt. It resists frost but not hard droughts, consequently it does not occur in arid or semiarid bioclimates.

Due to the climatic conditions and the fire dynamics of Mediterranean Basin, these three *Pinus* species use to regenerate after fire and they show early cone production

(Tapias et al. 2004), the total recovery and maturity (structurally and functionally) is estimated to require at least 20–30 years which has been known as immaturity risk (Trabaud 2000; Tsitsoni et al. 2004; Vega et al. 2010). Forest management should be prioritized depending on the vulnerability of the burned site. It should be character-ized taken into account vegetation characteristics before the fire, climate conditions, soils and landform characteristics. The rate of vegetation recovery is the combined result of plant reproductive strategy (seeder or resprouter species) plus physical factors, such as climate and aspect. Recruitment of Mediterranean serotinous pines could be stimulated by fire (Boydak 2006; Pausas 1999a; Vega et al. 2010), but in the middle term, ecosystem vulnerability will depend on the ability to persist without major changes in vegetation composition, structure and the relative cover and biomass of the species present. In this sense, ecosystems dominated by obligate seeder species have lower resilience in the young phase due to the immaturity risk (Keeley et al. 1999).

### 6.1.3  Composition of Natural Communities

The three serotinous pine species form natural stands in the previously mentioned geographical areas although they have been introduced by reforestation in almost all the Mediterranean Basin (within and outside their natural range).

They form monospecific or mixed forests, depending on the soil, altitude and bioclimate. These forests can be an intermediate step in the succesional series to broa-dleaved trees, such as some species of *Quercus*. However, in some areas they conform climacic communities as paraclimaxes (Barbero et al. 1998), composed of *Pinus* or other coniferous species, such as *Abies pinsapo* Boiss., *Tetraclinis articulata* (Vahl) Mast, *Juniperus thurifera* L., *J. phoenicea* L. and *J. oxycedrus* Sibth. & Sm.

The natural communities of the three Mediterranean pines include different shrub species depending on soil, bioclimates and forest degradation but the most common species which can also be found in shrubland formations (*maquis* and *garrigue*) are *Rhamnus lycioides* L., *Quercus coccifera* L. or *Q. calliprinos* Webb, and several species of the genus *Pistacia, Cistus, Phillyrea, Genista, Anthyllis, Rosmarinus, Myrtus, Erica, Arbutus* or *Calluna*.

### 6.1.4  Ecological and Socioeconomic Importance

The three pine species are very important due to their ecological role and economical uses in the Mediterranean Basin (Alia and Martin 2003; Fady et al. 2003). Aleppo and Brutia pines represent the only or main source of wood and forest cover in many Mediterranean countries, such as *P. brutia* in Turkey or *P. halepensis* in Tunisia. Their wood is used for many purposes such as construction, industry, carpentry, firewood and pulp. Even seeds are an important resource for pastry in several areas, mainly in North Africa. The importance of *P. pinaster* comes mainly from the uses related to wood and the production of high quality resin. It can be considered a fast-growing species in the Atlantic region, where it is used for pulp and paper production, construction, chipboards, floor boards and paletts.

Ecologically, *P. pinaster* has been considered, in its habitat region, the most important pine species for making recreational forests with high soil protection. On the other hand, Aleppo pine is the most important forest species in North Africa, and it has high ecological importance in southern France and Italy, especially at the urban-forest interface. The importance of *P. brutia* is similar in eastern Mediterranean Basin. In addition, the adaptation of the species to drought, poor soils and forest fires are the reasons by which they have been used in several afforestation pro-grammes throughout the Mediterranean Basin, both for wood production and soil protection (Alia and Martin 2003; Fady et al. 2003).

## 6.1.5   *Expansion or Regression of Populations*

An increasing number of pollen fossils and charcoal records from different areas in the Mediterranean Basin have related the presence and/or expansion of the three pine species to fire events (Quezel 2000; Vega et al. 2010), although high fire frequency would have limited the presence of *P. pinaster* and *P. halepensis* (Christakopoulos et al. 2007; Gil-Romera 2010).

Current trends indicate that in North Africa their distribution area decreased in the last decades due to human activity (land use change, overgrazing and overex-ploitation) whereas in northern part of the Basin their geographical extent is increasing due to the expansion after fire and European Union policies promoting afforestation (Barbero et al. 1998). Global warming can be a factor to be considered in the next years in relation to the dynamics of the distribution area of the serotinous pine species although this has not been studied.

## 6.1.6   *Fire Adaptions and Regeneration Strategies*

The landscapes in the Mediterranean Basin have been transformed in association with the effects of fire, being the first evidence of fires induced by human activity from about 10,000 years ago (Boydak et al. 2006). In this fire-prone habitat, as it has been above mentioned, one strategy has been developed by some Mediterranean pine species, which are obligate seeders: the seed storage on the canopy protected in the pine cones to take advantage of the conditions that taking into account after fire, delaying the opening of cones to release seeds after the fire (Saracino et al. 1997; Tsitsoni 1997).

Serotiny is a term pertaining to the retention of mature seeds in a canopy-stored seed bank for 1–30 years or even more (Lamont et al. 1991). According to Richardson

(1998), serotiny is analytically defined as a morphological feature of some pine cones (and reproductive structures in other plants) whereby the cones remain closed on the tree for one or more years after seed maturation; cones open rapidly when high temperatures melt the resin that seals the cone scales. Long-term seed storage is common in dominant plant families of fire-prone areas, e.g. *Proteaceae*, *Cupressaceae* and *Pinaceae* (Tapias et al. 2001). The degree of serotiny in pines varies considerably among species and populations of the same species, mainly in response to fire frequency as well as site productivity (Ne'eman et al. 2004). Mediterranean serotinous pine species have developed a dual strategy: post-fire obligate seeder (from serotinous cones) and an early colonizer (from non-serotinous cones) which attributes them high fire resistance. These pines regenerate easily after fires which trigger a massive release of seeds, but this also occurs in the absence of fires under suitable stand conditions (Nathan et al. 1999; Thanos 2000). Although the mature trees of the three considered pine species are tolerant to fire, life-history attributes such as a short juvenile period, large cone crops, fast growth and a massive regeneration capability after fire contribute considerably to fire resilience in these pine species (Thanos and Daskalakou 2000). Tapias et al. (2004) recorded a negative correlation between bark thickness and serotiny level in Iberian *P. pinaster* populations and linked the presence of serotinous cones in *P. pinaster* and *P. halepensis* to populations growing in areas with abundant scrub layer that are prone to very intense fires which frequently scorch or torch the crowns. In addition, they have developed the capacity for serotiny and are well adapted to fire prone habitats, mainly at low elevation along the Mediterranean Basin (Thanos and Daskalakou 2000).

However, serotiny is more than the delayed opening of pine cones since increases the probability of success to the recruitment (Moya et al. 2008c). The seeds released from serotinous cones reach better conditions for survival and germination (low competition) than those from non-serotinous cones and avoid excess predation (Saracino et al. 2004). In addition, serotiny produces a higher number of seeds that display the best biological qualities to be released after a forest fire (Daskalakou and Thanos 1996, 2004; Ferrandis et al. 2001; Leone et al. 1999; Saracino et al. 1997, 2004). These qualities include a higher number of sound seeds, greater weight and germination rate (Saracino et al. 1997) or higher heat insulation and resistance to unfavorable environmental conditions (Salvatore et al. 2010). In this way, serotiny in pine species maximizes the number of seeds available for the next generation by storing seed crops successfully and protecting them to ensure extensive and vigorous natural regeneration after fire or other disturbances.

Post-fire regeneration of *P. halepensis* has been extensively reviewed for the eastern and western Mediterranean Basin, with emphasis on community recovery, pine population reconstitution (Trabaud 2000), plant adaptive traits, pine forest communities and vegetation structure (Arianoutsou and Ne'eman 2000). The literature concludes that whole pine tree populations are usually burned by wildfires in *P. halepensis* or in *P. brutia* forests (Arianoutsou and Ne'eman 2000; Thanos and Doussi 2000). Taking into consideration that the three Mediterranean serotinous pines (*P. halepensis*, *P. brutia* and *P. pinaster*) are obligate seeders (Thanos and Daskalakou 2000; Tapias et al. 2001; Boydak 2006), the most critical point for natural regeneration and forest re-growth are the immaturity risk period and the amount of seeds stored in the canopy seed bank and the quantity of sound seeds available after fires (Thanos and Doussi 2000). Therefore, post-fire regeneration of these species is strongly related to the quality and quantity of the canopy seed bank, since soil seed banks have a transient character (Ferrandis et al. 1996, 2001; Boydak 2006). In addition, early flowering and cone production is a common characteristic for both *P. halepensis* and *P. brutia* species with a short juvenile period, which probably ends at the age of 3–6 years for several pine individuals or at 12–20 years for the whole population (Thanos and Daskalakou 2000), although the latter species seems to start flowering and produce cones a little later (Boydak 2006). In mature stands, serotinous cones remain closed for several years, thus forming the canopy seed bank that is responsible for post-fire regeneration. Additionally, in young reproductive Aleppo pine trees, a high percentage of serotinous cones has been observed, in which resources are allocated to seed production and thereby reduce the immaturity risk in the case of a second successive fire (Ne'eman et al. 2004).

Focusing on post-fire regeneration dynamics, massive seedling recruitment after fire has been extensively recorded (Boydak 2006; Daskalakou and Thanos 2004; Thanos and Marcou 1991; Tsitsoni et al. 2004). In many study cases, Aleppo pine seed germination, seedling emergence and establishment took place as a *single massive wave* shortly after the onset of the first post-fire rainy season which appears to be strongly related to rainfall and temperatures (Daskalakou and Thanos 2004), with no additional pine seedlings observed emerging during subsequent years. For *P. brutia* forests, densities of 10,000–30,000 biologically independent seedlings per hectare are proposed to be considered as successfully recruitment for natural regeneration after fire, whereas 3,000–10,000 seedlings are acceptable in certain parent materials and poor soils (Boydak 2006). On the other hand, mean post-fire seedling density (Thasos Island) was considerably high (20,000–60,000 seedlings ha$^{-1}$) at the end of the recruiting period. A significant drop in density was observed during the first summer and stabilization at 0.6–2.0 seedlings m$^{-2}$ was achieved 5 years after fire (Spanos et al. 2000).

Several studies reported the initial sapling density in the short term (3 years after the fire) and the survival rate, such as Brutia pine in Samos Island where the overall pine density was 0.15 saplings m$^{-2}$ 6 years after a fire (Thanos and Marcou 1991), while values of 10,600–12,000 stems ha$^{-1}$ at different sites were recorded in an artificial forest 20 years after fire (Tsitsoni et al. 2004). Moreover, sapling survival was found to be high (43%) in the sixth growing season after fire and was expressed by a negative exponential curve (Thanos and Marcou 1991). Sapling survival kinetics was described by a rectangular hyperbola, reaching a percentage of 28% 5 years after a fire (Spanos et al. 2000). For *P. pinaster,* an initial seedling density of about 8–12 seedlings m$^{-2}$ were found in Spain (Vega et al. 2010) and high variability for *P. halepensis* was found in eastern Spain, from about 10,000 to 60,000 seedlings/ha, depending on latitude (Moya et al. 2008a). However, a high recurrence of fire has a strong, negative effect on the ecosystems, increasing the risk of degradation and soil loss. The minimum time interval to avoid this negative effect has been estimated in 15 years (Eugenio and Lloret 2004; Eugenio et al. 2006).

## 6.2   Post-Fire Management in Serotinous Pine Forests

### 6.2.1   Current Practices and Management Objectives

In general, the three serotinous pine species have been under two contrasting management alternatives: under- or over-managed.

Excessive harvesting and unplanned exploitation, both past and present, have been destroying the original forest structure to promote monospecific stands where pine growth is maximized by decreasing interspecific competition with other trees (oaks or other coniferous species). Typical treatments include clear-cut operations applied over large areas and the establishment of monoculture afforestation. This management approach has been developed under the premise of maximizing the potential timberline in short rotation ages. Although this could be a correct management in high quality site areas, it is totally wrong in low productivity regions (Bravo et al. 2008a, b) and the result, in several cases, are highly degraded forests resulting from long periods of mismanagement (Çolak and Rotherham 2007).

In the last decades of twentieth century, the decrease of economic benefits from low productive areas for timber production lead to under-management, resulting in dense forests with almost null productivity rates and high fire hazard. The more common management practices after forest fires in the serotinous pine forests of Southern Europe are the salvage logging of the burned trees, usually carried out at early stages of succession (before 6 months after the fire). If the amount of seedlings is not enough to assure the natural regeneration, reforestation is usually carried out using the same main tree species. Stands naturally regenerated are not early managed and dense stands with lower productivity rates are developed.

In many cases, the current socio-economic and ecologic context in serotinous pine forests calls for a more sensitive and sustainable forest management. External benefits such as recreational uses, decrease of rural depopulation, soil protection, biodiversity, carbon sink capability, reducing fire risk should be taken into account. There is an opportunity to introduce the current scientific knowledge to promote good practices.

The management of burned pine forests should first establish the main objectives of actions. In most cases, the prioritized objectives should include soil protection and water regulation, the reduction of disturbance (fire) hazard as well as increased ecosystem and landscape resistance and resilience to disturbances (fires), and the promotion of mature, diverse, and productive forests (Vallejo and Valdecantos 2008). Foresters have to define whether they want to recover the pre-disturbance plant component of the ecosystem (i.e. a pure *Pinus halepensis* forest) or better moving the system towards alternative states (i.e. mixed pine-oak forest) following the classic restoration model proposed by Bradshaw and Chadwick (1980). The introduction of a larger and more diverse set of species in the restoration of pine forests with low regeneration may improve ecosystem resilience in the framework of current fire regimes (Valdecantos et al. 2009). However, both objectives may share common management tools and actions. An intermediate management goal must address avoiding future disturbances, such as fire, that convert the system to more degraded states. Selective clearing of obligate seeder shrubby vegetation has been shown to be an effective technique to transform a highly flammable, dense and continuous shrubland with a great amount of dead fuel, to grassland with sparse resprouting shrubs and regenerating pines and a discontinuous fuel load (Baeza et al. 2008). Mulching of the soil surface with brush-chipping of material from clearing greatly reduces the germination rates of obligate seeders (Tsitsoni 1997). In addition, the removal of standing vegetation by selective clearing may promote growth of remaining individuals and seedlings introduced by reforestation (Valdecantos et al. 2009).

## 6.2.2  *Management of Burned Trees and Soil Protection*

The first priority should be an emergency assessment of the burned area (Robichaud et al. 2009) (see Chap. 1). A multidisciplinary point of view is mandatory to the assessment of urgent threats and risks to human life, property and critical natural and cultural resources considered in an integrated plan. Following the procedures of BAER, Burned Area Emergency Responses (Napper 2006), the main topics to take into account should include forest fire severity, presence of water-repellent soils, mapping of effective soil cover, flood or debris risky areas, riparian stability assessment, potential erosion or sedimentation and water quality deterioration, and status of infrastructures. Immediately after forest fire soil erosion prevention may have to be carried out. When there is a low recovery rate of vegetation and/or where erosion is expected to occur, several management strategies might be applied. The application of an organic layer of mulch, either alone or combined with seeding native grasses, is a suitable rehabilitation option on steep slopes with low plant cover and high erosion risk, as it is aimed at reducing surface flow, rainsplash, soil crusting and compaction, thereby increasing infiltration. To be effective, this technique should be applied soon after the fire and before the heavy autumn rains, which means that areas vulnerable to erosion must be identified as soon as possible. In burned pine forests on steep slopes and soft soils, log dams or contour-felled barriers are also effective post-fire management practices for reducing physical soil degradation and erosion (Serrano et al. 2008).

In Aleppo pine stands, salvage logging has been checked to reduce the seedling density due to mechanical actions but also it induced the reduction of growth and the survival of pine seedlings due to the protection losses (Martínez-Sanchez et al. 1995). In addition, the burned wood promoted seedling recruitment due to microclimatic conditions which could be significant in poor site quality areas (Castro et al. 2010). These results are expanded and analyzed deeply in the study cases, Sect. 6.3.2.3.

## 6.2.3  *Plantation and Seeding*

Historically, extensive reforestation programs have been developed over the last two centuries in southern Europe, and more intensively in the last 50 years, resulting in expansion of the habitat distribution of these pines (Le Maitre 1998) due to the high availability of seeds, ease of seedling cultivation and success of survival in low fertility soils. In several countries, the reforested area is considerably larger than the original range. These plantations were undertaken for timber production and soil protection on degraded areas and wood production is currently an important economic resource on a local level.

In many reforestations, seed origin was not considered and this has become not only the main difficulty for clarifying migration pathways of the species, but more importantly, it may have drastically affected the genetic structure of populations with previous fire adaptive traits (Tapias et al. 2004; Vega et al. 2010), leading to difficulties in post-fire regeneration.

### 6.2.4 Management of Natural Regeneration

After the emergency treatments, the best option to support the restoration of serotinous pine forests is monitoring the burned area in order to record how natural regeneration progresses. After 1–3 years, a new evaluation should be carried out and a management plan can be implemented (Corona et al. 1998). Depending on the success of natural regeneration, the options are: no action, assisted natural regeneration or active restoration. Forest restoration treatments often include reforestation with planting or seeding (less frequent) to complete the recovery in the burned area. When fire recurrence is lower than the required period for the community to reach maturity and to increase the resilience to new fires, the young stands could be burnt before the production of mature seeds. In this situation, artificial regeneration (reforestation) should be carried out with seedlings coming from similar provenances than the burned area to reinsert populations with similar resilience to the previous one. The success could be improved by adding sewage sludge (and, if necessary, pine seedlings inoculated with mycorrhizas), mainly in arid and semiarid areas (Fuentes et al. 2007; Gonzalez-Ochoa et al. 2003).

Regeneration capacity of pine forests after a fire will determine the most appropriate post-fire management option. As mentioned above, usually serotinous pine forests show high recovery after a fire, showing a moderate to high resilience. In fact, pine seed germination after fire is often excessive, promoting high intra-specific competition and low tree growth rates and, as a consequence, high fuel load accumulation. Post-fire management should consider all of these points, including fire recurrence. Simulation models of vegetation dynamics in relation to fire frequency show that *Pinus* species dominate the plant community at medium fire recurrence, declining at intervals longer than 100 years (Pausas 1999b). However, when these forests are affected by recurrent fires, succession can be diverted. A single fire is enough to change serotinous pine forests into alternative stable shrubland, where the colonization of late-successional species is impeded (Rodrigo et al. 2004). Undoubtedly, the management options should consider the forest conversion after the fire approaching the ecological window to reintroduce the climax species.

Furthermore, post-fire recovery may not be homogeneous within the whole burned area of a serotinous pine forest (Pausas et al. 2004b). For instance, the fire severity will determine the regeneration depending on the availability of seeds and scorched needles protecting the soil surface (Pausas et al. 2003). Furthermore, plant cover and type influence soil protection by decreasing hydrophobicity and soil losses (Cerda and Doerr 2005). Land use history and soil type, properties that are often intimately related, also determine the type and characteristics of the regenerating community after a fire. Baeza et al. (2007) observed that the regeneration of a mixed forest of *Pinus halepensis* and *P. pinaster* 23 years after fire was mainly driven by the occurrence of a second wildfire, land use and soil type, releasing six different vegetation types from grasslands to dense shrubland with young regenerating pines. As a consequence, the landscape and the plant community are quite heterogeneous just after post-fire regeneration, suggesting that specific local management schemes are needed. In burned Aleppo pine forests in eastern Spain, sapling recruitment has been observed to be positively related to the presence of branches and trunks covering the soil surface, which create a favorable microclimate for germination and seedling establishment (Pausas et al. 2004a). At a later stage, planting seedlings is suggested for increasing site diversity or reintroducing species that have disappeared after the fire.

In some cases, high-density stands are opened by fires or silvicultural treatments which reduce understory fuel accumulations limiting crown fire propagation (Fernandes and Rigolot 2007) and increase seed production (Moya et al. 2008a). Dense stands which receive little treatment have high potential for destructive crown fires (Pausas et al. 2008), although *P. pinaster* stands are generally vulnerable to fire in any conditions (Fernandes and Rigolot 2007). Fuel treatments by means of prescribed fire in the understory or alternative techniques can make these stands less vulnerable to crown fire. The interaction between provenance, fire severity and post-fire management can be critical for seedling recruitment success. Some of these aspects are specifically addressed in other sections of this chapter.

Although *P. pinaster* re-establishment after a single fire is generally acceptable in part of its natural area, there are post-fire regeneration difficulties in some of these populations. In burned areas where *P. pinaster* and *P. halepensis* populations are in contact, *P. halepensis* can be more effective for seed dispersal than maritime pine due to its larger seed bank, since it is more thermophilous and tolerant to summer drought, and usually has a more abundant regeneration than maritime pine. More xeric conditions and more severe and frequent forest fires are expected, according to the climate change scenarios for the Mediterranean Basin over the coming decades (Fischlin et al. 2007; Moriondo et al. 2006). Consequently, it seems plausible to expect *P. pinaster* to present more difficulties in post-fire re-establishment than *P. halepensis* and for the geographic range of *P. pinaster* to be reduced (Benito-Garzón et al. 2008).

### 6.2.5 Herbivory

In general, grazing activity in post-fire serotinous pine forests has not been different to that of other pine forests in the Mediterranean Basin. The grazing load depends on the cattle species and the success of the natural regeneration (Nosetto et al. 2006). Although in the first 2 years a great cover of legumes and other pasture species cover the burnt areas, in the case of serotinous pine forests the advisable is to forbid it for 2 or 3 years after the fire to check how the forest regeneration is developing. If the natural regeneration is ensured, the maintenance of agricultural and pastoral activities can be used as tools for fire prevention.

The grazing rate depends on the plant community, the ecosystem response to fire (related to fire severity) and the livestock. Depending on the herbivore or lignivore and the livestock size (trampling effect), the shrub vegetation could be reduced and disappear mainly in areas bearing heavy and wild lignivores (McEvoy et al. 2006). Pine seedlings are compatible with horses and cattle but not with goats and sheeps which should be forbidden in regenerating areas after the fire (Mosquera et al. 2006). In other hand, because the source of more than 30% of the wildfires in southern Europe is agricultural and grazing activities, it is essential to consider the needs of rural communities who use fire for their livelihood, and how burning might help restore and maintain these ecosystems.

## 6.2.6 Pest and Diseases

Fire and insects are natural disturbance agents in many forest ecosystems, affecting succession, nutrient cycling and forest species composition. Abundance and species richness of herbivorous insects (xylophages and sap-feeders) decrease during succession whether or not major changes are found between the sites in terms of abundance or species richness of predators (Kaynaş and Gurkan 2008). Fire suppression policies implemented in the past (early 1900s) resulted in profound changes in biomass accumulation in forests, structure and species composition. Additionally, wildfires reduce some hosts that assist in the spread of pests and pathogens (McRae et al. 2001). Associated with these changes, there was an increasing vulnerability of forest stands to wildfires and damages during outbreaks of defoliating insects that are more frequent due to the increasing drought periods in the Mediterranean Basin. There is a gap in the knowledge of the relationship between serotinous pines and the outbreaks of the pest attacks; however, these pines are host of a large number of insects that can become pests (Mendel et al. 1985). The most common pest species in Mediterranean semiarid areas are bark beetles (*Carphoborus minimus, Hylurgus ligniperda, H. micklitzi, Hylastes linearis, Pityogenes calcaratus, Orthotomicus erosus, Tomicus destruens*), caterpillars (*Thaumetopoea pityocampa, T. wilkinsoni*) and other insects (*Neodiprion sertifer, Matsucoccus josephi, Palaeococcus fuscipennis*). Some studies reported that reforestation actions sometimes increase bark beetles attacks, mainly in semiarid areas. Also, some post-fire management measures, such as very early thinning, increase the sensitivity of these stands to pest attacks, such as *Pachyrhinus squamosus* in Spain (Gonzalez-Ochoa and de las Heras 2002). Outbreaks of *O. erosus* after the fire has been reported in *P. halepensis* stands in Israel (Mendel and Halperin 1982) and *Pinus pinaster* in South Africa (Baylis et al. 1986).

In *P. halepensis* and *P. brutia* stands the higher risk to become a pest was showed by Hymenoptera and Coleoptera families (Scolytidae, Buprestidae and Cerambycidae) in Greece (Markalas 1991). Pine engraver (*Ips sexdentatus*) and other scolytids (*Tomicus* spp, *Hylurgus piniperda*, *Hylastes* spp, *Pissodes notatus*) were the most frequent insects attacking fire-injured maritime pine trees in all burned areas in Spain (Vega et al. 2010).

### 6.2.7 Post-Fire Forest Conversion

In burned pine forests where reforestation is necessary or at least advised, a window of opportunity opens to introduce new forest structure and, even, new species.

As previously mentioned one common conversion is a change towards a mixed pine-oak forest, with gains in biodiversity and ecosystem resilience (Valdecantos et al. 2009; Moya et al. 2009). Conversion can also include changing land cover to create more fire resistant and resilient landscapes, increasing heterogeneity and landscape barriers or filters that inhibit the spread of fire (see Chap. 1).

To convert an area into serotinous pine stand, the selection of the main tree species is advisable, taking into account that the optimal seed provenance selection is mandatory to ensure genetic adaptation and promote resilience.

### 6.2.8 Climate Change

The three serotinous pine tree species present in Mediterranean forests could fail to develop adaptive strategies to increased aridity or habitat fragmentation as it has been recorded in the last two decades (Alig et al. 2002). Global change is also considered to increase the severity and recurrence of fires, the frequency of extreme events like hurricanes or windstorms, pest attacks and to promote invasions by alien species (Dale et al. 2001). Regarding European pines, and specifically the serotinous pines distributed in the Mediterranean Basin, there are some points in which climate change and its effects are impacting populations:

- Pine species which are not well adapted to intense drought or irregular weather, mainly during spring, are in jeopardy due to habitat reduction. In this way, *P. pinaster* distribution is becoming reduced, while altitudinal requirements are increasing and they are being replaced by other coniferous or broadleaf species (Resco et al. 2007).
- *P. halepensis* distribution is shifting upwards (up to 200 m) in the mountains close to the coast, replacing the lower altitudinal range of other species such as *P. sylvestris* in Southern France where the latter had lower productivity (Vennetier et al. 2005).

- Higher fire intensity and recurrence induced by climate change could reduce the period between two fire events, hampering the natural regeneration of the stand, altering the process of autosuccession (Initial Floristic Model) even in fire-prone ecosystems (Díaz-Delgado et al. 2002).
- Fire season is getting longer, which could decrease the resilience of communities and ecosystems adapted to recover after fire events in the drought season (Eugenio and Lloret 2004).
- The temperature and the longer activity period affect the survival of insects and the synchronization mechanism between hosts and herbivores. It alters diapauses, resulting in faster development and a higher feeding rate, increasing the risk of pest attack, mainly in the lower altitudinal areas, such as in the pine processionary moth (Battisti 2004).
- Higher aridity promotes an increase in respiration rate and decrease in Gross Primary Productivity (López-Serrano et al. 2009). It promotes carbohydrate use for growth and survival in drought years, reducing the investment in reproduction which increases the immaturity risk in young Aleppo and East Mediterranean (or Anatolian) pine stands, which in turn depends on the stored canopy seed bank for regeneration (Keeley et al. 1999).

The impacts from global climate change and its associated disturbances have created a growing demand for scientific research.

## 6.3 Case Studies

### 6.3.1 Early Post-Fire Management in Eastern Mediterranean Basin

#### 6.3.1.1 *Pinus brutia* Ten

A wildfire burned 1,664 ha in summer 1997 close to Thessaloniki, the North-western coast of the Aegean Sea (longitudes 22°57' to 23°04'E; latitudes 40°35' to 40°39'N). It was a reforestation coming from the fifties and resulted in a mixed coniferous forest composed mainly of Mediterranean pines and cypresses, although the climax vegetation in the area was an oak forest. The altitudes ranged between 85 and 560 m asl and the soil was shallow, rocky and intensely eroded with low productivity. It was an urban forest (about 60 years old) composed mainly of *P. brutia*.

After the fire, there was natural regeneration of *P. brutia* with high survival rates (72–90%) covering about 86% of the soil surface. The public administration decided to recover the stand as an urban forest managed for multiple use frameworks, such as flood prevention, soil protection and recreational use. An afforestation to increase diversity was carried out using mainly two coniferous and two deciduous tree species but deciduous showed high mortality rates (Grigoriadis et al. 2009).

The *P. brutia* tree density was very high and a thinning treatment was carried out 10 years after the fire. It improved tree growth, differentiation of young stands and shortened the age for cone production, which is very important due to the immaturity risk. In addition, the thinned stands showed higher mean diameter, mean height, vitality and faster developmental tendency compared to the unthinned stands. Autosuccession was occurring with the natural regeneration of *P. brutia* coming from restored areas, with similar results to those found in natural forests (Tsitsoni et al. 2004).

### 6.3.1.2  *Pinus halepensis* **Mill.**

A high recurrence of fires has been recorded in Kassandra peninsula (longitudes 25°25′ to 25°35′E; latitudes 39°90′ to 40°10′N), close to Chalkidiki (Northern Greece). The northern Kassandra peninsula is almost flat (100–250 m amsl). The soil is marly, covered by red-clays. Before fire, the forests were mainly managed for resin production. A large forest fire occurred in summer 1984 and the first plan to manage the burned area was no action and the short term monitoring of some plots to record the level of natural regeneration. Natural regeneration was supported by controlling grazing (goats and sheep were restricted to rangelands) and removing snags from late autumn to early winter in the year after the fire. On steep slopes (>50%), log dams were built with the burned branches to prevent soil erosion. In addition, new firebreak strips and roads were opened by removing the understory (Tsitsoni 1997) because a new fire within less than 10–15 years could be catastrophic for natural regeneration of *P. halepensis* (immaturity risk; Keeley et al. 1999).

The natural regeneration of *P. halepensis* in the first 8 years after the fire varied from 0.6 to 14.3 seedlings m$^{-2}$. The height of the Aleppo pine seedlings in their first year ranged from 1 to 30 cm (Tsitsoni 1997). Ten years after the fire, the plant composition in the burned area was following the Initial Floristic Model (Egler 1954). Shrub biomass accumulation was rapid during the early years and continued until the age of 10, reaching 11.5 t ha$^{-1}$. At this age the biomass was characterized by the high contribution of *Cistus* species (24.9%; Ganatsas et al. 2004).

A new fire occurred in summer 2006 in the same area, burning about 7,700 ha of Aleppo pine forest (Fig. 6.2). The emergency actions were to carry out salvage logging, using the deadwood to build log dams (148 km) and branch barriers (447 km) along the contours. Furthermore, in the second year after the fire, flood prevention tasks were planned and implemented using wooden dams (447 dams). Three years after the fire, a reforestation was carried out in places with no natural regeneration or very low tree density due to the high fire recurrence.

In nearby Penteli, in central Greece, two fires (summers 1995 and 1998) burned an Aleppo pine stand. The low fire intensity allowed for recovery of the composition of the vegetation but not the density. However, the main tree species, Aleppo pine, showed a very low natural regeneration in the monitored plots 3 or 4 years after the fire. Afforestation was necessary and the seedlings were planted in patches (2 × 2 m)

**Fig. 6.2** Burned area in Kassandra Peninsula, Northern Greece, after the large wildfire in 2006. Upper image shows the area a few weeks after the fire and the lower image the natural regeneration 3 years after the fire (summer 2009). (Source: Thekla Tsitsoni, Aristotle University of Thessaloniki)

using seeds from the same provenance, applying soil treatments (Tsitsoni 1997). Three restoration methods were used: two planting methods (paper-pot and bare-root seedlings), three seeding types (patches, strips and pits in lines) and no restoration action (Zagas et al. 2004). The results showed that all accelerated the rate of regeneration, while the most appropriate method to improve the regeneration process was to plant paper-pot or bare-root seedlings in pits which protected the seedlings.

The number of seedlings planted was reduced to a minimum (300 seedlings ha$^{-1}$), since soil protection is secured from the dense, broad-leaved evergreen vegetation (Zagas 1994). This choice has an additional economic advantage; planting seedlings in a similar density to mature stands implies no further silvicultural treatments should be needed if the seedling survival was successful.

## 6.3.2 Early Post-Fire management in Western Mediterranean Basin

### 6.3.2.1 *Pinus halepensis* Mill.

We studied three large fires which burned ca. 65,000 ha in July 1994, in Spain. In the burned forests, *P. halepensis* was the main tree species. All three areas naturally regenerated after the fire with very high initial seedling density, from about 7000 to 30,000 pine trees ha$^{-1}$. In the three areas, *P. halepensis* was the dominant regenerated species after fire (>90% of ground cover) and the soil was mainly carbonate substratum with a pH of about 8.5 and low slope (<5%).

Areas are described starting from the north and moving south:

- *Bages* in Barcelona (42°6′N, 2°1′E), 21,500 ha burned. Subhumid.
- *Yeste* in Albacete (38°20′ N, 2°20′ W), 14,000 ha burned. Dry.
- *Moratalla* in Murcia (38°16′, N1°38′W), 30,000 ha burned. Semiarid.

Immediately after the fires, several plots were set to monitor natural regeneration. Ten years later early silvicultural managements consisted of thinning the young Aleppo pine trees. In December 2003, a mixture of free and low thinning methods were applied, removing trees to control stand spacing and favour desired trees using criteria linked to the health appearance of individuals from the lower crown classes to favour those in the upper crown classes. The final pine tree density achieved was about 1,600 pine trees ha$^{-1}$. Following Moya et al. (2008a), Verkaik and Espelta (2006) and Espelta et al. (2008), in spring 2004 six experimental plots were randomly established per site, and a preliminary experiment was conducted to compare the size of the canopy seed bank stored (number, test of viability and germination of seeds were included). The pines in the monitored plots were tagged and the total height, basal diameter (30 cm above the soil to avoid irregularities) and crown coverage were recorded. In addition, the different cohorts of cones were identified and counted, according to their colour and position in the canopy (Daskalakou and Thanos 1996) and the serotiny level was calculated following Goubitz et al. (2004). The first reproductive year (age when a pine tree begins to bear cones) and the number of cone-bearing trees (reproductive trees) were also recorded to characterise the reproductive stages (juvenile and reproductive phase).

Growth and reproductive characteristics of trees from three sites were compared to observe the effects of the treatments on pine growth, improvement of cone yield and increase in canopy seed bank (Table 6.1). We found that thinning improved growth and the amount of reproductive trees (>10%) just 1 year after to apply the silvicultural treatments. Opened cones were not recorded, implying the serotiny level did not vary from 2004 to 2005. Across the decreasing site quality, a decreasing number of reproductive trees and an increasing serotiny level which was not influenced by thinning were found.

The first reproductive year was found in pines 4–7 years old in all the study plots. The cone yield was improved to the individual tree-level by thinning, in both mature and serotinous cones. However, when the values were recorded for unit of surface (ha), we checked that the final amount of cones born and the canopy seed bank stored in them was lower in thinned plots due to the lower pine tree density. The final canopy seed bank was $37,000 \pm 5,000$, $154,684 \pm 39,448$ and $16,710 \pm 4,592$ seeds ha$^{-1}$ in Bages, Yeste and Moratalla, respectively.

Initial tree density has been shown to have a direct relationship with growth and reproductive characteristics, mainly abortion and cone production, in different site quality areas (Moya et al. 2007, 2008b). Therefore, optimal management should be developed and designed to be flexible for diverse objectives specific to the area. Silvicultural treatments improve health and reproductive characteristics in these stands, although a drastic drop in pine tree density could require time to respond to the treatment. To improve the health of the stand and resilience to fire we recommend high intensity thinning in young stands with high tree density. These simple and cheap silvicultural treatments shape juvenile stands that regenerate after fires, improving the fertile seed production and thus fostering species resilience. At the same time, the treatment can be used as a prevention tool to reduce fire risk.

### 6.3.2.2   *Pinus pinaster* Aiton.

Two populations of *P. pinaster*, Nocedo (Lugo province) and Cabo Home (Pontevedra province) in north-western Spain were burned in the summers of 2000 and 2001, respectively (Table 6.2). Within each stand, five square plots ($30 \times 30$ m) were set in each area (border length between plots was at least 15 m). We found different fire severity level (estimated from crown damage):

- *High*: trees with combusted crowns (crown fire).
- *Medium*: trees scorched (surface fire)
- *Low*: not scorched crowns (surface fire)

The burned trees were harvested some months after the fires in Nocedo (September 2001) and Cabo Home (March 2002). The logging slash treatments applied were: *clearcutting + slash chopping* (tractor with a mechanical chopper) in Cabo Home and *clearcutting + slash windrowing* (manual) in Nocedo.

To evaluate seedling density and mortality, 16 subplots ($2 \times 2$ m) were set in each plot as a grid. On the first sampling date, all emergent *P. pinaster* seedlings were

**Table 6.1** Reproductive and epidometric characteristics in sites naturally regenerated after fire [silvicultural treatments were carried out 10 years after the fire (winter 2004) and the study period extended for 1 year long]

| SITE | Treatment | Density | Dbasal | H | CC | RP | MC | SC | SL |
|---|---|---|---|---|---|---|---|---|---|
| Bages | Control | 56404 | 2.4±0.2 | 212±9 | 0.4±0.1 | 52±2 | 0.22±0.05 | 2.20±0.31 | 59 |
| | Thinning | 1655 | 3.3±0.2 | 212±11 | 0.8±0.1 | 63±7 | 2.27±0.81 | 5.52±1.46 | |
| Yeste | Control | 24211 | 3.1±0.1 | 140±4 | 1.2±0.1 | 39±8 | 1.47±0.33 | 0.16±0.06 | 89 |
| | Thinning | 1600 | 4.4±0.2 | 177±4 | 1.5±0.1 | 49±9 | 5.34±0.79 | 3.64±1.52 | |
| Moratalla | Control | 62669 | 1.6±0.1 | 103±3 | 0.6±0.1 | 13±5 | 0.04±0.03 | 0.13±0.05 | 100 |
| | Thinning | 1688 | 1.9±0.1 | 111±2 | 0.8±0.1 | 26±8 | 0.14±0.09 | 0.16±0.12 | |

*Density* Aleppo pine tree density (pines ha$^{-1}$), *Dbcsa/* basal diameter recorded 30 cm above the soil (cm), *H* total height (cm), *CC* crown coverage of Aleppo pine trees (cm), *RP* reproductive pines (%) *MC* mature cones (cones tree$^{-1}$) *SC* serotinous cones (cones tree$^{-1}$), *SL* serotiny level (%)

**Table 6.2** Description of the two studied maritime pine stands (*Pinus pinaster*) burned by wildfires in 2000 and 2001

| Site | Fire date | Fire severity | BS | Age | SL | Climate | AR | MAT | $T_{max}$ | $T_{min}$ | Soil[a] |
|---|---|---|---|---|---|---|---|---|---|---|---|
| Nocedo | Summer 2000 | high-medium | 1050 | 45 | 0 | Mediterranean | 850 | 12.7 | 26.4 | 2 | Alumi-dystric Leptosols |
| Cabo Home | Summer 2001 | medium-low | 32 | 53 | 79 | Oceanic | 1565 | 14 | 24.1 | 8.2 | Dystric Cambisols |

*BS* burned surface (ha), *AGE* age of the stand before the fire (years), *SL* serotiny level (%), *AR* average rainfall (mm $yr^{-1}$), *MAT* mean annual temperature (°C), $T_{max}$ mean of absolute maximum temperatures (°C), $T_{min}$ mean of absolute minimum temperatures (°C)
[a]From Macías and Calvo (2001)

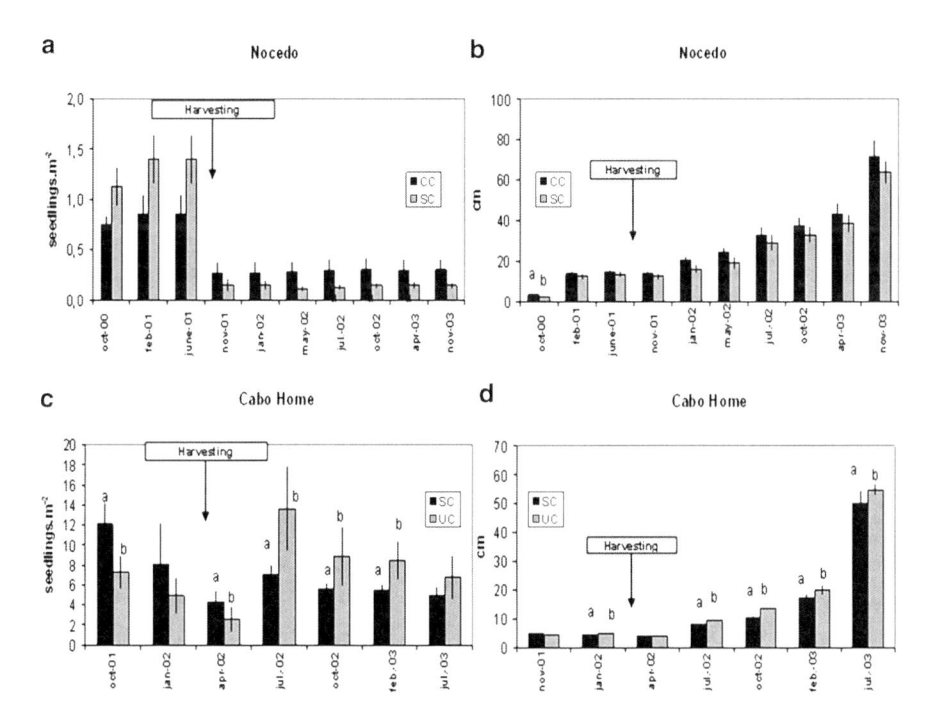

**Fig. 6.3** Seedling density (number of seedlings) and seedling height (in cm) in two burned and treated maritime pine stands from the first autumn after the fire to 2003. *CC* combusted crowns, *SC* scorched crowns, *UC* unaffected crowns. Within evaluation dates means values indicated with different letter are significantly different (P < 0.05). *Vertical bars* represent standard error. (Source: Cristina Fernandez, CIF Lourizan)

labelled and their heights were measured. In subsequent samplings, the number of dead seedlings was recorded and newly germinated seedlings were labelled. Seedling height was measured on each sampling date. In both stands, periodic measurements were carried out the first 2 years after the fire.

Seedling emergence began soon after the first rains in autumn in both sites. In both stands the initial seedling density was significantly higher in burned plots with lower fire severity. The salvage logging carried out 1 year after the fire induced a significant decrease in the initial seedling density. In Cabo Home, slash chopping favoured a new seedling cohort, especially in the unaffected crown plots, promoting an increase in seedling density the first spring after the fire (Fig. 6.3). Seedling height was higher in areas with lower fire severity in Nocedo, although not significantly. In Cabo Home, the mean seedling height was greater in the unburned plots (Fig. 6.3).

Scorched crown level exhibited higher seedling densities than trees with combusted crowns or unaffected crown in Nocedo and Cabo Home, respectively. This may be related to the fact that scorched trees dispersed seeds faster than trees with combusted crowns. In a study carried out in Cabo Home, Vega et al. (2010) found that seed release from scorched trees was faster than in non-scorched crowns. In burned *P. halepensis* and non-serotinous *P. pinaster* stands, the same recruitment pattern was observed (Martínez et al. 2002; Saracino et al. 1997).

The effects of harvesting on pine regeneration varied markedly depending on the stand and site characteristics. This suggests that it is not possible to design a common post-fire salvage logging schedule for all sites. For provenances with low serotiny, harvesting and logging must be planned and conducted more carefully. The combination of high temperatures and drought in the summer of 2001 in the south-facing slopes in Nocedo, and an early frost may have contributed to the sharp increase in mortality. By contrast, logging slash chopping in serotinous stands with unaffected crown trees can result in a new cohort of seedlings. This can be critical if the preceding one fails. Similar delayed pulse of germination was observed as a result of slash chopping in *P. pinaster* stands, by Canga et al. (2003) and Fernández et al. (2008).

There is no evident reason for higher seedlings in plots where fire severity, estimated through crown damage, was greater in Nocedo. The same conclusions were described for *P. halepensis* (Ne'eman 1997; Ne'eman et al. 2004; Pausas et al. 2003) and it was attributed to a greater nutrient availability and lower competition. Further research is necessary to clarify this point. Contrary to the results in Nocedo, in Cabo Home seedling height was greater in the unaffected crown plots than in scorched crown plots, but in this case damping off as well as harvesting and slash chopping effects might explain those results.

From a management point of view, even in the harsh environment of the Nocedo stand, seedling stocking was sufficient to ensure an acceptable level of pine recruitment after wildfire although logging slash carried out in summer seems inadvisable.

### 6.3.2.3 Post-Fire Salvage Logging Impact on *P. pinaster* and *P. halepensis* Natural Regeneration

A literature searching of the studies on the impact of post-fire salvage logging on *P. pinaster* and *P. halepensis* natural regeneration was carried out. Ten studies with more than one replicate, available means and variances and including harvesting activities were included (Table 6.3). The method to calculate the size-effect was the logarithmic response ratio, using variance to weight the importance of each study (Kopper et al. 2009; Kalies et al. 2010). Random effects model was used in the analysis to check significant differences in mean response among categorical variables (type of slash manipulation). Between groups heterogeneity (Q) for each categorical variable followed a chi-square distribution which allows developing a significance test of the null hypothesis.

The level of serotiny explained a significant portion of the variability of the *P. pinaster* seedling density after fire (Gil et al. 2009; Vega et al. 2008a). Slash manipulation after clearcutting affected significantly to seedling density after harvesting, without effects of harvest intensities and slash disposal techniques. In some

**Table 6.3** Summary of the studies used in the meta-analysis of impact of post-fire salvage logging on Mediterranean pine recruitment

| Species | Reference | Site | Fire date | TT | SL | Treatment |
|---|---|---|---|---|---|---|
| *Pinus halepensis* | Saracino et al. (1993) | Italy (Taranto) | 1988 | 0-24 | – | EXTRACTION |
| | Martínez-Sánchez et al. (1999) | Spain (Albacete) | 1994 | 10 | – | CONTROL |
| | | | | | | CLEARCUT+EXTRACTION |
| *Pinus pinaster* | Madrigal et al. (2007) | Spain (Cáceres) | 2003 | 2 | High | CONTROL |
| | | | | | | CLEARCUT+EXTRACTION |
| | | | | | | CLEARCUT+CHOPPING |
| | Fernández et al. (2008) | Spain (Ourense) | 2003 | 13 | 20–33 | CONTROL |
| | | | | | | CLEARCUT+WINDROW |
| | | | | | | CLEARCUT+CHOPPING |
| | Vega et al. (2008a) | Spain (Lugo) | 2000 | 11 | 0 | CLEARCUT+WINDROW (HIGH FIRE INTENSITY) |
| | | | | | | CLEARCUT+WINDROW (LOW FIRE INTENSITY) |
| | Vega et al. (2008b) | Spain (Guadalajara) | 2005 | 15 | 0 | CONTROL |
| | | | | | | CLEARCUT+WINDROW |
| | Gil et al. (2009) | Spain (Guadalajara) | 2005 | 12 | 32.8 | CLEARCUT+WINDROW (CONTROL SOIL) |
| | | | | | | CLEARCUT+WINDROW (ALTERED SOIL) |
| | | | | | 0.1 | CLEARCUT+WINDROW (CONTROL SOIL) |
| | | | | | | CLEARCUT+WINDROW (ALTERED SOIL) |
| | Madrigal et al. (2009) | Spain (Guadalajara) | 2005 | 12 | 0 | CLEARCUT+WINDROW |
| | Vega et al. (2010) | Spain (Guadalajara) | 2001 | 6 | 75-83 | CLEARCUT+CHOPPING (CROWN FIRE) |
| | | | | | | CLEARCUT+CHOPPING (SURFACE FIRE) |
| | Castro et al. (2010) | Spain (Granada) | 2005 | 8 | - | CONTROL |
| | | | | | | CLEARCUT+CHOPPING |
| | | | | | | CLEARCUT |

*TT* time to treatment (months), *SL* serotiny level (%), *WE* wood extraction, *CONTROL* no treatment was carried out cases, slash chopping favours a new seedling cohort, particularly in serotinous stands (Fernández et al. 2008; Vega et al. 2010), showing the least reduction of seedling stocking. Slash windrowing showed a significant negative effect, which could be influenced by the serotiny level although in some cases it is argued that this treatment could represent safe sites for pine installation (Castro et al. 2010).

Only a few studies found that salvage logging and slash manipulation could hamper pine regeneration, supporting that both ensured an adequate level of post-fire pine recruitment for the two species studied. It suggests that burned tree harvesting can be compatible with natural post-fire regeneration but low serotiny level and high water stress conditions can limit *P. pinaster* post-fire seedling establishment in many Mediterranean areas (Madrigal et al. 2010).

There is an obvious lack of information on the effect on salvage logging for serotinous pine species. Salvage logging might be detrimental or appropriated depending on serotiny level of the stand. For future studies we advise to include serotiny level, fire severity and a more appropriate control than the pre-treatment values in the sampling design.

Future studies should consider the origin of the stands because variability in provenance shows differences in serotiny level. Fire severity has also been shown to be a relevant variable in the regeneration process (Pausas et al. 2003; Broncano and Retana 2004; Vega et al. 2008a).

## 6.4   Key Messages

- Serotiny has been documented as, although not exclusively, a fire adaptation and is defined as a morphological feature whereby the cones remain closed for years after seed maturation and open after exposition to high temperatures. However, it has been proved to be a more complex trait increasing the survival of released seeds which showed high biological quality (sound seeds, weight and germination rate) and higher heat insulation and resistance to fire.
- The serotinous pines in the Mediterranean area of southern Europe are the obligate seeders *Pinus halepensis*, *Pinus brutia* and *Pinus pinaster,* covering a wide range, mainly in xeric and lower altitudinal areas.
- These forests have usually been under- or over-managed depending on the economic value. Regarding post-fire management, salvage logging has been usual and after they have been given up until to reach the natural recovering. In the last decades, the management objectives are including ecosystem protection, water regulation, hazard prevention and the promotion of mature, diverse, and productive forests.
- Emergency actions are a main topic to protect soils and prevent erosion. The monitoring of burned areas to check the success of natural regeneration should indicate if no action, assisted natural regeneration or active restoration had to be carried out.
- Climate change influences community structure, species composition and the resilience to fire, even for serotinous pine forests adapted to fire.

- Adaptive management should be developed to face up post-fire threats, such as soil loss, droughts, herbivore, pest attacks, etc., taking into account the variation induced by the predicted climate change.
- The case studies of post-fire management included in this chapter are focusing the forest management as a tool for restoration or assistance to natural regeneration. They show the effects of the main treatments used in burned serotinous pine stands in the Mediterranean area of southern Europe.

# References

Alía R, Martín S (2003) EUFORGEN Technical Guidelines for genetic conservation and use for Maritime pine (*Pinus pinaster*). International Plant Genetic Resources Institute, Rome, Italy, 6 p

Alig RJ, Adams DM, McCarl BA (2002) Projecting impacts of global climate change on the US forest and agricultural sectors and carbon budgets. For Ecol Mange 169:3–14

Arianoutsou M, Ne'eman G (2000) Post-fire regeneration of natural *Pinus halepensis* forest in the East Mediterranean Basin. In: Ne'eman G, Trabaud L (eds) Ecology, biogeography and management of *Pinus halepensis* and *P. brutia* forest ecosystems in the mediterranean basin. Backhuys Publishers, Leiden

Arnan X, Rodrigo A, Retana J (2007) Post-fire regeneration of Mediterranean plant communities at a regional scale is dependent on vegetation type and dryness. J Veg Sci 18(1):111–122

Baeza MJ, Vallejo VR (2008) Vegetation recovery after fuel management in Mediterranean shrublands. Appl Veg Sci 11:151–158

Baeza MJ, Valdecantos A, Alloza JA, Vallejo VR (2007) Human disturbance and environmental factors as drivers of long-term post-fire regeneration patterns in Mediterranean forests. J Veg Sci 18:243–252

Barbero M, Loisel R, Quezel P, Richardson DM, Romane F (1998) Pines of the mediterranean basin. In: Richardson DM (ed) Ecology and biogeography of *Pinus*. Cambridge University Press, New York

Battisti A (2004) Forests and climate change-lessons from insects. Forest 1(1):17–24

Baylis NT, De Ronde C, James DB (1986) Observations of damage of a secondary nature following a wild fire at the Otterford State Forest. South Afr For J 137:36–37

Benito-Garzón M, Sánchez de Dios R, Sainz-Ollero H (2008) Effects of climate change on the distribution of Iberian trees species. Appl Veg Sci 11(2):169–178

Boydak M, Dırık H, Calıkoglu M (2006) Forest stand dynamic, regeneration and fire in *Pinus brutia* ecosystems. In: Ogem-Vak (ed) Biology and silviculture of turkish red pine (*Pinus brutia* Ten.). Istanbul Universitesi Orman Fakultesi, Ankara

Bradshaw AD, Chadwick MJ (1980) The restoration of the land. University of California Press, Berkeley

Bravo F, Bravo-Oviedo A, Díaz-Balteiro L (2008a) Carbon sequestration in Spanish Mediterranean forests under two management alternatives: a modelling approach. Eur J For Res 127:225–234

Bravo F, Jandl R, Von Gadow K, Lemay V (2008b) Introduction. In: Bravo F (ed) Managing forest ecosystems: the challenge of climate change. Springer, Dordrecht

Broncano MJ, Retana J (2004) Topography and forest composition affecting the variability in fire severity and post-fire regeneration occurring after a large fire in the Mediterranean basin. Int J Wild Fire 13:209–216

Canga E, Rodríguez-Soalleiro R, Vega G (2003) Estudio de la regeneración natural de *Pinus pinaster* Ait. ssp. *atlantica* en el Noroeste de España. Cuad Soc Esp Cien For 15:101–106

Carrion JS, Navarro C, Navarro J, Munuera M (2000) The distribution of cluster pine (*Pinus pinaster*) in Spain as derived from palaeoecological data: relationships with phytosociological classification. Holocene 10:243–252

Castro J, Allen CD, Molina-Molares M, Marañón-Jiménez S, Sánchez-Miranda A, Zamora R (2010) Salvage logging versus the use of burnt wood as a nurse object to promote post-fire tree seedling establishment. Restor Ecol. doi:10.1111/j.1526-100X.2009.00619.x

Cerdà A, Doerr S (2005) The influence of vegetation recovery on soil hydrology and erodibility following fire: an eleven-year research. Int J Wild Fire 14(4):423–437

Christakopoulos P, Hatzopoulos I, Kalabokidis K, Paronis D, Filintas A (2007) Assessment of the response of a Mediterranean-type forest ecosystem to recurrent wildfires and to different restoration practices using Remote Sensing and GIS techniques. In: Proceedings of 6th international workshop of the EARSeL special interest group on forest fires, ,Thessaloniki, Greece, 27–29 Sep 2007

Çolak AH, Rotherham ID (2007) Classification of Turkish forests by altitudinal zones to improve silvicultural practice: a case-study of Turkish high mountain forests. Int For Rev 9(2):641–652

Corona P, Leone V, Saracino A (1998) Plot size and shape for the early assessment of post-fire regeneration in Aleppo pine stands. New Forest 16:213–220

Dale VH, Joyce LA, Mcnulty S, Neilson RP, Ayres MP, Flannigan MD, Hanson PJ, Irland LC, Lugo AE, Peterson CJ, Simberloff D, Swanson FJ, Stocks BJ, Wotton M (2001) Climate change and forest disturbances. BioScience 51:723–734

Daskalakou EN, Thanos CA (1996) Aleppo pine (*Pinus halepensis*) post-fire regeneration: the role of canopy and soil seed banks. Int J Wild Fire 6:59–66

Daskalakou EN, Thanos CA (2004) Post-fire regeneration of Aleppo pine – the temporal pattern of seedling recruitment. Plant Ecol 171:81–89

Díaz-Delgado R, Lloret F, Pons X, Terradas J (2002) Satellite evidence of decreasing resilience in Mediterranean plant communities after recurrent wildfires. Ecology 83(8):2293–2303

Egler FE (1954) Vegetation science concepts. I. Initial floristic composition-a factor in old-field vegetation development. Vegetatio 4:412–418

Espelta JM, Verkaik I, Eugenio M, Lloret F (2008) Recurrent wildfires constrain long-term reproduction ability in *Pinus halepensis* Mill. Int J Wild Fire 17:579–585

Eugenio M, Lloret F (2004) Fire recurrence effects on the structure and composition of Mediterranean *Pinus halepensis* communities in Catalonia (northeast Iberian Peninsula). Ecoscience 11:446–454

Eugenio M, Verkaik I, Lloret F, Espelta JM (2006) Recruitment and growth decline in *Pinus halepensis* populations after recurrent wildfires in Catalonia (NE Iberian Peninsula). For Ecol Manag 231:47–54

Fady B, Semerci H, Vendramin GG (2003) EUFORGEN Technical Guidelines for genetic conservation and use for Aleppo pine (Pinus halepensis) and Brutia pine (Pinus brutia). Int Plant Gen Res Inst, Rome

Fernandes PM, Rigolot E (2007) Fire ecology and management of maritime pine (*Pinus pinaster* Ait.). For Ecol Manag 241(1–3):1–13

Fernández C, Vega JA, Fonturbel T, Jiménez E, Pérez-Gorostiaga P (2008) Effects of wildfire, salvage logging and slash manipulation on *Pinus pinaster* Ait. Recruitment in Orense (NW Spain). For Ecol Manage 255:1294–1304

Ferrandis P, Martínez-Sánchez JJ, Herranz JM (1996) The role of soil seed bank in the early stages of plant recovery after fire in a *Pinus pinaster* forest in SE Spain. Int J Wild Fire 6:31–35

Ferrandis P, Las Heras J, Martínez-Sánchez JJ, Herranz JM (2001) Influence of a low-intensity fire on a *Pinus halepensis* Mill. forest seed bank and its consequences on the early stages of plant succession. Israel J Plant Sci 49:105–114

Fischlin A, Midgley GF, Price J, Leemans R, Gopal B, Turley C, Rounsevell M, Dube P, Tarazona J, Velichko A (2007) Ecosystems, their properties, goods, and services. In: Parry ML, Canziani OF, Palutikof JP, van der Linden PJ, Hanson CE (eds) Climate change 2007: impacts, adaptation and vulnerability. Contribution of working group II to the fourth assessment report of the inter-governmental panel on climate change. Cambridge University Press, Cambridge

Fuentes D, Disante K, Valdecantos A, Cortina J, Vallejo VR (2007) Response of *Pinus halepensis* Mill. seedlings to biosolids enriched with Cu, Ni and Zn in three Mediterranean forest soils. Environ Pollut 145:316–323

Ganatsas P, Zagas TD, Tsakaldimi MN, Tsitsoni TK (2004) Post-fire regeneration dynamics in a Mediterranean type ecosystem in Sithonia, northern Greece: ten years after the fire. In: Proceedings of 10th MEDECOS conference, Rhodes, Greece, 25 April–1 May 2004

Gil L, López R, García-Mateos A, González-Doncel I (2009) Seed provenance and fire-related reproductive traits of *Pinus pinaster* in central Spain. Int J Wild Fire 18:1003–1009

Gil-Romera G, Carrion JS, Pausas JG, Sevilla-Callejo M, Lamb HF, Fernandez S, Burjachs F (2010) Holocene fire activity and vegetation response in South-Eastern Iberia. Quaternary Sci Rev 29:1082–1092

González-Ochoa A, de las Heras J (2002) Effects of post-fire silviculture practices on Pachyrhinus squamosus defoliation levels and growth of *Pinus halepensis* Mill. For Eco Manag 167(1–3):185–194

González-Ochoa A, De las Heras J, Torres P, Sánchez-Gómez E (2003) Mycorrhization of *Pinus halepensis* Mill. and *Pinus pinaster* Aiton seedlings in two comercial nurseries. Ann For Sci 60:43–48

Goubtiz S, Nathan R, Roitemberg R, Shmida A, Ne'eman G (2004) Canopy seed bank structure in relation to: fire, tree size and density. Plant Ecol 173:191–201

Grigoriadis N, Galatsidas S, Takos I (2009) Post-fire regeneration in the urban forest of Thessaloniki, 10 years after fire. J Ecol Saf 3(I):75–82

Kalies EL, Chambers CL, Covington WW (2010) Wildlife responses to thinning and burning treatments in southwestern conifer forest: a meta-analysis. For Ecol Manage 259:333–342

Kaynaş BY, Gurkan B (2008) Species richness and abundance of insects during post-fire succession of a *Pinus brutia* forest in Mediterranean region. Pol J Ecol 56(1):165–172

Keeley JE, Zedler PH (1998) Life history evolution in pines. In: Richardson DM (ed) Ecology and biogeography of *Pinus*. Cambridge University Press, Cambridge, pp 219–251

Keeley JE, Ne'eman G, Fotheringham CJ (1999) Immaturity risk in a fire-dependent pine. J Med Ecol 1:41–48

Kopper KE, McKenzie D, Peterson DL (2009) The evaluation of Meta-analysis techniques for quantifying prescribed fire effects on fuel loadings. USDA Forest Service Research Paper PNW-RP-582

Lamont B, Le Maitre DC, Cowling RM, Enright NJ (1991) Canopy seed storage in woody plants. Bot Rev 57:277–317

Le Maitre DC (1998) Pines in cultivation: a Global view. In: Richardson DM (ed) Ecology and biogeography of *Pinus*. Cambridge University Press, Cambridge

Leone V, Borghetti M, Saracino A (1999) Ecology of post-fire recovery in *Pinus halepensis* in southern Italy. In: Trabaud L (ed) Life and environment in the Mediterranean. WIT Press, Southampton

López-Serrano FR, Rubio E, Andrés M, del Cerro A, García-Morote FA, de las Heras J, Lucas-Borja ME, Moya D, Odi M (2009) Efecto del cambio climático en los montes castellano-manchegos. In: Fundación General del Medio Ambiente (ed) Impactos del Cambio Climático en Castilla-La Mancha, Primer Informe, Toledo, Spain

Macías F, Calvo R (2001) Los suelos. Atlas de Galicia. Xunta de Galicia. Sociedade para o Desenvolvemento Comarcal de Galicia. Santiago de Compostela, Spain

Madrigal J, Hernando C, Guijarro M, Díez C, Gil JA (2007) Influencia de la corta a hecho y tratamiento de residuos en la supervivencia del regenerado natural post-incendio de *Pinus pinaster* Ait. en el monte "Egidos" Acebo (Cáceres, España). Proceedings of the 4th International Conference Wild Fire, Wildfire, Sevilla, Spain

Madrigal J, Vega JA, Hernando C, Fonturbel T, Díez R, Guijarro M, Díez C, Marino E, Pérez JR, Fernández C, Carrillo A, Ocaña L, Santos I (2009) Efecto de la corta a hecho y de la edad de la masa en la supervivencia de regenerado de *Pinus pinaster* Ait. tras el gran incendio del rodenal de Guadalajara. Proceedings of the 5° Congreso Forestal Español, Avila, Spain

Madrigal J, Hernando C, Guijarro M, Vega JA, Fonturbel T, Pérez-Gorostiaga P (2010) Smouldering fire-induced changes in a Mediterranean soil (SE Spain): effects on germination, survival and morphological traits of 3-year-old *Pinus pinaster* Ait. Plant Ecol 208:279–292

Markalas S (1991) Insects attacking burnt pine trees *Pinus halepensis, Pinus brutia,* and *Pinus nigra* in Greece. Anzeiger Fuer Schaedlingskunde 64(4):72–75

Martínez E, Madrigal J, Hernando C, Guijarro M, Vega JA, Pérez-Gorostiaga P, Fonturbel T, Cuiñas P, Alonso M, Beloso M (2002) Effect of fire intensity on seed dispersal and early regeneration in a *Pinus pinaster* forest. In: Viegas DX (ed) Proceedings of the 4th International Conference Forest Fire Research 2002, Wildland Fire Safety Summit, Millpress Science Publishers, Rotherdam, CD-ROM

Martínez-Sánchez JJ, Marín A, Herranz JM, Ferrandis P, Heras J (1995) Effects of high temperatures on germination of *Pinus halepensis* Mill. and *P. pinaster* Aiton subsp. *pinaster* seeds in southeast Spain. Vegetatio 116:69–72

McEvoy PM, McAdam JH, Mosquera-Losada MR, Rigueiro-Rodriguez A (2006) Tree regeneration and sapling damage of pedunculate oak Quercus robur in a grazed forest in Galicia, NW Spain: a comparison of continuous and rotational grazing systems. Agrofor Syst 66:85–92

McRae DJ, Duchesne LC, Lynham TJ (2001) Comparisons between wildfire and forest harvesting and their implications in forest management. Environ Rev 9:223–260

Mendel Z, Halperin J (1982) The biology and behavior of Orthotomicus erosus in Israel. Phytoparasitica 10:169–181

Mendel Z, Madar Z, Golan Y (1985) Comparison of the seasonal occurrence and behavior of 7 pine bark beetles (Coleoptera: Scolytidae) in Israel. Phytoparasitica 13:21–32

Moriondo M, Good P, Durao R, Bindi M, Giannakopoulos C, Corte Real J (2006) Potential impact of climate change on forest fire risk in Mediterranean area. Clim Res 13(31):85–95

Mosquera MR, McAdam J, Rigueiro A (2006) Silvopastoralism and sustainable land management. CAB International, UK

Moya D, Espelta JM, Verkaik I, López-Serrano F, De las Heras J (2007) Tree density and site quality influence on *Pinus halepensis* Mill. Reproductive characteristics after large fires. Ann For Sci 64:649–656

Moya D, De las Heras J, López-Serrano FR, Leone V (2008a) Optimal intensity and age of management in young Aleppo pine stands for post-fire resilience. For Ecol Manag 255:3270–3280

Moya D, Espelta JM, López-Serrano F, Eugenio M, De las Heras J (2008b) Natural post-fire dynamics and serotiny in 10-year-old *Pinus halepensis* Mill. Stands along a geographic gradient. Int J Wild Fire 17(2):287–292

Moya D, Saracino A, Salvatore R, Lovreglio R, De las Heras J, Leone V (2008c) Anatomic basis and insulation of serotinous cones in *Pinus halepensis Mill.* Trees–Struct Funct 22:511–519. doi:10.1007/s00468-008-0211-1

Moya D, De las Heras J, López-Serrano FR, Condes S, Alberdi I (2009) Structural patterns and biodiversity in burned and managed Aleppo pine stands. Plant Ecol 200(2):217–228

Napper C (2006) Burned Area Emergency Response (BAER) treatments catalog. United States Department of Agriculture Forest Service National Technology and Development Program Watershed, Soil, Air Management

Nathan R, Safriel UN, Noy-Meir I, Schiller G (1999) Seed release without fire in *Pinus halepensis*, a Mediterranean serotinous wind-dispersed tree. J Ecol 87:659–669

Ne'eman G (1997) Regeneration of natural pine forest-review of the work done after the 1989 fire in Mount Carmel, Israel. Int J Wild Fire 7:295–306

Ne'eman G, Goubitz S, Nathan R (2004) Reproductive traits of *Pinus halepensis* in the light of fire-a critical review. Plant Ecol 171:69–79

Nosetto M, Jobbágy E, Paruelo JM (2006) Carbon sequestration in semiarid rangelands: comparison of *Pinus ponderosa* plantations and grazing exclusion in NW Patagonia. J Arid Environ 67:142–156

Pausas JG (1999a) Mediterranean vegetation dynamics: modeling problems and functional types. Plant Ecol 140:27–39

Pausas JG (1999b) Response of plant functional types to changes in the fire regime in Mediterranean ecosystems: a simulation approach. J Veg Sci 10:717–722

Pausas JG, Ouadah N, Ferran A, Gimeno T, Vallejo R (2003) Fire severity and seedling establishment in *Pinus halepensis* woodlands, eastern Iberian Peninsula. Plant Ecol 169:205–213

Pausas JG, Bradstock RA, Keith DA, Keeley JE (2004a) Plant functional traits in relation to fire in crown-fire ecosystems. Ecology 85:1085–1100

Pausas JG, Ribeiro E, Vallejo R (2004b) Post-fire regeneration variability of *Pinus halepensis* in the eastern Iberian Peninsula. For Ecol Manag 203:251–259

Pausas JG, Llovet J, Rodrigo A, Vallejo R (2008) Are wildfires a disaster in the Mediterranean basin? – A review. Int J Wild Fire 17(6):713–723

Quezel P (2000) Taxonomy and biogeography of Mediterranean pines (*Pinus halepensis* and *P. brutia*). In: Ne'eman G, Trabaud L (eds) Ecology, biogeography and management of *Pinus halepensis* and *P. brutia* forest ecosystems in the mediterranean basin. Backhuys Publishers, Leiden, pp 1–12

Resco V, Fischer C, Colinas C (2007) Climate change effects on Mediterranean forests and preventive measures. New Forest 33:29–40

Richardson DM (1998) Glossary. In: Richardson DM (ed) Ecology and biogeography of *Pinus*. Cambridge University Press, Cambridge

Robichaud PR, Lewis SA, Brown RE, Ashmun LE (2009) Emergency post-fire rehabilitation treatment effects on burned area ecology and long-term restoration. Fire Ecol 5(1):115–128

Rodrigo A, Retana J, Pico X (2004) Direct regeneration is not the only response of Mediterranean forests to large fires. Ecology 85:716–729

Salvatore R, Moya D, Pulido L, Lovreglio R, Lopez-Serrano FR, De las Heras J, Leone V (2010) Morphological and anatomical differences in Aleppo pine seeds from serotinous and non-serotinous cones. New Forest 39(3):329–341

Saracino A, Corona P, Leone V (1993) La rinnovazione naturale del pino d'Aleppo (*Pinus halepensis* Miller) in soprassuoli percorsi dal fuoco. (II parte). Monti e Boschi 3:10–20

Saracino A, Pacella R, Leone V, Borghetti M (1997) Seed dispersal and changing seed characteristics in a *Pinus halepensis* Mill. Forest after fire. Plant Ecol 130:13–19

Saracino A, D'Alessandro CM, Borghetti M (2004) Seed colour and post-fire bird predation in a Mediterranean pine forest. Acta Oecol 26:191–196

Serrano L, Vallejo VR, Valdecantos A (2008) Forests and natural landscapes. In: LUCINDA. Land Care in Desertification Affected Areas. Booklet Series C, 1. http://geografia.fsch.unl.pt/lucinda/desertification_processes.html

Spanos IA, Daskalakou EN, Thanos CA (2000) Post-fire, natural regeneration of *Pinus brutia* forests in Thasos Island, Greece. Acta Oecol 21:13–20

Tapias R, Gil L, Fuentes-Utrilla P, Pardos JA (2001) Canopy seed banks in Mediterranean pines of south-eastern Spain: a comparison between *Pinus halepensis* Mill., *P. pinaster* Ait., *P. nigra* Arn. and *P. pinea* L. J Ecol 89:629–638

Tapias R, Climent J, Pardos JA, Gil L (2004) Life histories of Mediterranean pines. Plant Ecol 171:53–68

Thanos CA (2000) Ecophysiology of seed germination in *Pinus halepensis* and *P. brutia*. In: Ne'eman G, Trabaud L (eds) Ecology, biogeography and management of *Pinus halepensis* and *P. brutia* forest ecosystems in the mediterranean basin. Backhuys Publishers, Leiden

Thanos CA, Daskalakou EN (2000) Reproduction in *Pinus halepensis* and *P. brutia*. In: Ne'eman G, Trabaud L (eds) Ecology, biogeography and management of *Pinus halepensis* and *P. brutia* forest ecosystems in the mediterranean basin. Backhuys Publishers, Leiden

Thanos CA, Doussi M (2000) Post-fire regeneration of *Pinus brutia* forests. In: Ne'eman G, Trabaud L (eds) Ecology, biogeography and management of *Pinus halepensis* and *P. brutia* forest ecosystems in the mediterranean basin. Backhuys Publishers, Leiden, pp 291–301

Thanos CA, Marcou S (1991) Post-fire regeneration in *Pinus brutia* forest ecosystems of Samos island (Greece): 6 years after. Acta Oecol 12:633–642

Trabaud L (2000) Post-fire regeneration of *Pinus halepensis* forest in the West Mediterranean. In: Ne'eman G, Trabaud L (eds) Ecology, biogeography and management of *Pinus halepensis* and *P. brutia* forest ecosystems in the mediterranean basin. Backhuys Publishers, Leiden

Tsitsoni T (1997) Conditions determining natural regeneration after wildfires in the *Pinus halepensis* (Miller, 1768) forests of Kassandra Peninsula (North Greece). For Ecol Manage 92:199–208

Tsitsoni T, Ganatsas P, Zagas T, Tsakaldimi M (2004) Dynamics of post-fire regeneration of *Pinus brutia* Ten. In an artificial forest ecosystem of northern Greece. Plant Ecol 171:165–174

Vallejo VR, Valdecantos (2008) Fire. In: LUCINDA. Land Care in Desertification Affected Areas. Booklet Series B, 2. http://geografia.fsch.unl.pt/lucinda/desertification_processes.html

Valdecantos A, Baeza MJ, Vallejo VR (2009) Vegetation management for promoting ecosystem resilience in Fire-Prone Mediterranean shrublands. Rest Ecol 17(3):414–421

Vega JA, Fernández C, Pérez-Gorostiaga P, Fonturbel T (2008a) The influence of fire severity, serotiny, and post-fire management on *Pinus pinaster* Ait. recruitment in three burnt areas in Galicia (NW Spain). For Ecol Manage 256:1596–1603

Vega JA, Fonturbel T, Pérez JR, Fernández C (2008b) Proyecto Rodenal, Final Report

Vega JA, Fernández C, Pérez-Gorostiaga P, Fonturbel T (2010) Response of maritime pine (Pinus pinaster Ait.) recruitment to fire severity and post-fire management in a coastal burned area in Galicia (NW Spain). Plant Ecol 206(2):297–308

Vennetier M, Vila B, Liang EY, Guibal F, Ripert C, Chandioux O (2005) Impacts du changement climatique sur la productivité forestière et le déplacement d'une limite bioclimatique en région méditerranéenne française. Ingénieries 44:49–61

Verkaik I, Espelta JM (2006) Post-fire regeneration thinning, cone production, serotiny and regeneration age in *Pinus halepensis*. For Ecol Manage 231:155–163

Zagas T (1994) Studies for pilot project reforestations and silvicultural treatments in Hymettus. Workshop for the rehabilitation of the Urban Forest of Hymettus Mountain. Athens.

Zagas T, Ganatsas P, Tsitsoni T, Tsakaldimi M (2004) Post-fire regeneration of *Pinus halepensis* Mill. Stands in the Sithonia peninsula, northern Greece. Plant Ecol 171:91–99

# Chapter 7
# Post-Fire Management of Non-Serotinous Pine Forests

**Javier Retana, Xavier Arnan, Margarita Arianoutsou, Anna Barbati, Dimitris Kazanis, and Anselm Rodrigo**

## 7.1  Ecological Context

### 7.1.1  Short Definition/Justification of the Set of Forest Types Tackled in the Chapter

Nine *Pinus* species are found in the Mediterranean Basin (Barbéro et al. 1998). The respective forest types that are dominated by these pine species can be divided into four groups according to their ecological characteristics (EEA 2007):

- Thermophilous pine forests: Forests of *Pinus halepensis*, *P. pinaster*, *P. brutia* and *P. pinea* distributed at low elevations under thermo-Mediterranean climate.
- Black pine forests: Forests dominated by the various sub-species of *Pinus nigra* are regarded as typical of the mountainous zone of the Mediterranean Basin.
- Scots pine forests: *Pinus sylvestris* forests which are found across the higher, oro-Mediterranean altitudinal zone
- Alti-Mediterranean pine forests: Forests distributed near the timberline, with *Pinus heldreichii* and *P. peuce* as the dominant pine species.

J. Retana (✉) • X. Arnan • A. Rodrigo
CREAF and Unit of Ecology, Faculty of Biosciences, Campus UAB, Barcelona, Spain

M. Arianoutsou • D. Kazanis
Department of Ecology and Systematics, Faculty of Biology, School of Sciences, National and Kapodistrian University of Athens, Athens, Greece

A. Barbati
DIBAF, Dipartimento per la Innovazione nei Sistemi Biologici, Agroalimentari e Forestali, University of Tuscia, Viterbo, Italy

Although fire is generally recognized as an important ecological factor associated with pine-dominated forest ecosystems, not all pine forests share the same fire regime (Agee 1998). Pine species that have evolved under a fire regime characterized by frequent, high intensity events have serotinous cones in order to reassure their post-fire regeneration (see Chap. 6). This is the case of the thermophilous species *Pinus halepensis*, *P. pinaster* and *P. brutia*. The fourth thermophilous pine, *Pinus pinea*, does not have serotinous cones but it shows the highest resistance to fire in comparison to all the other Mediterranean pines due to the thickness of its bark (Fernandes et al. 2008).

All the other Mediterranean pine species are non-serotinous and can either persist fire if it is a surface, low intensity one or they are killed by extreme fire events and depend on seed dispersal from unburned sources in order to re-establish at the burned site (Arianoutsou et al. 2010a; Ordóñez et al. 2006). According to fire statistics, three out of the seven non-serotinous Mediterranean pine forest types (i.e. *Pinus pinea*, *P. nigra* and *P. sylvestris*) have suffered a considerable number of fire events during the last decades (Pausas et al. 2008; Rodrigo et al. 2007; Vilà et al. in press). This chapter is based on these three non-serotinous pine species.

## 7.1.2   Distribution of Non-Serotinous Pine Species

*Pinus nigra* (black pine) is as a typical Mediterranean pine species, with its natural geographic distribution ranging from Spain eastwards to Southern France, Italy and Austria, the Balkans and Turkey. It is also present in some localities of NW Africa, some of the Mediterranean islands (Corsica, Sicily, Cyprus and some Aegean islands) and the Crimean Peninsula in Black Sea (Fig. 7.1a). Its natural distribution also extends to southern Balkans and Anatolia. It is considered as a species capable to establish, grow and develop forest stands under a wide range of climatic and edaphic conditions, across an altitudinal zone that varies from 450 to 1,500 m.a.s.l (Dafis et al. 2001). For example, *P. nigra* is the only tree species of the meso-mountainous Mediterranean zone (800–1,600 m.a.s.l.) that forms forest ecosystems on ultramaphic rocks. *Pinus nigra* forests are considered as priority habitats under the Annex I of the 92/43 Directive of the European Union. This is the result of two main factors: firstly, the high genetic diversity of *P. nigra*, accounting for the large number of sub-species across its natural geographical distribution, and secondly, its overall pattern of sporadic occurrence across the European Mediterranean countries (Fig. 7.1a).

Scots pine (*P. sylvestris*) is widely distributed (Fig. 7.1b). Its native range includes Scotland, Scandinavia (excluding Denmark), northern Europe, and northern Asia. Scots pine has some populations that are very well adapted to the Mediterranean mountain environment, particularly in NE Spain, S. France, N. Italy and the Balkans, with the Greek populations of Macedonia marking the southernmost limits of its natural distribution at the Balkan Peninsula (Barbéro et al. 1998; Dafis 2010). It prefers siliceous substrates, where it can form either pure forest stands or mixed stands with other high-altitude tree species, such as *Betula pendula*, *Picea abies*, *Fagus sylvatica* and *P. nigra*.

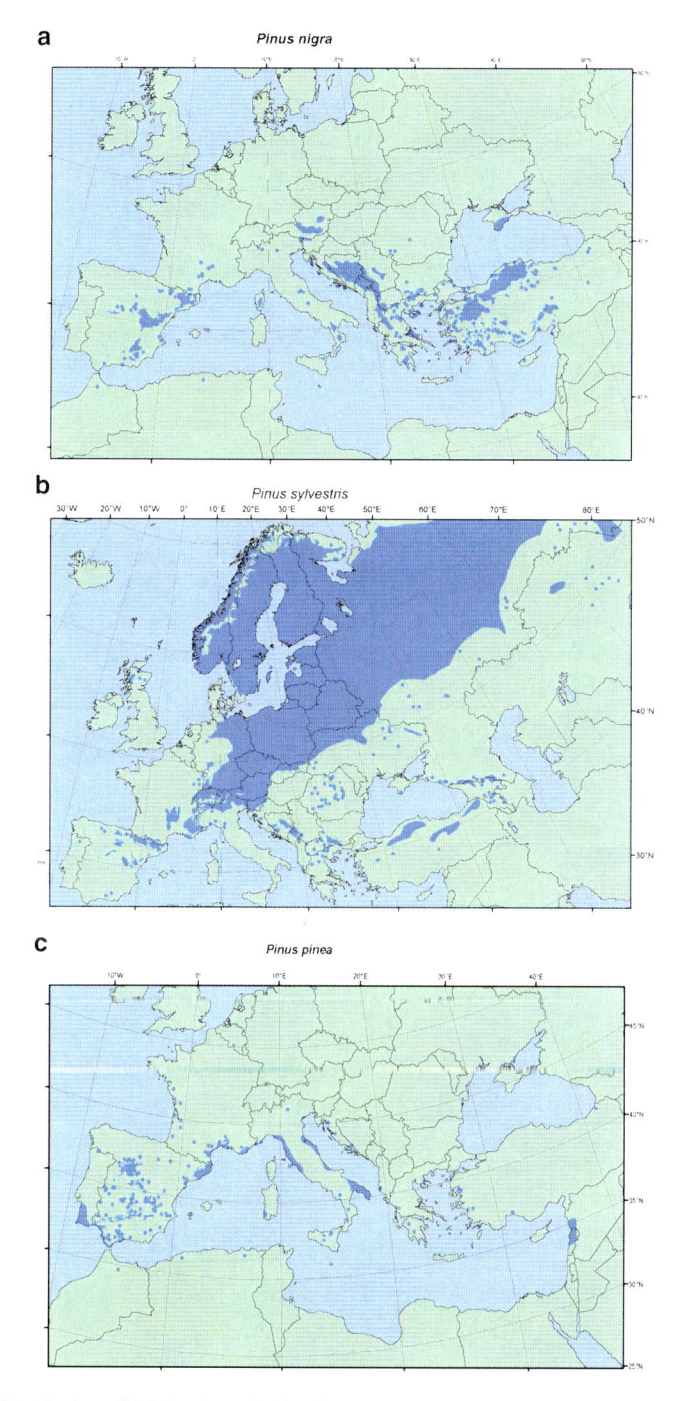

**Fig. 7.1** Distribution of (**a**) *P. nigra*, (**b**) *P. sylvestris* and (**c**) *P. pinea* in the Mediterranean basin. Data obtained from Euforgen.org

*Pinus pinea* is a Mediterranean pine whose natural range is difficult to define because it has been planted in the Mediterranean Basin so widely for so long (Mirov 1967), but different authors suggest that it may have occurred naturally in the whole Mediterranean Basin (Le Maitre 1998; Martínez and Montero 2004). It occupies ca. 320,000 ha (Barbéro et al. 1998), 75% of them in Spain (Montoya 1990). Its distribution has been traditionally restricted to coastal sandstone and siliceous low elevation ranges (Fig. 7.1c). However as a consequence of the profitable historical exploitation of the pinion of this species for human consumption, its present distribution in the Mediterranean Basin is considerably wider than these sandy habitats. Now, in the Mediterranean Basin, *P. pinea* is common at low and intermediate altitudes and occurs in scattered populations (Barbéro et al. 1998; Dafis 2010).

### 7.1.3   Vegetation Composition in Non-Serotinous Pine Forests

The composition and structure of the major natural communities for the three non-serotinous pine species are significantly different among them (Gracia and Ordóñez 2011a, b, c). The three species form monospecific or mixed forests depending on climate, topography and disturbances. *Pinus pinea* does not usually form monospecific stands, in fact it is quite common to find this species mixed with other tree species such as *Quercus ilex, Quercus suber* or *Quercus humilis* and also some pine species such as *Pinus halepensis* or *Pinus pinaster*. *Pinus nigra* more frequently forms monospecific stands, but also shares the overstory with other species such as *P. halepensis, P. sylvestris, Q. ilex, Q. humilis* and *Fagus sylvatica*. *Pinus sylvestris* also forms monospecific forests or mixed forests with few species. In this later case, the most frequent tree species are *Q. ilex, Q. humilis, F. sylvatica, Pinus uncinata* and *P. nigra*.

The understory of the natural communities of the three pine species also varies depending on the pine species considered. Thus, in the western Mediterranean Basin, the most common species in *P. pinea* forests are *Cistus salviifolius, Cistus monspeliensis, Erica arborea, Rhamnus alaternus, Pistacea lentiscus* and *Arbutus unedo* (Gracia and Ordóñez 2011a). In the eastern Mediterranean Basin, the understorey of *P. pinea* stands is occupied by typical species of evergreen sclerophyllous shrubs, species such as *Myrtus communis, Arbutus unedo, Erica arborea, Pistacia lentiscus* and *Phillyrea latifolia*, whereas whenever some gaps of the tree layer are encountered, seasonal dimorphic shrubs (primarily, *Cistus* spp.) are present (Dafis et al. 2001; Vassiliou 2007).

*Pinus nigra* forests in the western part of the Mediterranean have a diverse shrub and herbaceous layer dominated by *Buxus sempervirens, Juniperus communis, Crataegus monogyna, Prunus mahaleb, Thymus vulgaris* and *Viburnum lantana* (Gracia and Ordóñez 2011b). In the eastern part of the Mediterranean, a variety of factors, including climatic conditions, site pedology and topography and the stand history are expressed in the composition of the *P. nigra* forest understorey. For example, at Mt. Olympus, the understorey of *Pinus nigra* forests is dense, occupied by evergreen sclerophyllous shrubs such as *Quercus coccifera* and *Arbutus* spp., whereas at higher altitudes the presence of the woody vegetation is reduced and the commonest species are *Rhus coriaria* and *Staehelina uniflosculosa* (Strid 1980). Species of the genera *Erica, Juniperus, Crataegus* and *Quercus* are common in the understorey of *P. nigra* throughout Greece (Dafis et al. 2001; Kazanis et al. 2011).

The vegetation associated to *P. sylvestris* forests from the Iberian Peninsula is dominated by *Buxus sempervirens* and *Juniperus communis*, accompanied by species such as *Vaccinium myrtillus* or *Arctostaphyllos uva-ursi* in wet and mountain areas and by species of the genus *Quercus, Rhamnus* or *Erica* in drier ones (Gracia and Ordóñez 2011c). The commonest species in the understorey of *P. sylvestris* forests of Greece are *Pteridium aquilinum, Juniperus communis, Brachypodium sylvaticum, Rubus idaeus, Calamogrostis arundinacea, Fragaria vesca, Clinopodium vulgare* and *Rosa arvensis* (Dafis et al. 2001).

## 7.1.4  Ecological and Socioeconomic Importance

The ecological and socioeconomic importance of the different species of non-serotinous pines in the Mediterranean Basin is very high. The timber of *P. nigra* and *P. sylvestris* is of high technological quality, with both species having large and straight trunks. Their wood is used for many purposes such as construction, floor boards, saw logs, pulp and fuel. Additional products of these species are bark for gardening and resins. The timber of *P. pinea* is of mediocre quality and has short durability. The wood of this species is used as structural timber, sawn timber for light construction purposes, containers, as well as pulp for cellulose and paper (Giordano 1988). Although since the Roman period *P. pinea* has been artificially spread and cultivated also for timber, the pine nut of *P. pinea,* used in confectionery and food industries, has been an important commercial product from the ancient Egyptians and the Romans to present (Le Maitre 1998). However, it has been progressively replaced by pine nuts imported from other areas of the world, which involves low interest from private or public owners in carrying out plantations with this species in burned areas. Additionally, pinewoods of this species are able to provide other non-wood products such as resin, bark, honey and grazing for livestock.

Ecologically, *P. nigra* and *P. sylvestris* have been usually planted for erosion control and for reforesting burned sites, and they have been recommended for planting on strip-mined lands. Due to its tolerance to poor soil conditions, *P. sylvestris* is widely used in binding loose sands and in land reclamation programs in temperate zones. *Pinus pinea* tolerates sandy soils and, for this reason, has been traditionally used to consolidate coastal dunes and protect coastal agricultural crops. This species also has high interest in recreational forests due to its umbrella-like crown. Some varieties of these three non-serotinous pine species are frequently used in ornamental gardening.

## 7.1.5   Post-Fire Regeneration

In areas affected by understory fires, several adults of these three species (*P. nigra*, *P. sylvestris* and *P. pinea*) escape relatively unscathed even if these fires are intense, because their thick bark helps them to survive fire by insulating the cambium against lethal temperatures (Agee 1998; Fernandes et al. 2008). However, the survival of adults in these areas affected by understory fire does not promote an establishment of new seedlings of these species (Rodrigo et al. 2007 for *P. pinea*)

After crown fires, the natural regeneration of tree species lacking mechanisms to overcome the effects of fire (Tapias et al. 2004) is severely constrained (Retana et al. 2002; Rodrigo et al. 2004). Any of these three pines have serotinous cones (Escudero et al. 1997; Tapias et al. 2001, 2004), they open all the cones each year and they are not able to store seeds in closed cones, as *P. halepensis*, *P. brutia* or *P. pinaster* do (see Chap. 6). Moreover *P. nigra* and *P. sylvestris* show a similar seeding phenology, with cones maturing over 2 years and dispersal of seeds occurring in late winter to spring of the third year (Laguna 1993; Skordilis and Thanos 1997), just before their germination season. Then, most of the cones in *P. nigra* and *P. sylvestris* trees affected by summer wildfires (which are the most common ones in the Mediterranean region) are empty at that time and the seed canopy is exhausted. Seeds that germinate in late spring are burned as seedlings during summer fire (Retana et al. 2002) and the few seeds that remain in the soil over the ground are not able to resist the high temperatures of the fire (Habrouk et al. 1999). Thus recruitment into the burned area is difficult; in fact field data in different studies show an almost nil regeneration of *P. sylvestris* and *P. nigra* after fires (Espelta et al. 2002; Rodrigo et al. 2004; Vilà et al. in press). On a similar way, *P. pinea* cones ripen in spring over 3 years (Ganatsas et al. 2008; Tapias et al. 2001), while some authors consider that cone opening and seed release extend until autumn (Tapias et al. 2004). Although the cones are not serotinous and open at low temperatures (Tapias et al. 2001) some of these seeds can potentially survive summer wildfires, which may explain some field data documenting low but non-zero values of *P. pinea* establishment just after fires (Rodrigo et al. 2004, 2007). However mortality of these seedlings is high and does not allow a significant recruitment of *P. pinea* in burned areas (Rodrigo et al. 2007).

Moreover, when dispersed seeds arrive at the burned area, this does not necessarily imply a good seedling establishment. *Pinus nigra*, *P. sylvestris* and *P. pinea* seeds are consumed by mammals, birds and ants. These groups of animals, especially small mammals and birds, show a drastic decrease in their populations just after fire, but they recover few years later (Prodon et al. 1987; Torre and Díaz 2004; Brotons et al. 2005). As a consequence seed predation just after fire would be low. Moreover, in the case of *P. pinea* the harvesting of pine nuts for human consumption could limit seed availability in certain areas. Another limiting factor for recruitment is the competition between new seedlings and other vegetation regeneration in the burned area, especially for *P. sylvestris* and *P. pinea* that are more shade-intolerant than *P. nigra*. Therefore the seedlings established during the first years after fire, when vegetation cover is low, are those that have higher probability of survival (Ordóñez and Retana 2004). In consequence the regeneration in burned areas of these three pines is concentrated in a limited spatio-temporal window, when the predation of their seeds and plant cover are still low. The high growth and survival rates of these seedlings of *P. nigra* and *P. sylvestris* established during the first years after fire allows the recovery of these species in a narrow area close to unburned trees. On its hand, *Pinus pinea* shows high mortality rates when shrubby vegetation recovers, further limiting the recovery of this species (Rodrigo et al. 2007). This is even more evident if we consider that these pines achieve reproductive maturity around 15–20 years for *P. nigra* (Tapias et al. 2004), 15 years for *P. sylvestris* (Vilà et al. in press) and 10–20 years for *P. pinea* (Tapias et al. 2004). This time lag and the short dispersal distance of this species limit their overall regeneration rate in large burned areas.

### 7.1.6  Post-Fire Dynamics of Animal and Plant Communities in Non-Serotinous Pine Stands

It has been traditionally accepted that Mediterranean plant communities have a high resilience to fire, that is, the composition and structure of burned communities is restored very quickly, and the burned ecosystem cannot be distinguished from the predisturbance state after a few decades (Trabaud and Lepart 1980; Thanos 1999). However, recent studies (Retana et al. 2002; Rodrigo et al. 2004) indicate that Mediterranean basin forest communities and their dominant tree or shrub species show different responses after large fires. Thus, forests of seeder species that produce few seedlings after fire and have limited long distance dispersal (as it is the case for *P. nigra*) are replaced by other vegetation types. As trees have a key effect on the composition of the communities where they grow (Ne'eman et al. 1995), the composition and structure of whole plant communities after fire are directly related to the regeneration of the dominant tree species in the canopy (Arnan et al. 2007). In this context, *P. nigra* forests showed low regeneration after fire (Arnan et al. 2007). The lack of the main tree species (Retana et al. 2002) agrees with the pattern shown by the whole plant community, with low similarity composition values between burned and unburned plots. Although there were not great differences in dominance or diversity in unburned and burned *P. nigra* forest plots, the fact that the canopy does not re-establish completely or recovers to another vegetation type (Retana et al. 2002; Rodrigo et al. 2004) determines changes in the plant species present and causes large variation between burned and unburned plots (Arnan et al. 2007; Arianoutsou et al. 2010b).

Recently, a database was formed, including species that are known to inhabit mature *P. nigra* forests in Spain and Greece in order to realize the degree of our knowledge regarding their response to fire (Kazanis et al. 2011). The database consists of 131 taxa (with the pine species included), 47 woody and 84 herbaceous. It was proved that for most plant taxa of the database there was adequate knowledge supporting their ability to regenerate after a fire event. With the exception of 'annuals' and 'sub-shrubs', for all the other growth forms the commonest post-fire regeneration mode is resprouting. It should also be noted that it is among the 'perennials' that most species with remaining unknown response to fire are found. The percentage of taxa characterized by anemochorous and zoochorous seed dispersal is high. In the former group, taxa with small, light and numerous seeds are included, predominately of the Asteraceae and Poaceae families. In the later group, either taxa with fleshy fruits (endo-zoochorous dispersal, e.g. Rosaceae, Fagaceae) or taxa with fruits bearing spines (exo-zoochorous dispersal, e.g. Fabaceae, Poaceae) are included. Regarding early post-fire establishment, seeds of anemochorous are more prone to arrive at a burned stand from unburned sites, since for zoochorous taxa the improbable arrival of animals at the burned site would have been required. Nevertheless, for both categories the importance of unburned forest stands is once more highlighted.

Regarding the post-fire patterns of animal communities, they clearly parallel with those of overall plant community. For instance, in a study with ants conducted in the same vegetation types than those of Arnan et al. (2007), ant communities of *P. nigra* forests were the least resilient ones (Arnan et al. 2006), and this was attributed to the fact that *P. nigra* forests show little resilience to disturbance, and after fire they are replaced by completely different communities, either by coppices of resprouter species or by grasslands (Retana et al. 2002; Ordóñez and Retana 2004; Rodrigo et al. 2004). Other study found a persistent replacement of ant species in burned *P. nigra* forests (Rodrigo and Retana 2006), as it is also the case with vegetation. Such explanation is also applied to ground beetles communities, which in burned forests of *P. nigra* remain different from those in unburned areas along a long-term post-fire chronosequence (Rodrigo et al. 2008). In the case of birds, post-fire community is also strongly associated to vegetation recovery even when strong vegetation changes occurred due to non-direct regeneration of dominant forest tree species after a large fire. This is the case of *P. nigra* and *P. pinea* forests, where the non-direct regeneration process might create the appropriate habitats for open habitats species (Ukmar et al. 2007; Zozaya et al. 2011) and, thus, induce a turnover in species. As a deviation to this general trend, we are aware about a work with rodents where no differences in composition were detected between burned and unburned habitats of *P. nigra* few years after the fire (Ordóñez and Retana 2004).

## 7.2  Post-Fire Management: Issues and Alternatives

Ecological traits of non-serotinous pines considerably constrain the range of post-fire management options applicable in fire affected areas. The lack of seed and seedling survival of these species implies that their natural post-fire regeneration depends closely on the existence of *seed sources*, either isolated trees or groups of trees that survive fire both within the burned area and in the edges (Retana et al. 2002; Ordóñez and Retana 2004; Arianoutsou et al. 2010a). After the fire, survival and seed production of these trees determine the potential for colonization of the burned area.

*Pinus pinea* trees survive crown fire considerably better than other Mediterranean pines (Rodrigo et al. 2004), and their thick bark (Rigolot 2004; Mutke et al. 2005) make adult *P. pinea* trees relatively resistant to intense fire (Tapias et al. 2001). As a consequence, the presence of small groups of surviving trees of this species in areas affected by crown fire was relatively frequent. Rodrigo et al. (2007) found that 62.5% of *P. pinea* plots burned with crown fire in Catalonia had at least one *P. pinea* individual alive after fire. On the contrary *P. nigra* and *P. sylvestris* show lower survival of isolated trees in small groups of trees into the burned area (Rodrigo et al. 2004). But in the case of large fires often occur areas not affected by the fire, *green islands*, which can have different sizes and allow also the presence of groups of alive trees into the burned area. For *P. nigra*, Ordóñez et al. (2005) have found that trees, especially large trees, located in these small "green islands" produce more cones and more frequently than those in the edges or in large islands, and represent an important seed source for the post-fire regeneration of this species.

The success of post-fire regeneration from these seed sources depends, however, on the distance of seed dispersal. The distance at which pine seeds are dispersed is usually short. For *P. nigra* in a burned area in NE of Spain the distance observed was less than 50 m (Ordóñez et al. 2006), and reports of similar or even lower distances were obtained in Mt. Parnonas (S. Greece) (Arianoutsou et al. 2010b; Kakouros and Dafis 2009). For *P. sylvestris*, Vilà et al. (in press) have found that 90% of recruits were located at less than 25 m from the pines in the edge of the burned area. For *P. pinea* this dispersal distance is even shorter, with a maximum distance of 10–15 m from the crown (Montoya 1990; Rodrigo et al. 2007). The short dispersal in this species is related to seed morphology and weight: it is the heaviest seed of all Mediterranean pines, and has a rudimentary wing, considerably shorter than the seed itself (Tapias et al. 2001, 2004) which renders wind dispersal difficult. In areas affected by large fires where the proportion between burned area and unburned perimeter or green islands is very low and where most of the burned surface is far from seed sources, this type of regeneration is restricted to a little proportion of burned area or to small burned areas (Gracia et al. 2002). The competition of natural regeneration with resprouters and seeders (shrubs and oaks mainly) further limits possibilities of natural regeneration of these forest types after fires.

Under these conditions, successional trajectories after fire disturbance would most likely lead to the conversion of pre-fire non-serotinous pine dominated communities into other vegetation types (shrubland, grassland, oak-dominated woodlands), unless active restoration of pine woodland is implemented through reforestation measures (see Sect. 7.3). Post-fire management practices currently applied in non-serotinous pine forests reflect these issues. Table 7.1 reports the main practices applied in the management of non-serotinous pine forests affected by fire, as reported through a specific questionnaire by a sample of European countries in the framework of the FP0701 COST Action. Post-fire logging, mainly salvage logging, is widely and timely applied. The lack of any active post-fire response mechanisms explains well the widespread application of active restoration measures (planting or seeding) within 3 years after fires.

**Table 7.1** Example of post-fire management measures applied in the case of non-serotinous pine forest (Processed from COST Action FP0701 data)

| Post-fire management practices | Bulgaria | France | Greece | Italy | Latvia | Lithuania | Poland | Tunisia |
|---|---|---|---|---|---|---|---|---|
| Post-fire logging (Y/N) | Y | Y | Y | Y | Y | Y | Y | N |
| Post-fire logging timing (< 3 months after fire; 3–6 months; > 6 months) | 3–6 | 3–6 | 3–6 | 3–6 | < 3 | <3 | < 3 | – |
| Post-fire natural regeneration (Y/N) | Y | N | N | Y | Y | Y | N | Y |
| Active seeding or planting (Y/N) | Y | Y | Y | Y | Y | Y | Y | N |
| Active seeding/planting timing (<1 year after fire; 1–3 years; >3 years) | 1–3 | 1–3 | 1–3 | < 1 | 1–3 | 1–3 | 1–3 | – |

There is no paradigm or context to recommend active restoration to support the re-establishment of non-serotinous pines in fire affected areas. Alternative options exist and are mainly related to the conversion into more fire resilient communities, e.g. oak dominated woodlands (see Sect. 7.3.1). Reforestation option must be carefully evaluated being aware that restoration goals are often limited by the claims of other land uses, (see Sect. 7.3.1) and should be recognized as valuable by policy makers and public opinion (e.g., high conservation value, see Sect. 7.3.2). When planning restoration actions, land managers must be also aware that developing ecosystems, because of their low fire resilience, may undergo rapid transitions that would not result in the recovery of the components of the pre-fire forest community; thus, it is critically important to increase fire-resilience by enhancing the presence of resprouters, in order to promote self-regenerating forest types as independent as possible from further external subsidies. There are relevant experiences of post-fire management of Scots pine stands in Alpine environments: in such cases, the interventions are tuned to climatic conditions quite different from the Mediterranean ones (Beghin et al. 2010).

## 7.3  Case Studies

### 7.3.1  Early Post-Fire Management in P. nigra Forests in Central Catalonia (Western Mediterranean Basin)

#### 7.3.1.1  The Wildfires

Since 1990, large wildfires have destroyed more than 25% of the total area occupied by *P. nigra* forests in Catalonia (NE Spain) (Gracia et al. 2000). Central Catalonia was affected by two of the largest historically recorded wildfires: the Bages-Berguedà fire, which burned ca. 24,300 ha of forested land in July 1994, and the Solsonès fire, which burned ca. 14,300 ha in 1998. Prior to these fires, these areas had not burned for at least 70 years. The climate of the region is dry-subhumid Mediterranean (according to the Thornwaite index) with mean annual temperature between 10° and 13°, and annual precipitation between 550 and 750 mm. According to the data provided by the Ecological Forest Inventory of Catalonia (IEFC) carried out in 1993 (Gracia et al. 2000), natural *P. nigra* ssp. *salzmanii* forests were dominant before the fires occurred (78% of the burned surface), with *Q. ilex* and *Q. cerrioides* being extensively present in their understorey. *Pinus halepensis* forests were also represented in 14% of the burned area. Non-forested areas were mainly represented by croplands (Gracia et al. 2000). The resulting burned land was characterized by areas of high spatial heterogeneity in the distribution of burn severities. Thus, the fire left a mosaic of surviving green islands (10–15% of the total area affected by fire, Román-Cuesta 2002) immersed in a charred matrix.

### 7.3.1.2  Land Use Changes After Fire

An extensive survey of the entire burned area of the Bages-Berguedà wildfire, carried out through the comparison of aerial photographs prior (1993) and after (2005) the fire (Fig. 7.2), coupled with an extensive field sampling, revealed the existence of important land use changes (14% of the burned surface) (Espelta et al. 2002). Due to the regeneration failure of *P. nigra*, the area previously occupied by this species was extraordinarily reduced (form 15,700 ha to less than 100 ha). On the other hand, due to the vigorous resprouting of oaks, these species largely increased their presence in the landscape (from 1,048 ha to 12,450 ha). Transformation of previously forested areas to new croplands and rangelands accounted, respectively, for 1,005 (4.1%) and 2,365 ha (9.7%) of the 24,300 ha burned. However, a detailed analysis of land use changes points out that their distribution in the whole burned area is aggregated. Development of new croplands areas is positively related with the amount of previous croplands in the area and negatively related to the regeneration of resprouters. On the other hand, new rangeland areas are positively linked to the amount of surface where regeneration of resprouters is successful. These land use changes can be interpreted in the light of three major influences: (a) the need of landowners to find economic alternatives to forest logging, (b) the conviction of local authorities that favouring agricultural practices can decrease forest continuity and, thus, help to avoid the recurrence of large wildfires, and (c) the EU agricultural policy in the recent years, which has subsidized some crops and extensive livestock in the Mediterranean area.

### 7.3.1.3  Post-Fire Regeneration of *P. nigra* in the Study Area

Tree Survival and Cone Production

Tree survival and seed production were analyzed during 5 years as a function of tree size, crown damage, and tree location (i.e. location in edges or in islands)

1978 (MCA-1)                          2005 (MCSC-3)

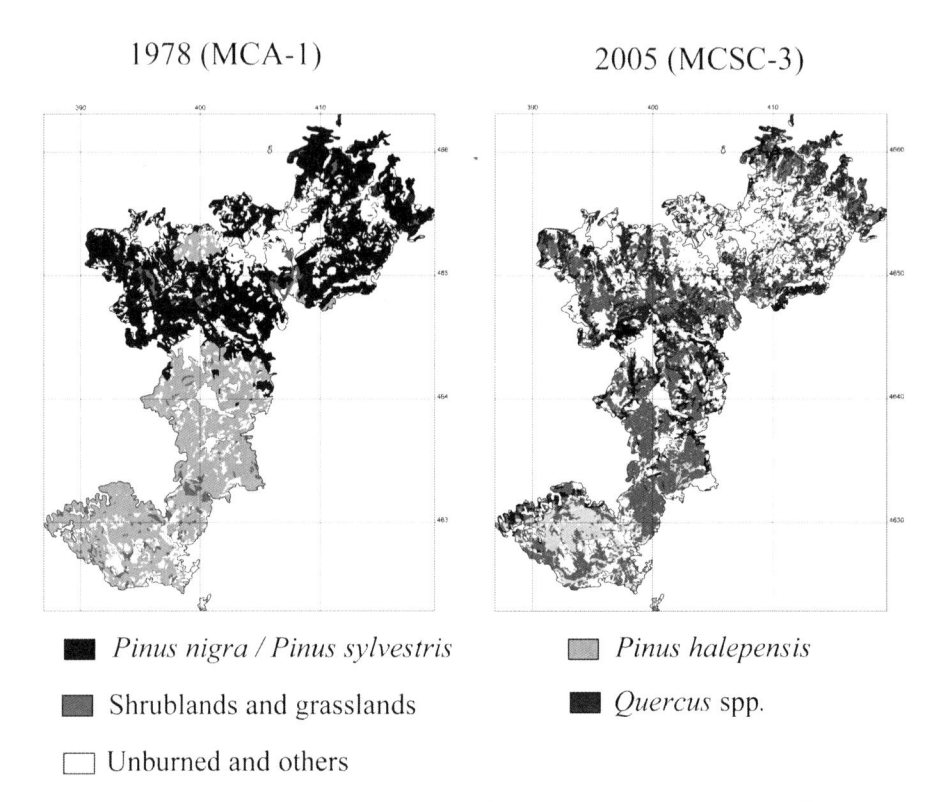

■ *Pinus nigra / Pinus sylvestris*          ☐ *Pinus halepensis*

■ Shrublands and grasslands                ■ *Quercus* spp.

☐ Unburned and others

**Fig. 7.2** Comparison of land covers in the area of the Bages and Berguedà regions affected by a large wildfire in 1994. *At left*, the map of land cover types in 1978 (Primer Mapa de cultivos y aprovechamientos, MCA-1). *At right*, the map of land cover types in 2005 (Tercer Mapa de Cobertes del Sòl de Catalunya, MCSC-3). Land cover types has been grouped in five categories. Source: Joanjo Ibañez, CREAF

(Ordóñez et al. 2005) in the area burned in 1998. Survival of *P. nigra* trees 5 years after fire increased with tree size (DBH > 20 cm) and decreased with crown damage. However, the response of trees with different fire damage varied with location, as less-affected trees showed higher survival in islands, while more affected ones performed better in the perimeter of the burned area. The main factor determining cone production during the first years after fire was tree size, because large trees produced more cones and more frequently than small ones. No differences were observed due to crown damage, but differences in cone production according to tree location were important. Thus, small trees produced cones more frequently in islands than on edges, while both cone production and the proportion of years that each tree produced cones decreased with island size. These results highlight the role of large trees and unburned islands as the main post-fire seed sources.

Seedling Establishment

The establishment of seedlings of *P. nigra* was examined under experimental controlled conditions of light and water as well as under natural conditions in the field (Ordóñez et al. 2004). The results suggest that seedling establishment after fire is scarce in field conditions and under a wide range of degrees of plant cover. However, *P. nigra* behaves as a species more shade tolerant than other pines but, given that this is not a common condition in recently burned areas, its regeneration is strongly influenced by fire.

The effects of seed predators (ants, rodents and birds) on post-dispersal seed removal and early seedling establishment of *P. nigra* were also evaluated by means of selective enclosure experiments limiting their access to seeds (Ordóñez and Retana 2004). The results indicated that the overall predation rates of *P. nigra* seeds by the three groups of predators were quite high, and the contribution of each group to overall predation showed seasonal variations. In the seedling establishment experiment, only in the exclusion treatment of the three predator groups there was initial establishment in all habitats, especially in the recently burned area. As in other species, seed predation can strongly limit population recruitment of *P. nigra* by reducing seed availability.

Post-Fire Regeneration Patterns

Post-fire regeneration patterns of *P. nigra* in the burned area of Bages and Berguedà were analyzed by developing a model of succession to predict medium-term changes in forest composition 30 years after fire (Rodrigo et al. 2004) from the regeneration monitored during the first years after fire (Retana et al. 2002). The results show that although *P. nigra* was the dominant species in the area before the fire, it almost disappears after the fire because its seedling density is almost nil. The highest proportion (76.7%) of plots originally dominated by pines changes after fire to communities dominated by oaks (*Q. ilex, Q. cerrioides*). Although a high proportion of the pre-fire oak seedlings and saplings present in the understory of *P. nigra* stands dies after the passage of fire, there is still a considerable percentage that sprouts vigorously and allows oak dominance in the future forest. There is also a considerable percentage of burned pine plots that change to shrublands or grasslands of *Brachypodium* spp. The transformation of these pine forests into shrublands or oak woodlands seems related to the length of time since abandonment, with the more recently abandoned lands becoming shrublands and the older areas, those providing time and a suitable habitat for oaks to establish (Lookingbill and Zavala 2000), becoming oak woodlands. Moreover, 7% of plots dominated by *P. nigra* before the fire showed a large post-fire regeneration of the serotinous pine *P. halepensis*, originating from seeds produced by the rare *P. halepensis* trees present in the canopy before the fire. Transformation into *P. halepensis* forests would be related to the proximity to seed sources of this pine or the fact that *P. nigra* can also form mixed forests with *P. halepensis*.

## Difficulties and Predictions for the Future

Three processes linked with the beginning of the recruitment, that is, 1) seed survival after fire, 2) seed predation, and 3) seedling survival, do not predict good perspectives for *P. nigra* forests after large wildfires. Thus, most seeds are dispersed in late winter and have already germinated in spring (Habrouk et al. 1999). Consequently, fire burns these seedlings, while the few seeds remaining in the soil are unable to withstand the high temperatures attained during intense summer wildfires (Habrouk et al. 1999). Moreover, post-dispersal seed predation by different animal groups is also high, as they consume many of the *P. nigra* seeds remaining on the ground. Finally, this low seed availability is particularly important taking into account that seedling survival is also very low during the first year after germination (Franco 2001). All these results suggest that the natural post-fire recovery of *P. nigra* in burned areas is difficult. Thus, natural regeneration of non-fire-prone seeder species after fire relies entirely on the arrival of propagules from seed sources, such as isolated trees present within the burned area or those at the unburned edges (Espelta et al. 2002). However, the distance at which pine seeds are dispersed is usually short, less than ca 50 m for *P. nigra* (Ordóñez et al. 2006), and larger trees (the main seed sources) are scarce in the study area. Under these constraints, colonization from the surrounding unburned landscape is likely to take many decades.

Ordóñez et al. (2006) developed a simulation model for predicting the recruitment response of *P. nigra* from unburned edges in the study area. The distribution of *P. nigra* seedlings at different distances from the unburned edges was simulated by integrating empirical field data for the different processes affecting seed (cone production, pre-dispersal cone predation, seed production per cone, seed dispersal, post-dispersal seed predation and seed germination) and seedling success. The model was successfully validated in old-burned areas. The simulated values of established seedlings 30 years after fire followed a normal distribution in the first 100 m, with a wide range of 2,000–25,000 seedlings/ha. The maximum dispersal distance showed a shorter range around 120 m. Thus, fires create large post-fire heterogeneity in seedling densities that depend on the distance to the seed source, but also on the characteristics of the forest before the fire. Plots with medium and large trees showed an increment in increasing seedling establishment with tree density, whereas plots dominated by small trees had very low regeneration. If we consider that *P. nigra* trees in the study area are concentrated in the smaller size classes, and that the forest front advanced in a much closed form, the natural post-fire recolonization of *P. nigra* in burned areas is very difficult.

## Post-Fire Management Practices

Under these circumstances a major interest grew up in the last decades about the best alternatives to restore *P. nigra* forests throughout suitable artificial reforestation programs, and to ameliorate the structure of mixed oak coppices (Espelta 1999).

An extensive experiment combining different methods of vegetation clearing (mechanical, controlled burning or grazing), soil preparation (ripping or planting holes) and reforestation methods (broadcast seeding, spot seeding and planting) was conducted in view of assessing the best alternatives to restore *P. nigra* forests (Espelta et al. 2003). These practices were compared in terms of seedling establishment. The results showed that the final establishment of *P. nigra* seedlings 2 years after the experiment onset ranged drastically from $7 \pm 4$ seedlings/ha in the broadcast seeding treatment to $610 \pm 40$ seedlings/ha in the plantations. The failure of the broadcast seeding assay points out that, although it may be recommended to wait a precautionary time to observe whether natural regeneration occurs after a wildfire (Espelta 1999), the fast recovery of ground vegetation, as well as that of some animal groups (especially seed-harvesting ants), severely threatens the success of broadcast seeding treatments. Seedling establishment after sowing was very poor and not influenced by vegetation clearing. In plantations, seedling survival was higher in the ripper treatment than in planting holes for all vegetation clearing treatments except the control one. Although the establishment of seedlings of *P. nigra* obtained in the plantation experiences is low in comparison to success reported in temperate and boreal forests, it is in the range of other experiences carried out in Mediterranean environments (Vallejo and Alloza 1998).

### 7.3.2   Early Post-Fire Management in P. nigra Forests in Southern Greece (Eastern Mediterranean Basin)

#### 7.3.2.1   The Wildfire

Mt. Parnonas (1,935 m) lies across the south-eastern part of the Peloponnese District, Southern Greece. It is characterized by a variety of habitats and a rich plant and animal diversity, which justifies its inclusion in the Natura 2000 network (GR2520006). An extended zone of coniferous forests, formed by *P. nigra* and *Abies cephalonica*, lies between 700 m and 1,700 m. For both species, the populations of Mt. Parnonas (together with those of the neighboring Mt. Taygetos) are on the southern-most edges of their natural geographical distribution. On the 23rd of August 2007, 1,921 ha of *P. nigra* forest were burned, i.e. about 36% of the total *P. nigra* forest cover on Mt. Parnonas.

#### 7.3.2.2   Management Objectives

Taking into consideration the fact that *Pinus nigra* is not adapted to high intensity fire events and the importance of this specific pine population of Mt. Parnonas, a pilot action plan for the post-fire restoration of the burned forest was proposed and funded under the LIFE + European Union Programme, titled "Restoration of *Pinus nigra* forests on Mt. Parnonas (GR2520006) through a structured approach".

The current LIFE + programme started in January 2009 and is expected to be completed in June 2013. Still, some preliminary results have been presented for the various scheduled actions. The actions scheduled in the context of the project were as follows (Kakouros 2009, http://www.parnonaslife.gr/en):

- Impact assessment of 2007 fire
- Demonstration of a structured approach for the restoration of *P. nigra* forests
- Implementation of restoration measures
- Monitoring and evaluation of the restoration

The impact assessment of fire was based on the detailed mapping of the burned areas and the evaluation of the fire impacts through the use of remote sensing, geographical information systems and fieldwork. The assessment showed that across the burned area, several forest stands remained unburned, accounting for a total area of 420 ha (Kakouros et al. 2009). As mentioned earlier, the importance of such stands acting as seed sources is very high.

The program involved the development and demonstration of a structured approach for the restoration of *P. nigra* forests that will help to define priorities for the restoration of the affected areas. Of major importance is the description of a step-by-step process for prioritizing and selecting the most suitable areas for restoration. Prioritization and selection is achieved by applying exclusion criteria, eligibility criteria in terms of abiotic parameters and technical criteria (Kakouros and Dafis 2010). In the case of Mt. Parnonas, the exclusion criteria were (i) the potential of natural regeneration and (ii) the potential of low survival of planted individuals. Following the first exclusion criterion all sites where pine seedling density exceeded or was expected to exceed 1 ind/m$^2$ were excluded. This was the case of sites near unburned stands or where fire burned only the understory. The second exclusion criterion corresponds to sites found outside the altitudinal limits that are regarded as the best for *P. nigra* growth, thus reducing the success of artificial reforestation.

The areas remaining after the application of the exclusion criteria have been ranked according to (1) the representativeness (sensu Annex I of the Habitats Directive) of the habitat type (i.e. areas with high representativeness before fire should have higher priority for restoration), (2) the inclusion of sites under conservation status (e.g. Natura 2000 sites or other protected areas), (3) the presence of important species (burned sites where rare or endemic species are known to be present should have higher priority for restoration), (4) the re-establishing forest connectivity (priority should be given to the restoration of forest stands that promote connectivity), and (5) the abiotic variables of the prospective areas (mainly soil depth and aspect), in order to select those with the higher potential of reforestation success. The final step is the consideration of the available resources (financial, personnel, seed stock) and the cost per hectare and per restoration method (seeding or planting) to determine the total area that will be restored.

After taking into consideration all the above mentioned criteria, the pilot restoration of 290 ha of *P. nigra* forests consisted of planting 464,000 *P. nigra* seedlings at 19 forest stands (Simadi 2010). The seedlings have been produced from seeds collected from cones of the Mt. Parnonas pine population in 2007 and reforestation begun in December 2010.

A crucial element of a restoration project, especially in cases where sites of the Natura 2000 network are involved, is the installation of a monitoring system for the evaluation of the restoration effectiveness. On Mt. Parnonas 33 permanent plots have been established, 13 plots for the monitoring of natural regeneration and 20 plots for the monitoring of artificial restoration (Kakouros and Dafis 2009). Valuable data are expected to be produced from the frequent sampling of these plots.

## 7.4  Key Messages

- Non-serotinous pine species do not regenerate after fire because there are no seeds available.
- The lack of active regeneration makes post-fire natural recovery mainly dependent upon seed dispersal from unburned patches.
- Natural succession trajectories after fire would most likely lead to the conversion of pre-fire non-serotinous pine forests into other vegetation types.
- This explains the widespread application of active restoration measures (planting or seeding) during the first years after fire.

## References

Agee JK (1998) Fire and pine ecosystems. In: Richardson DM (ed) Ecology and biogeography of *Pinus*. Cambridge University Press, Cambridge

Arianoutsou M, Christopoulou A, Tountas T, Ganou E, Kazanis D, Bazos I, Kokkoris, I (2010a) Effects of fire on high altitude coniferous forests of Greece. In: Viegas DX (ed) Book of proceedings of the VIth international conference on forest fire research, Coimbra (electronic edition)

Arianoutsou M, Kazanis D, Bazos I, Kokkoris I, Christopoulou A, Constantinidis-Georgiou P (2010b) Biological indices of the conservation status of burned communities at mountainous forest ecosystems of the Peloponnese. Final report for the WWF Hellas Project "Actions for the restoration of Peloponnese burned forests and the conservation of unburned patches", Athens (in Greek)

Arnan X, Rodrigo A, Retana J (2006) Post-fire recovery of Mediterranean ground ants follows vegetation and dryness gradient. J Biogeo 33:1246–1258

Arnan X, Rodrigo A, Retana J (2007) Post-fire regeneration of Mediterranean plant communities at a regional scale is dependent on vegetation type and dryness. J Veg Sci 18:111–122

Barbéro M, Loisel R, Quézel P, Richardson D, Romane F (1998) Pines of the Mediterrranean basin. In: Richardson DM (ed) Ecology and biogeography of *Pinus*. Cambridge University Press, Cambridge

Beghin R, Lingua E, Garbarino M, Lonati M, Bovio G, Motta R, Marzano R (2010) *Pinus sylvestris* forest regeneration under different post-fire restoration practices in the northwestern Italian Alps. Ecol Eng 36:1365–1372

Brotons L, Pons P, Herrando S (2005) Colonization of dynamic Mediterranean landscapes: where do birds come from after fire? J Biogeo 32:789–798

Dafis S (2010) The forests of Greece. Goulandris Natural History Museum, Kifissia (in Greek)

Dafis S, Papastergiadou E, Lazaridou E, Tsiafouli M (2001) Technical guide for the identification, description and mapping of the habitat types of Greece. Greek Center of Biotopes and Wetlands, Thessaloniki (in Greek)

EEA (2007) European forest types. Categories and types for sustainable forest management reporting and policy. European Environmental Agency report, No 9/2006, Copenhagen

Escudero A, Barrero S, Pita JM (1997) Effects of high temperatures and ash on seed germination of two Iberian pines (*Pinus nigra* ssp *salzmannii*, *P. sylvestris* var *iberica*). Ann Sci For 54:553–562

Espelta JM (1999) La reconstrucció de paisatges forestals afectats per grans incendis. Projecte pilot a l'incendi del Bages-Berguedà del 1994. Silvicultura 27:6–9

Espelta JM, Rodrigo A, Habrouk A, Meghelli N, Ordóñez JL, Retana J (2002) Land use changes, natural regeneration patterns, and restoration practices after large wildfire in NE Spain: challenges for fire ecology landscape restoration. In: Trabaud L, Prodon R (eds) Fire and biological processes. Backhuys Publishers, Leiden

Espelta JM, Retana J, Habrouk A (2003) An economic and ecological multi-criteria evaluation of reforestation methods to recover burned *Pinus nigra* forests in NE Spain. For Ecol Manag 180:185–198

Fernandes PM, Vega JA, Jiménez E, Rigolot E (2008) Fire resistance of European pines. For Ecol Manag 256:246–255

Franco S (2001) Efecto de las características del microhábitat en el establecimiento de plántulas de Pinus nigra después de grandes incendios forestales. Ms dissertation, Autonomous University of Barcelona, Barcelona

Ganatsas P, Tsakaldimi M, Thanos C (2008) Seed and cone diversity and seed germination of *Pinus pinea* in Strofylia Site of the Natura 2000 Network. Biodivers Conserv 17:2427–2439

Giordano G (1988) Tecnologia del legno. UTET 3:897–898

Gracia M, Ordóñez JL (2011a) Manuals de gestió d'hàbitats. Les pinedes de pi pinyer. Diputación de Barcelona, Barcelona

Gracia M, Ordóñez JL (2011b) Manuals de gestió d'hàbitats. Les pinedes de pinassa. Diputación de Barcelona, Barcelona

Gracia M, Ordóñez JL (2011c) Manuals de gestió d'hàbitats. Les pinedes de pi roig. Diputación de Barcelona, Barcelona

Gracia C, Burriel JA, Ibáñez JJ, Mata T, Vayreda J (2000) Inventari ecològic i forestal de Catalunya, Regió forestal IV. Centre de Recerca Ecològica i Aplicacions Forestals, Barcelona

Gracia M, Retana J, Roig P (2002) Mid-term successional patterns after fire of mixed pine-oak forests in NE Spain. Acta Oecol 23:405–411

Habrouk A, Retana J, Espelta JM (1999) Role of heat tolerance and cone protection of seeds in the response of three pine species to wildfires. Plant Ecol 145:91–99

Kakouros P (2009) Proposal to restore the black pine forests that have been affected by fires on Mount Parnonas (GR2520006). Greek Biotope-Wetland Centre, Thermi (in Greek)

Kakouros P, Dafis S (2009) Establishment of the monitoring system for the restoration of black pine forests on Mount Parnonas (GR2520006). Greek Biotope-Wetland Centre, Thermi (in Greek)

Kakouros P, Dafis S (2010) Guidelines for restoration of *Pinus nigra* forests affected by fires through a structured approach. Greek Biotope-Wetland Centre, Thermi (in Greek)

Kakouros P, Apostolakis A, Dafis S (2009) Report on the assessment of impacts to the habitat type "Mediterranean pine forests with endemic black pines" on Mount Parnonas (GR2520006). Greek Biotope-Wetland Centre, Thermi (in Greek)

Kazanis D, Pausas JG, Vallejo R, Arianoutsou M (2011) Fire related traits of *Pinus nigra* plant communities in the Mediterranean: a data base for Greece and Spain. In: Book of abstracts of Medpine 4th international conference, Avignon, 6–10 June 2011

Laguna M (1993) Flora forestal española. Galicia Editorial S.A, La Coruña

Le Maitre DC (1998) Pines in cultivation: a global view. In: Richardson DM (ed) Ecology and biogeography of *Pinus*. Cambridge University Press, Cambridge

Lookingbill TR, Zavala MA (2000) Spatial pattern of *Quercus ilex* and *Quercus pubescens* recruitment in *Pinus halepensis* dominated woodlands. J Veg Sci 11:607–612

Martínez F, Montero G (2004) The *Pinus pinea* L. woodlands along the coast of South-western Spain: data for a new geobotanical interpretation. Plant Ecol 175:1–18

Mirov NT (1967) The genus *Pinus*. The Ronald Press Co., New York

Montoya JM (ed) (1990) El pino piñonero. Agroguias mundi-prensa, Madrid

Mutke S, Sievanen R, Nikinmaa E, Perttunen J, Gil L (2005) Crown architecture of grafted Stone pine (*Pinus pinea* L.): shoot growth and bud differentiation. Trees 19:15–25

Ne'eman G, Lahav H, Izhaki I (1995) Recovery of vegetation in a natural east Mediterranean pine forest on Mount Carmel, Israel as affected by management strategies. For Ecol Manag 75:17–26

Ordóñez JL, Retana J (2004) Early reduction of post-fire recruitment of *Pinus nigra* by post-dispersal seed predation in different time-since-fire habitats. Ecography 27:449–458

Ordóñez JL, Franco S, Retana J (2004) Limitation of the recruitment of *Pinus nigrat* subsp. *salzmanii* in a gradient of post-fire environmental conditions. Ecoscience 11:296–304

Ordóñez JL, Retana J, Espelta JM (2005) Effects of tree size, crown damage, and tree location on post-fire survival and cone production of *Pinus nigra* trees. For Ecol Manag 206:109–117

Ordóñez JL, Molowny-Horas R, Retana J (2006) A model for the recruitment of *Pinus nigra* from unburned edges after large wildfires. Ecol Model 197:405–417

Pausas JG, Llovet J, Rodrigo A, Vallejo R (2008) Are wildfires a disaster in the Mediterranean basin? -a review. Int J Wildland Fire 17:713–723

Prodon R, Fons R, Athias-Binche F (1987) The impact of fire on animal communities in the Mediterranean area. In: Trabaud L (ed) The role of fire in ecological systems. SPB Academic, The Hague

Retana J, Espelta JM, Habrouk A, Ordóñez JL, Solà-Morales F (2002) Regeneration patterns of three Mediterranean pines and forest changes after a large wildfire in North-eastern Spain. Ecoscience 9:89–97

Rigolot E (2004) Predicting post-fire mortality of *Pinus halepensis* Mill and *Pinus pinea* L. Plant Ecol 171:139–151

Rodrigo A, Retana J (2006) Post-fire recovery of ant communities in Submediterranean *Pinus nigra* forests. Ecography 29:231–239

Rodrigo A, Retana J, Picó X (2004) Direct regeneration is not the only response of Mediterranean forests to large fires. Ecology 85:716–729

Rodrigo A, Quintana V, Retana J (2007) Fire reduces *Pinus pinea* distribution in the northeastern Iberian Peninsula. Ecoscience 14:23–30

Rodrigo A, Sardá-Palomera F, Bosch J, Retana J (2008) Changes of dominant ground beetles in black pine forests with fire severity and successional age. Ecoscience 15:442–452

Román-Cuesta RM (2002) Human and environmental factors influencing fire trends in different forest ecosystems. Ph.D. thesis, Autonomous University of Barcelona, Barcelona

Simadi P (2010) Reforestation of 290 ha on the burnt forest of Parnonas – extended summary. Programme LIFE 07 NAT/GR/000286, Forest Service of Sparti, Sparti

Skordilis A, Thanos CA (1997) Comparative ecophysiology of seed germination strategies in the seven pine species naturally growing in Greece. In: Ellis RH, Murdoch AI, Hong TD (eds) Basic and applied aspects of seed biology. Kruwer Academic Publishers, Dordrecht

Strid A (1980) Wild flowers of mountain Olympus. Goulandris Natural History Museum, Kifissia

Tapias R, Gil L, Fuestes Utrilla P, Pardos JA (2001) Canopy seed banks in Mediterranean pines of south-eastern Spain: a comparison between *Pinus halepensis* Mill., *P. pinaster* Ait., *P. nigra* Arn. and *P. pinea* L. J Ecol 89:629–638

Tapias R, Climent J, Pardos JA, Gil J (2004) Life histories in Mediterranean pines. Plant Ecol 171:53–68

Thanos CA (1999) Fire effects on forest vegetation, the case of Mediterranean pine forests in Greece. In: Eftichidis G, Balabanis P, Ghazi A (eds) Wildfire management. Algosystems SA and European Commission DGXII, Athens

Torre I, Díaz M (2004) Small mammal abundance in Mediterranean post-fire habitats: a role for predators? Acta Oecol 25:137–142

Trabaud L, Lepart J (1980) Diversity and stability in garrigue ecosystems after fire. Vegetation 43:49–57

Ukmar E, Battisti C, Luiselli L, Bologna MA (2007) The effects of fire on communities, guilds and species of breeding birds and control pinewoods in central Italy. Biodivers Conserv 16:3287–3300

Vallejo VR, Alloza JA (1998) The restoration of burned lands: the case of eastern Spain. In: Moreno JM (ed) Large forest fires. Backhuys Publishers, Leiden

Vassiliou C (2007) Evaluation of the regeneration potential of Pinus pinea at Koukounaries (Skiathos) Natura 2000 site. Diploma thesis, University of Athens, Athens (in Greek)

Vilà A, Rodrigo A, Martínez-Vilalta J, Retana J (in press) Lack of regeneration and climatic vulnerability to fire of Scots pine may induce vegetation shifts at the southern edge of its distribution. J Biogeo

Zozaya EL, Brotons L, Vallecillo S (2011) Bird community responses to vegetation heterogeneity following non-direct regeneration of Mediterranean forests after fire. Ardea 99:73–84

# Chapter 8
# Post-Fire Management of Mediterranean Broadleaved Forests

**Josep Maria Espelta, Anna Barbati, Lídia Quevedo, Reyes Tárrega, Pablo Navascués, Consuelo Bonfil, Guillermo Peguero, Marcos Fernández-Martínez, and Anselm Rodrigo**

## 8.1 Ecological Context

Broadleaved trees are those species that have wide and flat leaves in contrast with the needle-like ones typical of conifers. Indeed, this categorization is helpful to make a coarse and general distinction of trees (e.g. most broadleaved trees are angiosperms) but it includes a heterogeneous group of species with contrasting functional traits and ecological requirements. For example, Mediterranean broadleaved trees include deciduous and evergreen ones, and even in that last group, there is the sub-category of sclerophyllous trees with hard- and leather-like leaves (Quézel and Médail 2003). According to the European forest-type categories established by the EEA (2006), Mediterranean broadleaved forest mostly corresponds to: broadleaved evergreen forests (category 9), thermophilous deciduous forests (category 8) and to a lesser extent some mountain beech forests (category 6). In spite of being all

J.M. Espelta (✉) • G. Peguero • M. Fernández-Martínez • A. Rodrigo
CREAF and Unit of Ecology, Faculty of Sciences, Campus UAB, Barcelona, Spain
e-mail: josep.espelta@uab.es

A. Barbati
DIBAF, Dipartimento per la Innovazione nei Sistemi Biologici, Agroalimentari e Forestali, University of Tuscia, Viterbo, Italy

L. Quevedo
CREAF and Unit of Ecology, Faculty of Sciences, Campus UAB, Barcelona, Spain

Ajuntament d'Esparraguera. Plaça de l'Ajuntament, Esparreguera, Spain

R. Tárrega
Fac Biol, Area Ecol, Univ Leon, Leon, Spain

P. Navascués
Bureau of Forest Fire Prevention (OTPMIF), Diputació de Barcelona, Barcelona, Spain

C. Bonfil
Dept. de Ecología y Recursos Naturales, UNA, Distrito Federal, Mexico

F. Moreira et al. (eds.), *Post-Fire Management and Restoration of Southern European Forests*, Managing Forest Ecosystems 24, DOI 10.1007/978-94-007-2208-8_8,
© Springer Science+Business Media B.V. 2012

broadleaved, tree species included in these forest types have very different sensitivity to water stress, soil fertility, and resilience to disturbances. At a worldwide scale, the distribution of broadleaved deciduous and evergreen trees is mainly driven by macroclimate conditions, such as temperature and thermal range (Chabot and Hicks 1982). Notwithstanding this general trend, their coexistence is frequent in transition zones – as the Mediterranean Basin – between temperate and tropical shrublands (Terradas 1999). At this more regional and local scales, other factors such as microclimate, aspect, topography, geomorphology, disturbance or historical events may overlay the macroclimatic conditions and influence the distribution of broadleaved species (Damesin et al. 1998; Gracia et al. 2000). From a historical and biogeographically point of view, broadleaved trees in Southern Europe include a rich and diverse ensemble of species linked to the complex paleo-history of this region, including elements from the Paleocene to the Holocene (Quézel and Médail 2003).

In the Mediterranean Basin the general trend is that broadleaved sclerophyllous-evergreens are located in warmer and drier zones, while deciduous species are more common in cooler areas, and thus are usually found at higher altitudes within the same latitude or, for a similar altitude, in sites with northern aspects (Quézel and Médail 2003). Edaphic variables may also partly control their distribution, as evergreens have been presumed to dominate nutrient-poorer habitats, while deciduous species would be more competitive in high nutrient habitats (Aerts and van der Peijl 1993). Moreover, the ancient human occupation in the Mediterranean Basin has been also claimed to play an important role in determining the vegetation structure and the distribution of some broadleaved species (Acacio et al. 2010). For example in the case of oaks (*Quercus* sp.), a similar proportions of evergreen and deciduous species were found in NE Spain 5000 B.P. but the former spread more recently into the area of deciduous forests owing to increasing aridity and anthropogenic disturbances (Riera-Mora and Esteban-Amat 1994). This hypothesis relies on the fact that evergreen species are better adapted to harsh environments than deciduous due to the lower resource-loss ratios of the former (Aerts 1995). Thus, critical conditions during post-disturbance regeneration (e.g., water stress or high temperatures) would have a more negative impact in deciduous-broadleaved than in evergreen-sclerophyllous species (see for oaks, Mazzoleni and Spada 1992).

All these things considered, and having in mind the high diversity of Mediterranean broadleaved tree species, a question arises: Why all broadleaved species – excepting *Q. suber* are included in a single chapter in this book? Two pieces of evidence support this rationale. First, most – if not all – broadleaved Mediterranean tree species share in common the same regeneration mechanism after fire, i.e. resprouting (see next section in this chapter). Second, despite the high diversity of the group, species of a single genus (*Quercus* spp., oaks) – either evergreen or deciduous – are the most prevalent and dominant broadleaved trees throughout the Mediterranean Basin and they are the keystone and most relevant floristic element of evergreen and deciduous forests (Quézel and Médail 2003). Moreover, oak forests have been – and they still are – the Mediterranean broadleaved forests with the highest ecological and socioeconomic importance to obtain a high variety of: (i) primary forest goods (wood, firewood, charcoal), (ii) secondary products (e.g. mushrooms, livestock feeding) and (iii) other alternative uses (e.g. hunting). By far much less abundant, other broadleaved trees we can find in this region belong to the genus *Fagus, Castanea, Corylus, Ilex, Sorbus, Alnus, Populus, Fraxinus, Ceratonia, Olea* and other species in the halfway between shrubs and tree growth forms as *Arbutus, Laurus, Phyllirea* or *Rhamnus*. Even though these species are much less abundant than oaks, they are of great importance for the ecosystem services of Mediterranean-type forests, owing to they may play an important role in trophic webs (e.g. pollen for insects or fleshy fruits for frugivorous animals) and they contribute to maintain the high richness and diversity of forest Mediterranean communities (Garcia-Villanueva et al. 1998; Hulme 1997; Myers et al. 2000). Moreover, in recent years, a greater incidence of fire in non-fire-prone Mediterranean areas, where some of these non-dominant species mostly occur, has been detected and it is becoming a matter of concern (Rodrigo et al. 2004).

The abundance and socioeconomic importance of oaks has probably conditioned that the post-fire regeneration of these species has been better studied than that of other broadleaved trees (Espelta et al. 1999). Thus, in this chapter, we will mostly focus on recommendations arisen after the study of the post-fire regeneration and management of Mediterranean oak forests, although we will try to provide as much information as possible from other species.

## 8.2   Fire Adaptation and Regeneration Strategies

Most, if not all, broadleaved species inhabiting Mediterranean-type forests are able to resprout – i.e. to produce sprouts from buds located on pre-existing plant organs – after disturbance, and this process, turns into their main regeneration strategy after fire (Fig. 8.1). The ability to resprout requires both the presence of a protected bud bank (Klimešová and Klimeš 2007) and the presence of stored reserves to sustain initial regrowth (Chapin et al. 1990). The onset of resprouting after disturbance (e.g. fire) occurs by the activation of a dormant bud bank located in the stump, root collar or roots (Pascual et al. 2002; Verdaguer et al. 2001) – also in the branches of *Quercus suber* (see Chap. 9) – probably driven by a combined effect of changes in hormonal levels and physical conditions (temperature, moisture, light) after the destruction of the aerial parts (Champagnat 1989). As resprouting is the main post-fire regeneration mechanism of most Mediterranean broadleaved species this process will be on the heart of most remarks we will make concerning the influence of several factors on the success of the natural regeneration of these species after fire and also the basis to design the best alternatives to post-fire management.

Resprouting is an efficient life-history trait by which woody plants can recover lost biomass after disturbance (Bellingham and Sparrow 2000). This response is observed in many taxa around the world, but it has been in Mediterranean-climate regions where it has probably received more attention, as one of the main regeneration mechanisms of plant communities subjected to fire (see the seminal work by Keeley and Zedler 1978). In fact, the resprouting ability of many Mediterranean shrubs

**Fig. 8.1** Basal resprouts of
*Quercus cerriodes* 3 months
after fire

and trees has been one of the most important keystones to build up the paradigm of
the resilience (sensu Westman 1986) and autosuccessional nature (Hanes 1971) of
Mediterranean-type communities after fire.

Nevertheless, this conspicuous evidence does not allow us to unequivocally
consider resprouting of many broadleaved Mediterranean species as a fire adaptation
(sensu Bond and Keeley 2005). Indeed, resprouting is an ancient and widespread
trait among many plants, even in those not regularly exposed to fire (Axelrod 1989).
Thus, it might be a pre-adaptive trait of damage tolerance to other selective forces,
such as grazing or drought (Bradshaw et al. 2011, see for dry-tropical forests Janzen
and Martin 1982). Therefore, resprouting must be considered a remarkable characteristic of some Mediterranean plants to withstand fire but it is an attribute potentially evolved in response to multiple or different selective forces. In the case of
many Mediterranean broadleaved species (e.g. *Quercus* spp., *Phyllirea* spp., *Arbutus*
spp.) resprouting is usually accompanied by other characteristic life history traits,
such as longevity, slow growth and some degree of shade tolerance during the
juvenile stages.

The success of resprouting after fire will vary depending on: (i) the species involved, (ii) the intensity, season and frequency of the fire events, (iii) characteristics related with site quality (topography, soil, climate) and, (iv) individual traits, such as size (and probably age) of the trees affected.

## 8.2.1 Species-Specific Characteristics

Albeit most Mediterranean broadleaved species are able to resprout after fire, large inter-specific differences in resprouting vigor have been observed (Calvo et al. 2003; Catry et al. 2010; Espelta et al. 2003; Lopez-Soria and Castell 1992; Quevedo et al. 2007). This pattern may be linked to inter-specific differences *per se* in the size of the bud bank (Bonfil et al. 2004), the degree of bud protection (Catry et al. 2010) or the amount of stored resources in belowground organs (burl, taproot). However, they may also depend on each species particular sensitivity to other environmental factors influencing the resprouting process, such as resource availability (Cruz et al. 2002), topography and soil characteristics (Lopez-Soria and Castell 1992) or the season when fire occurs (Bonfil et al. 2004). For example, after a fire event in 1994 in the upper part of the Montseny Massif (NE, Spain) both *Fagus sylvatica* – a species not regularly exposed to fire – and *Quercus ilex* – a species highly resilient to fire – were able to resprout vigorously from the stump (respectively, 80% and 98% of trees; JM Espelta unpubl. data). However, a drought episode in 1998 killed most resprouted beeches but had no influence on oaks dramatically enlarging the difference in the survival rate of the two species (respectively, 21% and 95%). See also Catry et al. 2010 for a similar study. Unfortunately, there is yet not enough information to establish a comprehensive rank of the resprouting ability after fire of all Mediterranean broadleaved species. We have tried to resume in Table 8.1 the information available according to several published sources. Clearly, future studies should be aimed to provide additional information on this point.

## 8.2.2 Fire Regime

Many characteristics of the particular fire event (e.g. intensity, duration and season) and of the fire regime of the area (e.g. fire frequency) may influence the success of resprouting. Numerous studies have observed that more intense wildfires decrease resprouting ability, as they probably cause the physical destruction of part of the bud-bank, damage superficial roots or affect the physiology of the stump (Obón 1997). In addition, the particular moment of the year when fire occurs can also critically influence resprouting (Bonfil et al. 2004). Certainly, most wildfires in Southern Europe occur during the hottest and driest season of the year (summer). Nevertheless, whether a fire occurs at the beginning or at the end of summer may be critical. However, there is no general consensus in the sign of these effects.

**Table 8.1** Post-fire tree mortality of some common broadleaved tree and shrub species found in Southern Europe according to different studies

| Species | Post-fire mortality (%) | Source |
|---|---|---|
| *Acer opalus* | 0 | Quevedo et al. (2007)[a] |
| *Acer campestre* | 6 | Quevedo et al. (2007) |
| *Acer monspessulanum* | 7 | Quevedo et al. (2007) |
| *Amelanchier ovalis* | 0 | Quevedo et al. (2007) |
| *Castanea sativa* | 20–83 | Catry et al. (2010)[b] |
| *Crataegus monogyna* | 6 | Quevedo et al. (2007) |
| *Crataegus monogyna* | 0–7 | Catry et al. (2010) |
| *Cornus sanguinea* | 3 | Quevedo et al. (2007) |
| *Fagus sylvatica* | 20–79 | J.M. Espelta (unpubl. results)[b] |
| *Fraxinus angustifolia* | 0–0 | Catry et al. (2010) |
| *Ilex aquifolium* | 26 | Quevedo et al. (2007) |
| *Olea europaea sylvestris* | 0–0 | Catry et al. (2010) |
| *Prunus spinosa* | 7 | Quevedo et al. (2007) |
| *Quercus cerrioides* | 1 | Bonfil et al. (2004) |
| *Quercus coccifera* | 0–10 | Catry et al. (2010) |
| *Quercus faginea* | 2–14 | Catry et al. (2010) |
| *Quercus ilex* | 1 | Bonfil et al. (2004) |
| *Quercus ilex* | 2–5 | J.M. Espelta (unpubl. results) |
| *Sorbus aria* | 17 | Quevedo et al. (2007) |
| *Sorbus domestica* | 0 | Quevedo et al. (2007) |
| *Sorbus torminalis* | 0 | Quevedo et al. (2007) |
| *Viburnum lantana* | 0 | Quevedo et al. (2007) |
| *Viburnum tinus* | 0 | Quevedo et al. (2007) |

[a] These results may underestimate mortality values as they were not obtained after a summer wildfire but after a prescribed burning in late winter

[b] The first value is survival 1 year after fire while the second value is survival after 4 years

Thus, while some studies have reported the most negative consequences negative for plants when fires occur in at the beginning of summer (see DeBano et al. 1998) others have reported that "late season fires" – fires occurring towards the end of summer – may constrain the growth of new resprouts (Fig. 8.2, Bonfil et al. 2004, see also Ducrey and Turrel 1992). This negative effect of "late season fires" has been also observed in other species living in seasonal climates with a harsh season (e.g. see for dry tropical forests Peguero and Espelta 2011) and it has been attributed to the worsening of the plant water status and the consumption of stored reserves in order to sustain metabolic activity as summer progresses (Bonfil et al. 2004; Bowen and Pate 1993; Cruz et al. 2002; Hodgkinson 1992). Indeed, studies that have analyzed the mobilization of nutrients after resprouting have reported that plants burned at the end of summer show lower N and P levels in their reservoir organs than those burned earlier (Peguero and Espelta 2011). This suggests that, rather than starch, N and especially P availability may be responsible of this "phenological" constrain of a lower resprouting vigour after late season fires (Canadell and López-Soria 1998; Saura-Mas and Lloret 2009). Especially in the case of P, limitation for

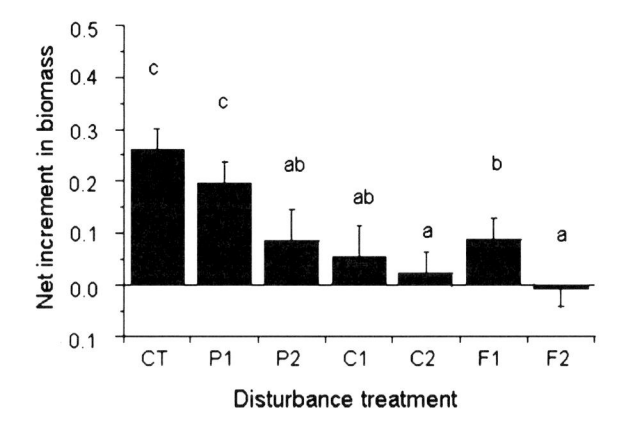

**Fig. 8.2** Net increment of biomass (once the effects of size have been accounted for) of Mediterranean oaks (*Q. ilex* and *Q. cerrioides* individuals) 1 year after different types of disturbances (CT = control, P = clipping, C = cutting and F = Burning) applied just before (*1*) and just after (*2*) summer. Notice that individuals disturbed after summer (*2*) show lower recovery than those disturbed at the onset of summer and this response is specially evident in burned individual (compare *F1* and *F2*). Different letters indicate significant differences among treatments according to the Fisher-LSD test (Modified after Bonfil et al. 2004)

resprouting could arise from the need to carry out a massive mobilisation of this nutrient from woody reservoirs towards leaves to improve water-use efficiency during summer drought conditions (Sardans and Peñuelas 2007). Interestingly, this negative effect of late summer fires for resprouting might be magnified in the near future with the lengthening of the summer drought period, as predicted by several future scenarios of climate change (IPCC 2007).

Not just the characteristics of a particular fire event (intensity, seasonality) but also the frequency of fire occurrence will condition resprouting. Indeed, a high frequency of fire increases mortality and decreases the resprouting vigor of surviving individuals (Díaz-Delgado et al. 2003). For example, *Q. ilex* trees twice burned in 1994 and 1998 exhibited three times less growth rates than those only burned in 1998 (Bonfil et al. 2004). The negative effect of repeated fires is explained by a progressive destruction and exhaustion of the bud bank and the depletion of stored resources to sustain re-growth.

Concerning the potential role of resources, some controversy still exists. Thus, while some studies have proven carbohydrates to be mobilised during resprouting (Bowen and Pate 1993), and to constrain re-growth after burning (Peguero and Espelta 2011), others have suggested nitrogen and phosphorus to be the most limiting resources (Canadell and López-Soria 1998; Miyanishi and Kellman 1986) and, even some studies have claimed for no effects of stored resources on resprouting vigour (Cruz et al. 2003a). Whatever the type of resources involved, and potentially exhausted, a decrease in resprouting vigour (survival, growth) after repeated fires has been observed in monitoring surveys or experimental studies (Bonfil et al. 2004).

## 8.2.3  Site Quality

Resprouting likelihood has been also observed to be conditioned by soil properties (Peguero and Espelta 2011) and meteorological conditions following the disturbance event (Riba 1998). In one of the most comprehensive studies of the influence of environmental factors on resprouting after fire, Lopez Soria and Castell (1992) reported a higher percentage of individuals resprouting in deep vs. shallow soils and north vs. south slopes, which was supposed to be related with higher soil moisture in the former conditions. In line with these results, the precipitation regime following fire has also been considered to have a crucial influence on the resprouting success of many Mediterranean species: i.e. resprouting vigor increases if repeated rainfall episodes occur after the disturbance event (Riba 1998).

## 8.2.4  Individual Characteristics

In addition to species-specific differences, fire regime characteristics and site quality, resprouting vigor is also closely linked with the size of the individual (genet) prior to fire (Bonfil et al. 2004; Catry et al. 2010). Thus, post-fire survival tends to be higher in larger sized individuals, probably because they have a larger bud-bank and more reserves in their below-ground organs (Catry et al. 2010; Peguero and Espelta 2011; Quevedo et al. 2007). Similarly, the number and the growth of the new sprouts are also positively linked with the previous size of individuals (Espelta et al. 2003). Interestingly, in comparison to the large amount of information we have about the importance of size for resprouting, almost nothing is known about the potential effect of age. The only information we have about age effects on resprouting is that of studies conducted after clear-cutting where a negative effect of ageing on two basic features of the resprouting process as the total number of resprouts and mean resprout height has been suggested (Ducrey and Boisserie 1992). This process has been attributed to the ageing and senescence of the bud bank and the poorest functioning of the root system with tree age.

## 8.3  Post-Fire Management Issues and Alternatives in Mediterranean Broadleaved Forests

### 8.3.1  Current Post-Fire Management Practices

Management practices currently implemented in fire affected areas are a straightforward consequence of the ability of Mediterranean broadleaved forests to resprout after fire. Table 8.2 summarizes key post-fire management practices as reported through a specific questionnaire by a sample of European countries in the

**Table 8.2** Example of post-fire management measures applied in Mediterranean broadleaved forests (processed from COST Action FP0701 data)

| Post-fire management practices | France | Greece | Israel | Italy | Portugal | Spain | Switzerland[*] | Tunisia |
|---|---|---|---|---|---|---|---|---|
| Post-fire logging (Yes/No) | Yes | Yes | No | Yes | Yes | No | Yes | No |
| Post-fire logging timing (< 3 months after fire; 3–6 months; > 6 months) | 3–6 | 3–6 | – | > 6 | > 6 | – | 3–6 | – |
| Reliance upon post-fire natural regeneration through resprouting (Yes/No) | Yes | Yes | Yes | Yes | Yes | Yes | Yes | Yes |
| Active seeding or planting (Yes/No) | Yes | No | No | No | No | No | Yes | No |
| Active seeding/planting timing (< 1 year after fire; 1–3 years; > 3 years) | 1–3 | – | – | – | – | – | < 1 | – |

[*]Mainly represented by chestnut dominated forest

framework of the FP0701 Cost Action. Harvesting of damaged trees is fairly widespread; it is implemented, mainly through salvage logging. Active restoration techniques (planting or seeding) are applied in a few countries just in case of low resprouting success. Thus, reliance upon post-fire natural regeneration by resprouting is by far the most widely applied approach to recover fire affected areas.

## 8.3.2   Post-Fire Management Alternatives

After bud onset, vertical growth of resprouts increases quickly and they may reach considerable lengths just in the first growing season (Cañellas et al. 2004). Resprouts also show a higher specific leaf area and nutrient content (e.g. N, P, K) in leaves and stems (Peña-Rojas et al. 2005) compared to the values exhibited by branches of adult trees (Castell et al. 1994) or young seedlings (Espelta et al. 2005). This advantageous leaf morfometry and nutrient content in resprouts has been assumed to occur because the large unbalance between above- and below-ground biomass in resprouting individuals, favors a greater amount of resources available and thus higher photosynthetic rates (Cutini et al. 1996; El Omari et al. 2003, see also Fleck et al. 2010).

The advantage of resprouts over seedlings results in a very important recommendation when examining the most suitable post-fire management practice in these areas: hardly ever, if there is a moderate density of individuals, burned areas dominated by resprouting species will require plantation or seeding practices (Moreira et al. 2009, see also Chap. 1). Of course, a critical point is to determine whether the density of resprouting individuals will be enough to recover the forest. It is difficult to give a general recommendation but we have observed that burned oak forests with 400–600 individuals (genets) per hectare were able to attain a continuous forest cover in 20–25 years. Even though reforestation – seeding or planting – may not be necessary, a wise management of post-fire natural regeneration should be recommended. Indeed, in Mediterranean broadleaved forests, the rapid recovery of vegetation cover through resprouting after fire helps to prevent soil erosion (Calvo and Cerdà 1994; Calvo et al. 2003), reduce nutrient losses (Trabaud 1987), provide shelter and suitable environmental conditions for the recovery of the fauna (Garcia-Villanueva et al. 1998; Jaquet and Prodon 2009) and enhance the rapid return of some crucial ecosystem services (e.g. $CO_2$ fixation). The rapid recovery of vegetation cover may be also of paramount importance to diminish the risk of ecological damage if logging of burned trees is conducted.

Notwithstanding the advantages of resprouting as a post-fire regeneration mechanism, the fact that the actual fire regime in the Mediterranean Basin is characterized by more intense and large wildfire events, coupled with the low revenue obtained from traditional forestry practices (e.g. charcoaling, livestock) and the new requests of a modern society (e.g. spaces of leisure), raises doubts on the suitable alternatives to manage extensive forest landscapes arisen through resprouting (Cotillas et al. 2009). Indeed, extensive burned Mediterranean forests and shrublands regenerated through resprouting result in large coppices-like structures with a high homogeneity in their structure (Terradas 1999). These "young" coppices are characterized by high-density stands of multi-stemmed stumps with relatively small resprouts, slow vertical growth and low production rates (Calvo et al. 1999; Espelta et al. 2003, 2006). Certainly, once resprouts have started to develop, interaction and competition among them will progressively result in the mortality of the weakest ones. (Espelta et al. 1999). As observed by Castell et al. (1994) for *A. unedo*, this process is mostly owned to great differences in light availability between dominant and suppressed resprouts (aerial competition) rather than to differences in their water relations (below ground competition). During ageing, mortality of resprouts (self-thinning) and the growth of those surviving will progressive drive forests stands to be composed by individuals with few stems (ca 1–5) of greater size. However, this process may take long and, in the meanwhile, dense young coppices with a great vertical and horizontal continuity will provide little social and economic revenues and may be particularly sensitive to the spread of new fires.

### 8.3.3 *Assisting Natural Regeneration*

According to the scientific and technical literature, the best alternative to manage and ameliorate post-fire stands of resprouted (multi-stemmed) individuals appears to be their gradual conversion into either *stored coppices* (i.e. coppices in which there remains only one or two stems per stool) or eventually, if sexual reproduction may be encouraged, to *high forests* (Serrada et al. 1996). This process involves the elimination and selection of resprouts (i.e. cleaning of the stumps) in order to reduce competition among the selected ones, raise the forest canopy (Cotillas et al. 2009) and encourage sexual reproduction (Sánchez-Humanes and Espelta 2011). Moreover, this practice has been argued to increase the potential of forest for wood production, livestock grazing, hunting and other alternative uses, while preserving their protective function (Cañellas et al. 1996) and reducing the risk of wildfires.

When considering the possibility to conduct a selection thinning practice in a broadleaved forest regenerated through resprouting three major questions arise: (i) Which is the best regeneration age to conduct this treatment? (ii) What is the best intensity of thinning to apply? and (iii) How to control the second wave of resprouts that will appear after thinning?

### 8.3.3.1   At What Age Should We Apply Selective Thinning?

Ideally, to perform the selection thinning treatment we should wait until some degree of apical dominance and asymmetry in the size of resprouts has developed. From a practical point of view, this is when dominant and suppressed resprouts can be distinguished. If selection thinning is conducted too early the potential benefits will be suppressed by the new wave of young resprouts appearing after thinning. The age when dominant resprouts start to be distinguished may vary depending on the species considered, the size of individuals and the quality of the site. For both deciduous and evergreen oaks (*Quercus cerrioides* and *Quercus ilex*) and also *Arbutus unedo* in NE Spain this has been observed around 10–12 years after post-fire resprouting onset (Espelta et al. 2002). At this regeneration age, stump cleaning increases absolute and relative growth in diameter, height and canopy cover of the retained resprouts (Fig. 8.3) and it promotes the onset of acorn production (Sánchez-Humanes and Espelta 2011). Moreover, thinning may help to improve growth especially under negative environmental conditions (e.g. drought episodes). Thus, as shown in Fig. 8.3 thinned individuals had more similar growth rates in a normal (2004) and in a drought year (2005) than non-thinned ones.

In contrast to the benefits obtained from thinning when it is conducted at an appropriate age, in thinning experiments conducted too early (e.g. in oak forests only 5 years after the fire event), no significant effects where observed in the growth of retained resprouts and, just in a year, the new wave of new resprouts produced meanwhile attained a similar size (Espelta et al. 2008). One practical drawback of carrying out thinning at a suitable time is that it is cheaper to carry out this operation at an earlier post-fire regeneration than in older forests. All things considered, and as a rule of thumb, thinning of resprouts in post-fire regenerated areas is recommended to be carried out once they are 10–15 years old.

### 8.3.3.2   What Is the Best Thinning Intensity to be Applied?

When planning the thinning treatment, a second major concern for forest managers is to decide the most suitable thinning intensity to be applied. This intensity must be viewed as a compromise between two processes: it should be strong enough to

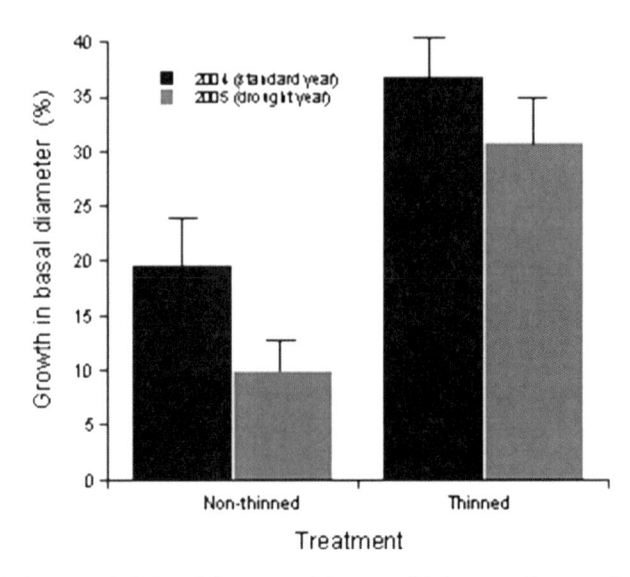

**Fig. 8.3** Relative growth in basal diameter of 9 years old *Quercus ilex* post-fire resprouts in non-thinned (control) and thinned individuals in two consecutive years: 2004 (standard year) and 2005 (drought year). Notice that thinned individuals better overcame the negative consequences of drought: i.e. growth in 2004 and 2005 was much more similar than in non-thinned individuals (JM Espelta unpubl. data)

stimulate the growth of the retained resprouts but not so much to promote the appearance of a vigorous second wave of resprouts.

In experiments which have applied different thinning intensities to post-fire regenerated mixed evergreen and deciduous oak forests, it has been observed that cleaning of stumps increases to a great extent the height and basal diameter of the selected resprouts, with low differences between the two cleaning intensities applied: one or three resprouts retained per stump (Espelta et al. 2003). However, the number and height of the new wave of resprouts which appeared after thinning was greater in those individuals with a single resprout retained. This suggests that a moderate intensity of thinning (i.e. leaving 2–4 resprouts per stump) may render better results considering both the growth of the retained resprouts and the inhibition of new resprouts. Indeed, the vigor of this second wave of resprouts (number, growth) in *Arbutus unedo* thinned individuals has been observed to depend on the number and size of the resprouts retained and the basal area of the resprouts thinned (see Fig. 8.4 for the effects of the number of resprouts retained). If the tinning intensity is appropriate, resprouts will increase their growth and new resprouts will remain under control (Fig. 8.5). In contrast with the reported benefits of selective thinning to ameliorate the structure of post-fire regenerated broadleaved forests, the use of pruning in these young coppices has been reported to produce barely any significant effect (Espelta et al. 2003).

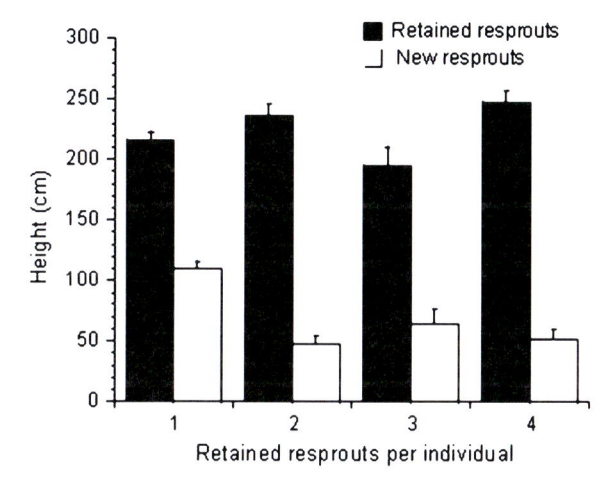

**Fig. 8.4** Height of resprouts (retained and new) according to the number of resprouts retained (from 1 to 4) in thinned 10 years old *Arbutus unedo* individuals resprouting after fire (JM Espelta unpubl. data)

**Fig. 8.5** If wisely conducted, selection thinning of post-fire resprouting broadleaved forests may enhance their structure and ameliorate their growth: Post-fire resprouted oaks 1 year after selective thinning (upper-left), post-fire resprouted oaks 7 year after selective thinning (*downer-left*) , and post-fire resprouted oaks 20–30 years after selective thinning (*upper-right*). Note that the three pictures are for illustrative purposes and do not correspond to the same individuals

### 8.3.3.3 How to Control the New Wave of Resprouts After Thinning?

As mentioned above, one of the main concerns after selective thinning is conducted is how to control the new wave of resprouts that will appear. On the one hand, this may be achieved if a moderate intensity of thinning is applied (e.g. leaving several resprouts per stump). On the other hand, the growth of these resprouts can be controlled by moderate livestock grazing (Espelta et al. 2006). However, this treatment should only be applied if retained sprouts have already reached a size that allows them to escape damage from livestock grazing (e.g. more than 2–3 m of height in the study of Espelta et al. 2006).

## 8.4 Pests and Diseases

Mediterranean broadleaved forests regenerated through resprouting after fire are rarely sensitive to suffer catastrophic events of pests, attack or diseases. Among the most important pests, especially in oaks, there are two Lepidoptera that may cause intensive defoliation episodes: *Lymantria dispar* and *Tortrix viridiana*. These species tend to appear in post-fire regenerated areas after some years after resprouting onset (i.e. 8–10 years, J.M. Espelta, personal observation) but these episodes are very much linked to specific meteorological situations and they rarely cause tree mortality.

## 8.5 Climate Change

One of the main concerns for forest conservation in the Mediterranean Basin is the potential increase in aridity due to climate change. Indeed, a progressive trend towards warming has been detected over the last century (Pausas 2004; Piñol et al. 1998) and this situation may turn even worst during the current century, when a rise in mean temperature of between 2.2 and 5.1°C is expected to occur (IPCC 2007). Concerning precipitation, potential changes are more uncertain but include a 4–27% reduction of rainfall and important changes in the precipitation patterns (IPCC 2007). This new climatic scenarios are very likely to lead to reduced soil moisture, to promote recurrent drought episodes (Douville et al. 2002; Wang 2005) and to increase wildfire occurrence (Pausas 2004).

Whether this new climatic scenario may constrain resprouting ability of Mediterranean broadleaved trees after disturbance is a research topic starting to receive increasing attention. Thus, in experimental studies that simulated a reduction in rainfall (15%) in post-fire regenerated mixed oak forests it has been demonstrated that this moderate drought increase may reduce the growth of resprouts of the deciduous *Q. cerrioides* while no effects were observed in the growth of the evergreen *Q. ilex* (Cotillas et al. 2009). Nevertheless, despite the lack of effects in the growth of the evergreen oak, increased drought delayed reproduction -the onset of flower production and increased acorn abortion – in the resprouts of this species

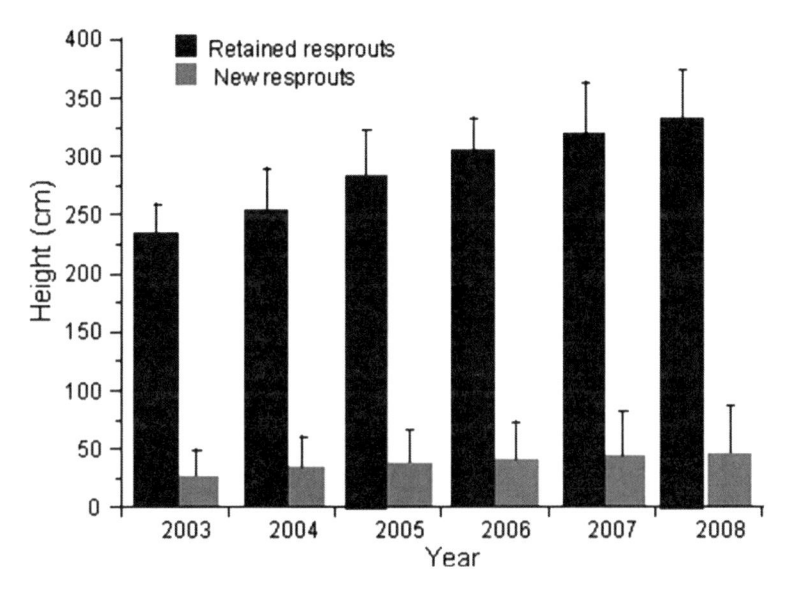

**Fig. 8.6**  Mean ± SE relative growth rate (RGR) in height of resprouts in thinned and non-thinned *Quercus ilex* individuals resprouting after fire in plots subjected to rainfall exclusion. Notice that the difference between non-thinned and thinned individuals quickly decreases and finally disappears at the fourth year (Modified from Cotillas et al. 2009)

(Sánchez-Humanes and Espelta 2011). In these two oak species, selective thinning of resprouts could partly mitigate the negative effects of increasing drought on resprouts growth, although this positive effect was very transient and disappeared 2–3 years after treatment (see Fig. 8.6, Cotillas et al. 2009; Sánchez-Humanes and Espelta 2011). This suggests that the traditional thinning intensity applied (removing 30–40% of basal area, see Retana et al. 1992; Cotillas et al. 2009) may not be intense enough under upcoming drier climatic scenarios. In addition, to the consequences of increased drought, occurrence of more intense and frequent fire events can constrain the post-fire natural regeneration of Mediterranean broadleaved species. Clearly, this warns about the urgent need to reconsider the intensity and the frequency of this traditional management practice.

## 8.6  Case Studies

### 8.6.1  Post-Fire Management of Burned Forests Dominated by Arbutus unedo in Montserrat (NE Spain)

This study was conducted in the Baix Llobregat area (41° 35′ N; 1° 52′ E, Catalonia, NE Spain) in a mountain range where wildfires in 1985, 1986, 1994 burned a total of ca 5,000 ha. Altitude ranges from 390 to 500 m and climate is Mediterranean dry

**Fig. 8.7** Post-fire resprouting individual of *A. unedo* 1 year after selective thinning has been applied. Notice the three dominant resprouts retained and the new wave of resprouts appeared after thinning

sub-humid (according to the Thornthwaite index). Average temperature is 14°, and mean annual rainfall is 675 mm. Before the occurrence of wildfires, forests in the area were predominantly dominated by *Pinus halepensis* with abundant *A. unedo* individuals in the understory.

The main objective of post-fire management in this site was to explore the possibility to ameliorate the structure (i.e. increase diameter and height growth and the raise of the canopy layer) of burned areas dominated by *Arbutus unedo* resprouting individuals (Fig. 8.7). Indeed, *A. unedo* is an abundant species in many forests in the Northern rim of the Mediterranean Basin. This is an interesting species – somehow between a tree and a shrub in its growth form – that vigorously resprouts after fire. Unfortunately, while many studies have analyzed the possibility to manage burned oak forests and to convert them towards *high forest* through selective thinning of resprouts (see Espelta et al. 2003, 2008; Bonfil et al. 2004), barely any study has focused on these other Mediterranean broadleaved species. Therefore, it remains unknown whether the same practices applied to post-fire regenerated oak forests

**Table 8.3**  Mean ± SE growth in length (cm) of retained resprouts of *Arbutus uned*

| Treatment | 2008 | 2009 | 2010 | 2007–2010 |
|-----------|------|------|------|-----------|
| Control | 5.0 ± 0.2 | 4.7 ± 0.2 | 5.4 ± 0.3 | 17.5 ± 0.6 |
| T | 7.6 ± 0.3 | 7.0 ± 0.3 | 7.7 ± 0.4 | 22.0 ± 0.7 |
| T + M | 7.9 ± 0.3 | 7.4 ± 0.4 | 7.8 ± 0.4 | 22.8 ± 0.7 |

*Control* control, *T* thinning, *T + M* thinning + mechanical clearing

could be applied to *A. unedo* to enhance the structure of these communities and even to provide some revenues.

In the present study three thinning intensities were applied in the winter of 2007: (i) selective thinning (hereafter T) of resprouts (1–5 resprouts retained), (ii) selective thinning of resprouts plus mechanical clearing (hereafter T + M) of all surrounding vegetation (shrubs) in the area and (iii) control (no thinning, hereafter C). These treatments were applied randomly in 50 m × 50 m plots in four blocks distributed through the study area. After the experimental treatments were applied, height of retained resprouts was measured in 2008, 2009 and 2010 and so was basal diameter in 2007 and 2010. Mortality of resprouts was recorded in 2010 (3 years after the onset of the experiment). Concerning the new wave of resprouts that appeared after thinning (Fig. 8.7), the number of resprouts, their dominant height and surface cover was measured in 2008, 2009 and 2010.

Thinning (T) and thinning plus mechanical clearing (T + C) enhanced the growth in height and basal diameter of retained resprouts during all the 3 years monitored in comparison to non-thinned (control) individuals (Table 8.3).

Yet, individuals in the T and T + C plots exhibited a similar length at the end of the experiment. The number and height of the new resprouts appearing after thinning, were similar for the two treatments and the slightly higher surface cover of the new resprouts observed in T + C individuals during the first year of monitoring, finally disappeared at the end of the experiment. These results suggest that selective thinning of resprouts may enhance the growth of post-fire resprouted individuals of *Arbutus unedo*. Conversely, clearing of surrounding vegetation in the area does not provide additional benefits. Considering the high resprouting vigor of this species, attention must be paid to the number of resprouts retained and a number higher than one is highly recommended. Indeed, as shown in Figure 8.8 for a similar proportion of basal area removed during the thinning process, the new wave of resprouts attains higher height in those individuals with only 1 resprout retained.

### 8.6.2  Post-Fire Management of Burned Mixed Oak Forests (Q. ilex and Q. cerrioides) After the Bages-Berguedà Wildfire (NE Spain)

This experiment was conducted in mixed coppices of *Q. cerrioides* and *Q. ilex* located in central Catalonia (NE Spain), a region affected by one of the largest historically recorded wildfires in NE Spain: the Bages-Berguedà fire, which burned

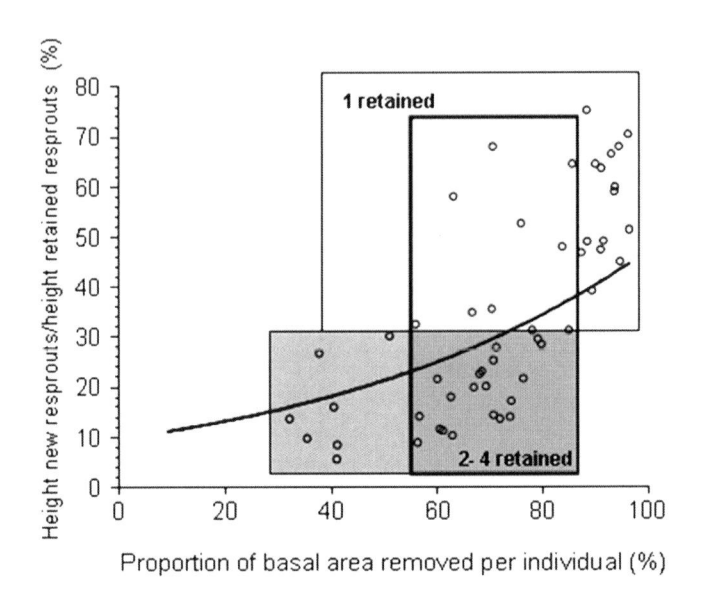

**Fig. 8.8** Relationship between the height of the new wave of resprouts and that of retained resprouts according to the proportion of basal area removed per individual after thinning in post-fire resprouting *Arbutus unedo* individuals. Individuals with *1* or *2* to *4* resprouts retained are differently highlighted. Note that for a similar proportion of basal area removed, the new wave of resprouts attains a higher height in individuals with only *1* resprout retained (JM Espelta unpubl. data)

ca. 24,300 ha in July 1994. Before this fire event, theses forests had not burned for at least the previous 80 years. The climate of the region is dry-subhumid Mediterranean (according to the Thornwaite index), with mean annual temperature of 10–13°C and an annual precipitation of 600–700 mm. According to the data provided by the Ecological Forest Inventory of Catalonia (IEFC) carried out in 1993, *Pinus nigra* forests were dominant before the fires (78% of the burned surface), with *Q. ilex* and *Q. cerrioides* being extensively present in their understory. After the fires, and due to the lack of *P. nigra* regeneration (Espelta et al. 2002), resprouting of *Quercus* transformed these areas into mixed *Quercus* forests, with the typical structure of a coppice woodland.

The main objective of this management and scientific experiment was to explore the response of these two oaks species to different treatments of selective thinning, taking into account both the growth of retained shoots and the possible appearance of new resprouts after thinning. This trial, that was initially planned as a scientific experiment to be conducted at a local scale, but was adopted and applied to a much larger surface (ca 3,000 ha) by foresters of local administrations as the Oficina Tècnica Municipal de Prevenció d'Incendis Forestals (OTMPIF, Diputació de Barcelona). In the experiment, four treatments of resprout selection and pruning were applied 9 years after the fire event to stumps of *Q. ilex* and *Q. cerrioides*: (i) selection of the largest resprout (S1), (ii) selection of the largest resprout and pruning of 40% of its lower height (S1P), (iii) selection

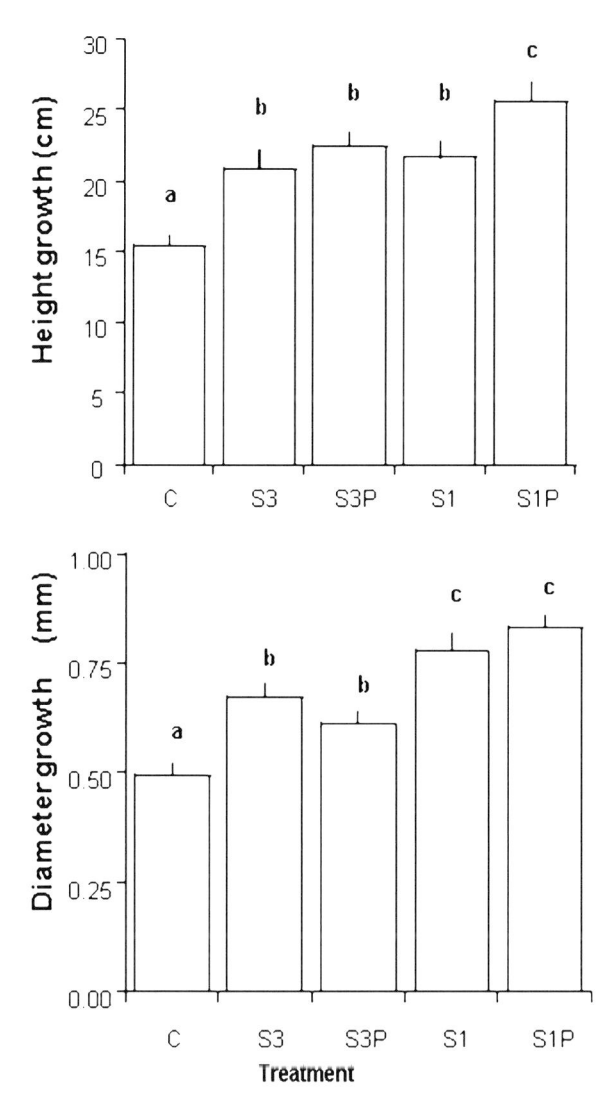

**Fig. 8.9** Height growth (**a**) and base diameter growth (**b**) of *Quercus ilex* and *Q. cerrioides* resprouts in individuals: (i) non-thinned (C), (ii) with the the three largest resprouts retained (*S3*), (iii) selection of the three largest resprouts and pruning of 40% of their height (*S3P*), (iv) selection of the largest resprout (*S1*) and selection of the largest resprout and pruning of 40% of its height (*S1P*). Different letters indicate significant differences among treatments according to the Fisher-LSD test (Modified from Espelta et al. 2008)

of the three largest resprouts (S3), (iv) selection of the three largest resprouts and pruning of 40% of their lower height (S3P), and (v) control (C), with neither selection nor pruning.

One year later, we observed that resprout selection and pruning increased growth rate in height and diameter of the dominant resprout: the highest growth was

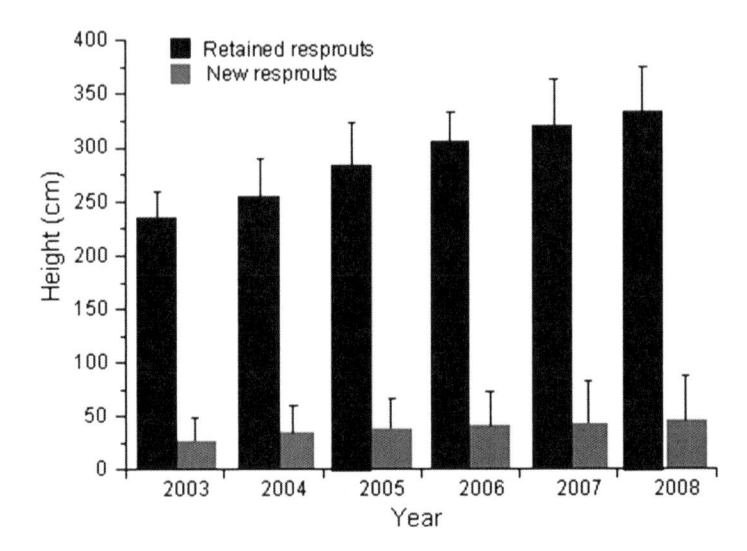

**Fig. 8.10** Mean ± SE height of resprouts (retained and new) in *Quercus ilex* individuals resprouting after a fire in 1994 with three retained resprouts (*S3*). Notice that the difference in height between retained and new resprouts progressively increases with time (JM Espelta unpubl. results)

observed in stumps with a single resprout and the lowest growth was found in control individuals (Fig. 8.9). Both pruned and non-pruned individuals showed a fairly similar response in terms of height and diameter growth. These results highlight the interest of selective thinning of oak resprouts in post-fire regenerated areas to enhance tree growth while they show that pruning does not provide additional benefits in this case. Moderate thinning (retaining three resprouts per stump) decreased the appearance of new resprouts and better controlled their growth (Fig. 8.10).

## 8.7 Key Messages

- Either evergreen or deciduous, most Mediterranean broadleaved species are able resprout after disturbances such as fire. Resprouting onset occurs after the activation of a protected bud bank located in the stump, root collar or roots.
- Resprouting is an important natural post-fire regeneration mechanism. However, forest managers must take into account that this ability may vary depending on the species considered, the intensity and recurrence of fire events, site quality and individual characteristics (e.g. plant size and probably age).
- As a consequence of the quick recovery of vegetation through resprouting, hardly ever, if there is a moderate tree density, burned areas dominated by resprouters require plantation or seeding practices.
- The rapid recovery of vegetation cover through resprouting helps to prevent soil erosion, reduce nutrient losses, and provide shelter for the recovery of the fauna.

Yet, a high density of multi-stemmed individuals in these young forests involves low growth rates and, sometimes, a high risk of the spread of new fires.

- According to different studies, the best alternative to manage burned forests regenerated through resprouting is selective thinning of resprouts to promote the gradual conversion of these forests into either *stored coppices* (i.e. coppices with a very few stems per stump) or ultimately, if sexual reproduction may be encouraged, to *high forests*.

- Selective thinning increases the growth in diameter, height and canopy cover of the resprouts retained, and it may help to shorten the reproduction onset. These benefits may even increase under harsh climatic conditions (e.g. drought episodes).

- The positive effects of selective thinning may be offset by the production of a new wave of resprouts after thinning. To control this, it is highly recommended to conduct selective thinning when dominance of some resprouts has appeared (ca. 10–12 years after post-fire resprouting) and to apply it with a moderate intensity (e.g. removing half of the basal area per individual and leaving up to 2–3 retained resprouts).

- When retained resprouts have attained a secure size, grazing by domestic livestock may help to control the new wave of resprouts.

**Acknowledgments**   We are grateful to Francisco Moreira and Filipe Catry for their comments on an earlier draft of this manuscript. Part of the research presented in this chapter is currently funded by the projects, MCIIN (CGL2008-04847-C02-02), Consolider-Ingenio Montes (CSD2008-00040) and the Bureau of Forest Fire Prevention (OTPMIF)- Diputació de Barcelona. This chapter has also benefited from discussion in the framework of the FPS COST Action FP0701 "Post-Fire Forest Management in Southern Europe".

# References

Acácio V, Holmgren M, Moreira F, Mohren GMJ (2010) Oak persistence in Mediterranean landscapes: the combined role of management, topography, and wildfires. Ecol Soc 15:40
Aerts R (1995) The advantages of being evergreen. Trees 10:402–407
Aerts R, van der Peijl MJ (1993) A simple model to explain the dominance of low-productive perennials in nutrient-poor habitats. Oikos 66:144–147
Axelrod DI (1989) Age and origin of chaparral. In: Keeley SC (ed) The california chaparral paradigms reexamined. Natural history museum of Los Angeles, Los Angeles, pp 7–19
Bellingham PJ, Sparrow AD (2000) Resprouting as a life history strategy in woody plant communities. Oikos 89:409–416
Bond WJ, Keeley JE (2005) Fire as a global 'herbivore': the ecology and evolution of flammable ecosystems. Trends Ecol Evol 20:387–394
Bonfil C, Cortés P, Espelta JM, Retana J (2004) The role of disturbance in the co-existence of the evergreen *Quercus ilex* and the deciduous *Quercus cerrioides*. J Veg Sci 15:423–430
Bowen BJ, Pate JS (1993) The significance of root starch in post-fire shoot recovery of the resprouter *Stirlingia latifolia* R. Br. (Proteaceae). Ann Bot 72:7–16
Bradshaw SD, Dixon KW, Hopper SD, Lambers H, Turner SR (2011) Little evidence for fire-adapted plant traits in Mediterranean climate regions. Trends Plant Sci 16:69–76
Calvo A, Cerdà A (1994) An example of the changes in the hydrological and erosional response of soil after forest fires, Pedralba (València). In: Sala M, Rubi JL (eds) Soil Erosion as a consequence of forest fires. Geoforma Ediciones, Logroño, pp 99–110

Calvo L, Tárrega R, Luis E (1999) Post-fire succession in two Quercus pyrenaica communities with different disturbance histories. Ann For Sci 56:441–447

Calvo L, Santalla S, Marcos E, Valbuena L, Tárrega R, Luís E (2003) Regeneration after wildfire in communities dominated by *Pinus pinaster*, an obligate seeder, and others dominated by *Quercus pyrenaica*, a typical resprouters. For Ecol Manag 184:209–223

Canadell J, López-Soria L (1998) Lignotuber reserves support regrowth following clipping of two Mediterranean shrubs. Funct Ecol 12:31–38

Cañellas I, Montero G, Bachiller A (1996) Transformation of Quejigo oak (*Quercus faginea* Lam.) coppice into high forest by thinning. Ann Ist Sperimentale Selvicultura 27:143–147

Cañellas I, del Río M, Roig S, Montero G (2004) Growth response to thinning in *Quercus pyrenaica* Willd. coppice stands in Spanish central mountain. Ann For Sci 61:243–250

Castell C, Terradas J, Tenhunen JD (1994) Water relations, gas exchange, and growth of resprouts and mature plant shoots of *Arbutus unedo* L. and *Quercus ilex* L. Oecologia 98:201–211

Catry FX, Rego F, Moreira F, Fernandes PM, Pausas JG (2010) Post-fire tree mortality in mixed forests of central Portugal For. Ecol Manag 260:1184–1192

Chabot BF, Hicks DJ (1982) The ecology of leaf life spans. Annu Rev Ecol Syst 13:229–259

Champagnat P (1989) Rest and activity in vegetative buds of trees. Ann Sci For 46(suppl):9–26

Chapin FS, Schultze ED, Mooney HA (1990) The ecology and economics of storage in plants. Annu Rev Ecol Evol Syst 21:423–447

Cotillas M, Sabaté S, Gracia C, Espelta JM (2009) Growth response of mixed Mediterranean oak coppices to rainfall reduction: could selective thinning have any influence on it? For Ecol Manag 258:1677–1683

Cruz A, Pérez B, Quintana JR, Moreno JM (2002) Resprouting in the Mediterranean type shrub *Erica australis* affected by soil resource availability. J Veg Sci 13:641–650

Cruz A, Pérez B, Moreno JM (2003) Resprouting of the Mediterranean-type shrub *Erica australis* with modified lignotuber carbohydrate content. J Ecol 91:348–356

Cutini A, Mascia V (1996) Silvicultural treatment of holm-oak (Quercus ilex L.) coppices in southern Sardinia: effects of thinning on water potential, transpiration and stomatal conductance. Ann Ist Sperimentale Selvicultura 27:47–53

Damesin C, Rambal S, Joffre R (1998) Co-occurrence of trees with different leaf habit: a functional approach on Mediterranean oaks. Acta Oecologica 19:195–204

DeBano LF, Neary DG, Ffolliott PF (1998) Fire's effects on ecosystems. Wiley, New York

Díaz-Delgado R, Lloret F, Pons X (2003) Influence of fire severity on plant regeneration by means of remote sensing imagery. Int J Remote Sens 24:1751–1763

Douville H, Chauvin F, Planton S, Royer JF, Salas-Melia D, Tyteca S (2002) Sensitivity of the hydrological cycle to increasing amounts of greenhouse gases and aerosols. Clim Dyn 20:45–68

Ducrey M, Boisserie M (1992) Recrû naturel dans des taillis de chêne vert (Quercus ilex L.) a` la suite d'explotations partielles. Ann Sci For 49:91–109

Ducrey M, Turrel M (1992) Influence of cutting methods and dates on stump sprouting in holm oak (*Quercus ilex* L.) coppice. Ann Sci For 49:449–464

EEA (2006). European forest types: categories and types for sustainable forest management reporting and policy. EEA technical report no 9, European Environmental Agency, Rome

El Omari B, Aranda X, Verdaguer D, Pascual G, Fleck I (2003) Resource remobilization in *Quercus ilex* L. resprouts. Plant Soil 252:349–357

Espelta JM, Sabaté S, Retana J (1999) Resprouting dynamics. In: Rodà F, Retana J, Gracia CA, Bellot J (eds) Ecology of Mediterranean evergreen oak forests. Springer-Verlag, Berlin, pp 61–73

Espelta JM, Rodrigo A, Habrouk A, Meghelli N, Ordoñez JL, Retana J (2002) Land use changes, natural regeneration patterns and restoration practices after a large wildfire in NE Spain: challenges for fire ecology and landscape restoration. In: Trabaud L, Prodon R (eds) Fire and biological processes. Backhuys Publishers, Leiden, pp 315–324

Espelta JM, Retana J, Habrouk A (2003) Resprouting patterns after fire and response to stool cleaning of two coexisting Mediterranean oaks with contrasting leaf habits on two different sites. For Ecol Manag 179:401–414

Espelta JM, Cortés P, Mangirón M, Retana J (2005) Differences in biomass partitioning, leaf nitrogen content and water use efficiency (δ13C) result in a similar performance of seedlings of two Mediterranean oaks with contrasting leaf habit. Ecoscience 12:447–454

Espelta JM, Retana J, Habrouk A (2006) Response to natural and simulated browsing of two Mediterranean oaks with contrasting leaf habit after a wildfire. Ann For Sci 63:441–447

Espelta JM, Arnan X, Verkaik I (2008) Evaluación ecológica de diferentes tratamientos silvícolas de mejora de la regeneración natural en zonas afectadas por incendios y sequías extremas. – In: Modelos silvícolas en bosques privados Mediterraneos. Colección_Documentos de Trabajo, Serie_Territorio, 5. Diputación de Barcelona, Barcelona, pp. 151–179

Fleck I, Hogan KP, Llorens L, Abadía A, Aranda X (2010) Photosynthesis and photoprotection in *Quercus ilex* resprouts after fire. Tree Physiol 18:607–614

Garcia-Villanueva J, Ena V, Tàrrega R, Mediavilla G (1998) Recolonization of two burned *Quercus pyrenaica* ecosystems by Coleoptera. Int J Wildland Fire 8:21–27

Gracia C, Burriel JA, Ibáñez JJ, Mata T, Vayreda J (2000) Inventari ecològic i forestal de Catalunya. Regió forestal IV. Centre de Recerca Ecològica i Aplicacions Forestals, Barcelona

Hanes T (1971) Succession after fire in the chaparral of southern California. Ecol Monogr 41:27–42

Hodgkinson KC (1992) Water relations and growth of shrubs before and after fire in a semi-arid woodland. Oecologia 90:467–473

Hulme PE (1997) Post-dispersal seed predation and the establishment of vertebrate dispersed plants in Mediterranean shrublands. Oecologia 111:91–98

IPCC (2007) Climate change 2007: the physical science basis. Contribution of working group I to the fourth assessment report of the intergovernmental panel on climate change, Cambridge University Press, Cambridge

Janzen DH, Martin PS (1982) Neotropical anachronisms: the fruits the *Gomphotheres* ate. Science 215:19–27

Jaquet K, Prodon R (2009) Measuring the post-fire resilience of a bird-vegetation system: a 28-year study in a Mediterranean oak woodland. Oecologia 161:801–811

Keeley JE, Zedler P (1978) Reproduction of chaparral shrubs after fire: a comparison of sprouting and seedling strategies. Am Midl Nat 99:142–161

Klimešovà J, Klimeš L (2007) Bud banks and their role in vegetative regeneration – a literature review and proposal for simple classification and assessment. Perspect Plant Ecol Evol Syst 8:115–129

López-Soria L, Castell C (1992) Comparative genet survival after fire in woody Mediterranean species. Oecologia 53:493–499

Mazzoleni S, Spada F (1992) Deciduous broadleaved versus evergreen sclerophyllous forest. Disturbance and local shifting dominance in Mediterranean environments. In: Teller A, Mathy P, Jeffers JNR (eds) Responses of forest ecosystems to environmental changes. Elsevier, Science Publishers, Amsterdam, pp 839–842

Miyanishi K, Kellman M (1986) The role of root nutrient reserves in regrowth of two savanna shrubs. Can J Bot 64:1244–1248

Moreira F, Catry F, Lopes T, Bugalho M, Rego F (2009) Comparing survival and size of resprouts and planted trees for post-fire forest restoration in central Portugal. Ecol Eng 35:870–873

Myers N, Mittermeier RA, Mittermeier CG, da Fonseca GAB, Kent J (2000) Biodiversity hotspots for conservation priorities. Nature 403:853–858

Obón B (1997) Recuperació de la vegetació després del gran incendi del estiu de 1994 a Gualba (Vallès Oriental). Master dissertation, University of Lleida, Lleida

Pascual G, Molinas ML, Verdaguer D (2002) Comparative anatomical analysis of the cotyledonary region in three Mediterranean Basin *Quercus* (Fagaceae). Am J Bot 89:383–392

Pausas JG (2004) Changes in fire and climate in the eastern Iberian Peninsula (Mediterranean Basin). Clim Change 63:337–350

Peguero G, Espelta JM (2011) How disturbance intensity and seasonality affect the resprouting ability of the dry tropical tree *Acacia pennatula*: do resources stored below-ground matter? J Trop Ecol 27:539–546

Peña-Rojas K, Aranda X, Joffre R, Fleck I (2005) Leaf morphology, photochemistry and water status changes in resprouting *Quercus ilex* during drought. Funct Plant Biol 32:117–130

Piñol J (1998) Climate warming wildfire hazard and wildfire occurrence in coastal Eastern Spain. Clim Chang 38:345–357

Quevedo L, Rodrigo A, Espelta JM (2007) Post-fire resprouting ability of 15 non-dominant shrub and tree species in Mediterranean areas of NE Spain. Ann For Sci 64:883–889

Quézel P, Médail F (2003) Écologie et biogéographie des forêts du bassin méditerranéen. Elsevier SAS, Paris

Retana J, Riba M, Castell C, Espelta JM (1992) Regeneration by sprouting of holm oak (*Quercus ilex*) stands exploited by selection thinning. Vegetation 99(100):355–364

Riba M (1998) Effects of intensity and frequency of crown damage on resprouting of *Erica arborea* L. (Ericaceae). Acta Oecol 19:9–16

Riera-Mora S, Esteban-Amat A (1994) Vegetation history and human activity during the last 6000 years on the central Catalan coast (north-eastern Iberian Peninsula). Veg Hist Archaebot 3:7–23

Rodrigo A, Retana J, Picó X (2004) Direct regeneration is not the only response of Mediterranean forests to large fires. Ecology 85:716–729

Sánchez-Humanes B, Espelta JM (2011) Increased drought reduces acorn production in *Quercus ilex* coppices: thinning mitigates this effect but only in the short term. Forestry 84:73–82

Sardans J, Peñuelas J (2007) Drought changes phosphorus and potassium accumulation patterns in an Evergreen Mediterranean forest. Funct Ecol 21:191–201

Saura-Mas S, Lloret F (2009) Linking post-fire regenerative strategy and leaf nutrient content in Mediterranean woody plants. Perspect Plant Ecol Evol Syst 11:219–229

Serrada R, Bravo A, Sánchez I, Allué M, Elena R, San Miguel A (1996) Conversion into high forest in coppices of *Quercus ilex* subsp. *ballota* L. in central region of Iberian Peninsula. Ann Ist Sperimentale Selvicultura 27:149–160

Terradas J (1999) Holm oak and holm oak forests: an introduction. In: Rodà F, Retana J, Gracia CA, Bellot J (eds) Ecology of Mediterranean evergreen oak forests. Springer-Verlag, Berlin, pp 3–14

Trabaud L (1987) Natural and prescribed fire: survival strategies of plants and equilibrium in mediterranean ecosystems. In: Tenhunen JD, Catarino FM, Lange OL, Oechel WC (eds) Plant response to stress. Functional analysis in Mediterranean ecosystems. NATO ASI series, ecological sciences, 15. Springer-Verlag, Berlin, pp 607–621

Verdaguer D, García-Berthou E, Pascual G, Puigderrajols P (2001) Sprouting of seedlings of three Quercus species (Q. humilis Miller, Q. ilex L., and Q. suber L.) in relation to repeated pruning and the cotyledonary node. Aust J Bot 49:67–74

Wang G (2005) Agricultural drought in a future climate: results from 15 global climate models participating in the IPCC 4th assessment. Clim Dyn 25:739–753

Westman WE (1986) Resilience: concepts and measures. In: Dell B, Hopkins AJM, Lamont BB (eds) Resilience in mediterranean-type ecosystems. Dr W Junk Publishers, Dordrecht, pp 5–19

# Chapter 9
# Post-Fire Management of Cork Oak Forests

**Filipe X. Catry, Francisco Moreira, Enrique Cardillo, and Juli G. Pausas**

## 9.1 Ecological and Socio-economic Context

Cork oak (*Quercus suber* L.) forests are defined here as the range of habitats from open savanna-like woodland formations to dense forests. According to the European forest type's nomenclature (EEA 2007) these ecosystems are included in the 'broadleaved evergreen forest' class and in the 'Mediterranean evergreen oak forest' type. This forest type is dominated by the evergreen sclerophyllous oak species *Q. suber*, *Q. ilex*, *Q. rotundifolia* and *Q. coccifera*, constituting the main natural forest formation of the meso-Mediterranean vegetation belt (EEA 2007). However, cork oak has a unique characteristic that makes it different from all the other Mediterranean broadleaved species: an outer insulating coat consisting of a corky bark, up to 30 cm thick, made of continuous layers of *suberized* cells that may have evolved as an adaptation to fire, and that has been used by people for millennia (Natividade 1950; Pausas et al. 2009). Periodical bark harvesting of cork oak trees makes them more vulnerable to external agents including wildfires. This is why cork oak forests are treated separately in this book.

F.X. Catry (✉) • F. Moreira
Centre for Applied Ecology, School of Agriculture (ISA), Technical University of Lisbon, Lisbon, Portugal
e-mail: fcatry@isa.utl.pt

E. Cardillo
Institute for the Wood, Cork and Vegetable Coal (IPROCOR), Polígono Industrial El Prado, Badajoz, Spain

J.G. Pausas
Desertification Research Centre, Spanish National Research Council (CIDE-CSIC), Valencia, Spain

F. Moreira et al. (eds.), *Post-Fire Management and Restoration of Southern European Forests*, Managing Forest Ecosystems 24, DOI 10.1007/978-94-007-2208-8_9,
© Springer Science+Business Media B.V. 2012

Nowadays, cork oak ecosystems cover nearly 2.5 million hectares of land in the western Mediterranean Basin. They can be found in southern Europe and North Africa, from the Iberian Peninsula and Morocco to the western rim of the Italian Peninsula (Fig. 9.1), occurring in a wide range of ecological conditions (APCOR 2009; Pausas et al. 2009). Cork oak trees show a high ecological plasticity. This species is well adapted to Mediterranean type climate, with mild, wet winters and dry, hot summers, occurring from more continental regions to coastal areas with Mediterranean and Atlantic influence. It grows well with mean annual precipitation of 600–1,000 mm, but stands up to 2,000 mm, 500 mm being the minimum usually considered for a balanced tree development (Natividade 1950; Pereira 2007). The optimum mean annual temperature is in the range 13–16°C, although the species can also occur in environments with up to 19°C. Cork oak grows from sea level to 2,000 m of altitude, but optimum growth occurs below 600 m. The species is tolerant to a variety of soils with the exception of calcareous and limestone substrates. It may grow on poor and shallow soils, with low nitrogen and organic matter content and it allows a pH range between 4.8 and 7.0. However, cork oak occurs preferentially in siliceous and sandy soils, preferring deep well aerated and drained soils, being very sensitive to compaction and water logging (Bernal 1999; Pereira 2007).

Most of the present distribution and physiognomy of cork oak forests is the result of an ancient anthropogenic alteration by clearance, coppicing, fires and overgrazing (e.g. EEA 2007), but also reforestation (plantation or seeding). A characteristic physiognomy of these ecosystems in the Iberian peninsula, found also locally elsewhere (Balearic islands, Sardinia), are savanna-like formations (known as *montado* in Portugal and *dehesa* in Spain) in which crops, pasture land or shrublands are shaded by a fairly closed to very open tree canopy (EEA 2007; Fig. 9.2). More rarely, denser cork oak forests can also be found, particularly in steep slopes and mountainous regions (Fig. 9.8).

Cork oak ecosystems play a very important ecological, economic and social role in several Mediterranean countries (e.g. Pereira and Fonseca 2003; Bugalho et al. 2011). Due to their uniqueness, these ecosystems are recognized as habitats of conservation value listed in the Habitats Directive: Habitat 6310 – *Dehesa*s with evergreen *Quercus* spp. and Habitat 9330 – *Quercus suber* forests (EEC 1992).

Cork oak ecosystems support a large variety of animal, plant and fungi species, including many endemisms (e.g. Bernal 1999). They have remarkable ecological value, providing habitat for several threatened species such as the Imperial eagle *Aquila adalberti*, the black vulture *Aegypius monachus* or the critically endangered Iberian lynx *Linx pardinus* (IUCN 2010).

Plant species composition depends on the ecological characteristics of each region and anthropogenic interventions. In southern Europe, and particularly in the Iberian Peninsula, mixed forests of cork oak and other oaks (*Q. rotundifolia*, *Q. ilex*, *Q. faginea*, *Q. robur*, *Q. pyrenaica*, *Q. canariensis*, *Q. coccifera* and *Q. lusitanica*) can be found. In France and Italy, other oak species, such as *Q. pubescens* and *Q. cerris* can be found; noteworthy are also savanna-like formations of *Q. suber* and *Q. congesta* in Sardinia (EEA 2007).

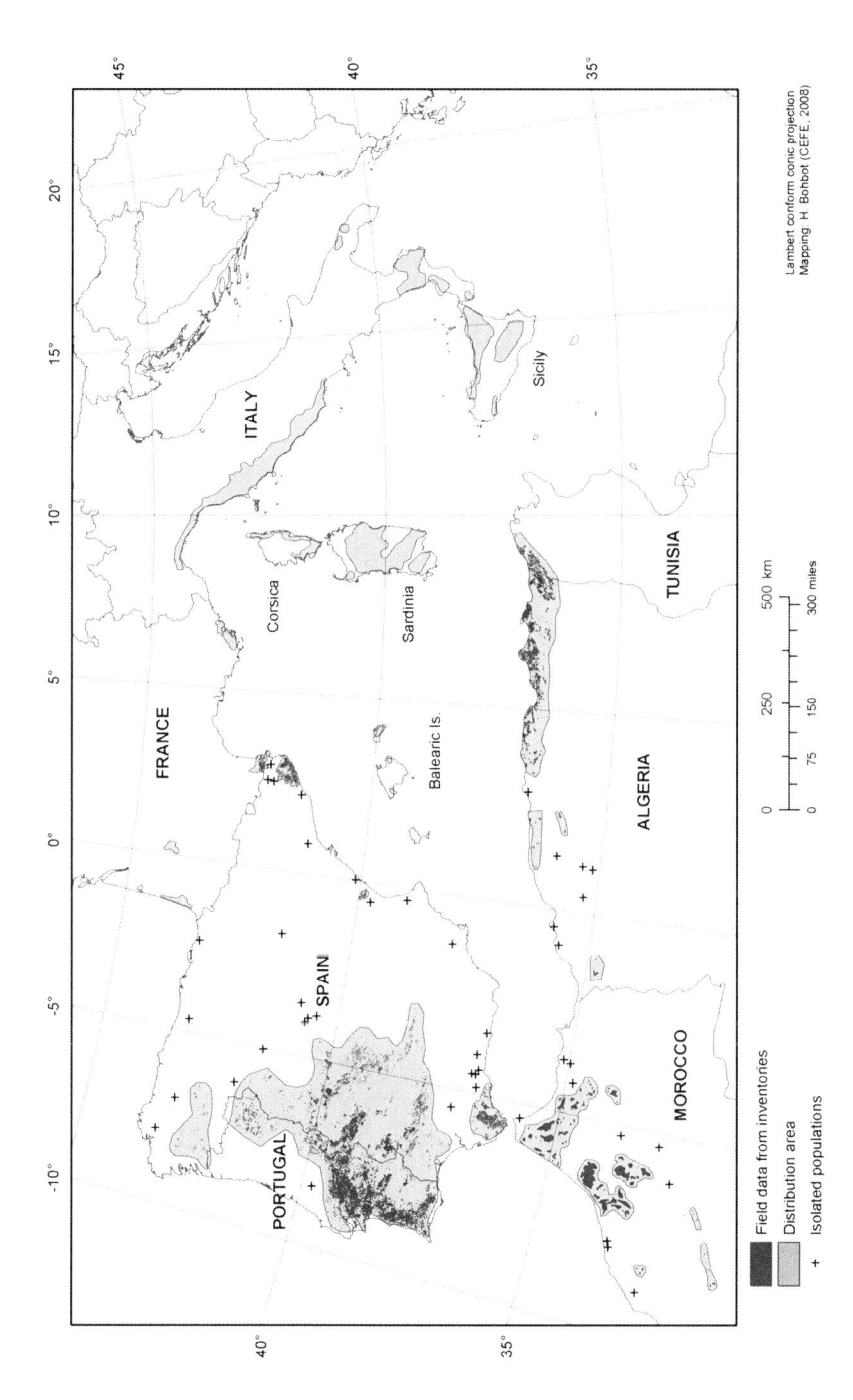

**Fig. 9.1** Actual distribution of cork oak forests in the western Mediterranean Basin. Reproduced from Pausas et al. (2009) (copyright © 2009 Island Press; reproduced by permission of Island Press, Washington, DC)

**Fig. 9.2** Cork oak forest (savanna-like formation, *left*) and detail of the main trunk and bark of a virgin cork oak tree (*right*) (Photos: F. Catry)

The structure of the more preserved cork oak forests includes a very dense tree cover up to 20 m high, often mixed with other Mediterranean broadleaved species (sometimes with conifers), and with shade-tolerant herbaceous species in the understory (Bernal 1999). Besides oaks, other small trees and large shrubs can coexist in these forests, such as *Arbutus unedo*, *Myrtus communis*, *Olea europaea* var. *sylvestris*, *Pistacia lentiscus*, *P. terebinthus*, *Crataegus monogyna*, *Viburnum tinus*, *Phillyrea angustifolia*, *P. latifolia*, *Rhamnus alaternus* and *Erica arborea*, among others. In more open forests, subjected to over-grazing, wildfires or with poor soils, other plants appear more often, such as species of the genera *Cistus*, *Cytisus*, *Erica*, *Genista*, *Ulex*, *Lavandula* and *Rosmarinus*, among others. Dominance of herbaceous species, such as *Agrostis*, *Brachypodium* or *Festuca*, is characteristic of more degraded woodlands (Bernal 1999).

Cork is a renewable natural resource constituting a valuable and versatile raw material for industry used for a large variety of products. Because of its economical value, cork oak silviculture is usually oriented towards periodical cork harvesting (Pereira 2007). Currently, cork is the second most important marketable non-wood forest product in the western Mediterranean, and the world cork market exports represent near US$2 billion annually (Mendes and Graça 2009; APCOR 2009). This species is particularly important in the Iberian Peninsula, which holds about 55% of the world's cork oak area and 82% of the world's cork production, representing thousands of jobs (Silva and Catry 2006).

However, despite of their value, several factors such as pests and diseases, over-harvesting, over-grazing and land use changes, are endangering *Q. suber* forests. These threats, exacerbated by climate change, affect tree health and increase vulnerability to wildfires (e.g. WWF 2007).

**Fig. 9.3** After a wildfire (photos *above*) cork oak often starts regenerating quickly; photos *below* show totally charred trees with crown (epicormic) regeneration three months after fire (*left*), and 16 months after fire (*right*; the tree in background resprouted after fire but died some months later) (Photos: F. Catry)

## 9.2  Post-Fire Cork Oak Regeneration Strategies

Most of Mediterranean broadleaved species have the capacity to resprout after disturbances, including wildfires, and most of them resprout from basal buds when stems or crowns are severely damaged. Similarly to other oaks, post-fire cork oak recovery occurs mainly through vegetative regeneration. However, cork oak is the only European tree with the capacity to resprout from epicormic buds (i.e. buds positioned underneath the bark) high on the tree (Fig. 9.3), a feature shared with many *Eucalyptus* species and the Canary Island pine (*Pinus canariensis*) but otherwise rare (Pausas et al. 2009). The insulating bark of cork oak, when sufficiently thick (see Sect. 9.3), protects the epicormic buds, permitting trees to resprout quickly and effectively from stem and crown buds after fire. Because of this feature, cork oak is undoubtedly one of the tree species best adapted to persist in recurrently burned ecosystems. The post-fire tree survival is often high and the regeneration of cork oak-dominated landscapes is remarkably quick (Silva and Catry 2006). The fact that cork oak can regenerate after fire from epicormic buds gives this species a competitive advantage over coexisting woody plants. Together with its socio-economic importance and cultural significance, this extraordinary resprouting capacity makes

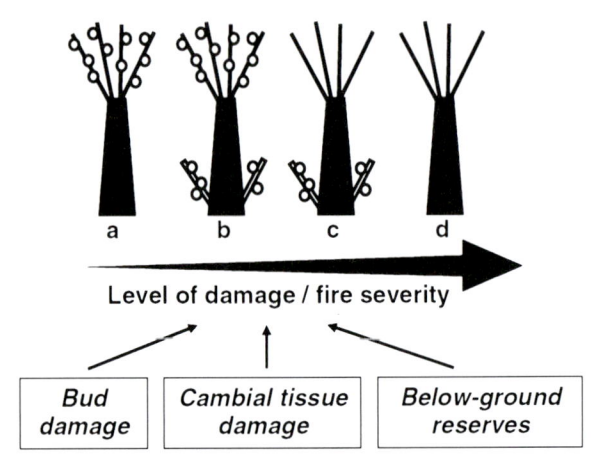

**Fig. 9.4** A conceptual model of post-fire responses of a sprouting tree that suffered total crown consumption (combustion of leaves and twigs during a wildfire) in relation to a gradient of increasing damage/fire severity: (**a**) crown resprouting, (**b**) resprouting from both crown and base, (**c**) basal resprouting, (**d**) plant death (reproduced from Moreira et al. 2009)

**Fig. 9.5** Jay (*Garrulus glandarius*) (*left*) is the main natural dispersal agent of cork oak acorns (*right*) (Photos: F. Catry)

the cork oak a very good candidate for reforestation programs in fire-prone areas (Pausas et al. 2009).

The post-fire cork oak responses are usually a function of the level of damage (fire severity). A conceptual model of vegetative tree responses was proposed by Moreira et al. (2009). At low levels of damage, a tree is expected to resprout from crown buds that survive the fire. At increasing levels of damage, the individual will resprout from both crown and base, just from the base, or will die (Fig. 9.4).

Cork oaks can also regenerate through seeds (acorns) during the inter-fire period (Pons and Pausas 2007), but rarely just after wildfires as acorns are usually destroyed. However, an increase in oak recruitment may occur not long after fire in areas where jays (*Garrulus glandarius*), the main oak dispersal agent, are abundant (Fig. 9.5). Post-fire conditions are suitable for jays to disperse acorns before the soil is covered by shrubs. A pair of jays may scatter and hoard several thousand acorns in a single season (Cramp 1994).

## 9.3   Factors Affecting Post-Fire Cork Oak Responses

### 9.3.1   Influence of Bark Thickness, Bark Exploitation and Tree Size

Previous research showed that bark thickness is a main driver of cork oak responses after fire (Catry et al. 2009, 2010a, b; Moreira et al. 2007, 2009; Pausas 1997). Tree vulnerability to fire significantly decreases with increasing bark thickness until bark reaches about 4 cm thick. Cork oak trees with bark more than 3–4 cm thick are well protected against heat injury having a very low probability of dying or suffering stem mortality (Fig. 9.6). Particularly in what concerns stem mortality it is noteworthy that for bark thickness lower than 3 cm, cork oak is apparently more fire resistant than other Mediterranean broadleaved species (Catry et al. 2010a). This can be explained by the high thermal insulating provided by cork, due to its high proportion of air and low density (Pereira 2007).

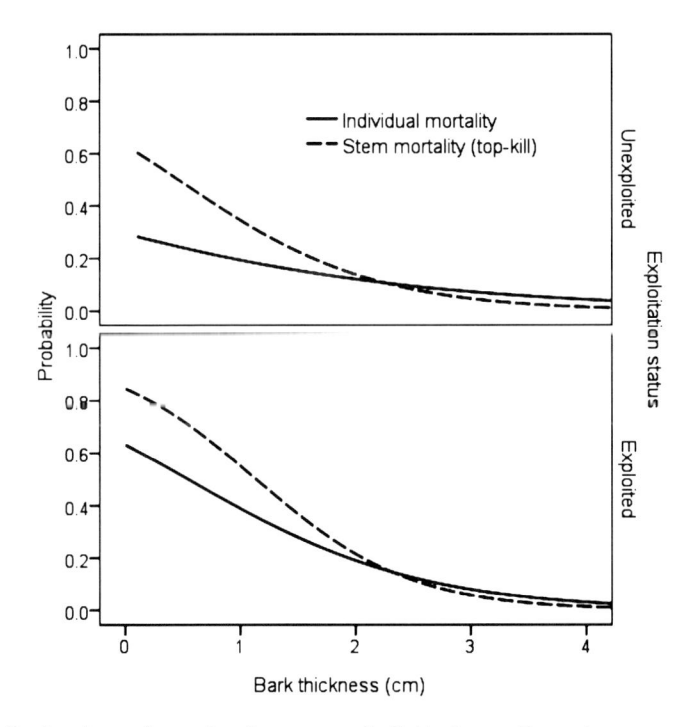

**Fig. 9.6** Predicted post-fire cork oak responses (individual mortality and stem mortality) as a function of bark thickness and cork management status (exploited for cork *versus* unexploited) for trees with 20 cm d.b.h. (diameter at breast height) (Catry et al. unpublished)

Cork harvesting does not only drastically reduce bark thickness, but it also has additional effects. Cork exploitation per se has been found to significantly increase tree vulnerability to fire (Fig. 9.6), with mortality being up to 40% higher on exploited trees, even for individuals with the same bark thickness (Catry et al. unpublished, Moreira et al. 2007). Debarking is a major stress factor for trees and has been associated to vigor loss (e.g. Natividade 1950). Bark extraction leads to considerable water losses through the stripped trunk surface which may negatively affect the trees photosynthetic activity and productivity (Correia et al. 1992). The injuries caused by cork harvesting operations can also be associated to loss of tree vigor (Costa et al. 2004). In fact, wounded trees were found to be less fire-resistant than undamaged trees (Catry et al. unpublished). Wounded trees are more vulnerable because bark is usually absent or much thinner near wounds making the trunk more heat-sensitive and more vulnerable to other external agents (Miller 2000). Wounding is also likely to reduce tree vigor, both because of the energy resources needed for cicatrisation and because the active xylem killed reduces the rate of water absorption (Rundel 1973). Additionally, the changes induced by stress reduce the trees ability to defend themselves from insect or fungi attacks (Wargo 1996).

Previous studies also indicate that larger trees (higher d.b.h., and usually older) are more vulnerable to fire damages than smaller trees (Catry et al. 2009; Moreira et al. 2009). Lower fire resistance of larger trees can be explained by the fact that older individuals were debarked more times during its life and were probably subjected more often to poor management practices (e.g. deep ploughing, excessive pruning or stripping damages), thus being less vigorous (Natividade 1950). For example in Sardinia, Barberis et al. (2003) reported that cork oaks stripped more often had higher post-fire mortality (~37%) than trees debarked only once (~17%).

## 9.3.2 Influence of Fire Regime and Local Factors

The fire regime, particularly fire intensity, severity, frequency and fire season, can also exert determinant effects on post-fire tree responses. The first two components can be evaluated through potential indicators of fire injury, such as the char height, char depth or the crown volume damaged. Previous studies showed that cork oak vulnerability to fire significantly increases with increasing char height (Catry et al. 2009; Moreira et al. 2007), as it happens with other species (e.g. Catry et al. 2010a).

There is very few information available on the effects of fire frequency and returning intervals on cork oak, but it is expected that increasing fire frequency will negatively affect tree resistance to fire, as suggested in a study in southern France (Curt et al. 2010). Similarly, the effects of fire season on post-fire cork oak responses were rarely evaluated. In a recent study (Catry et al. unpublished), trees burned earlier in the summer were found to be more likely to die than those burning later, which could be explained by seasonal variations in plant phenology. In spite of contradictory reports in the literature, several studies showed that plants are more vulnerable to fire damage when they are flowering or actively growing

(DeBano et al. 1998). Although cork oak is an evergreen species, the main growing and flowering periods occurs during spring and early summer, with the maximum stomatal conductance and transpiration rates occurring from March to June (Oliveira et al. 1992). Thus, the tree carbohydrate reserves are expected to be at a low level during this period, and the actively growing tissues are more susceptible to heat damages, which may increase fire vulnerability.

Previous studies also found that trees located in southern aspects are more vulnerable to fire (Catry et al. 2009; Moreira et al. 2007). In the Mediterranean, south-facing aspects are typically dryer and warmer and have less vegetation cover and a thinner soil layer (Kutiel and Lavee 1999; Sternberg and Shoshany 2001), being also more vulnerable to soil erosion (Marques and Mora 2003). Additionally, some of the more important insects and diseases affecting cork oak have been reported to have higher incidence on south-facing slopes (Du Merle and Attié 1992; Moreira and Martins 2005). All these unfavorable conditions are likely to increase tree stress and consequently increase vulnerability to wildfires.

## 9.4 Post-Fire Management Issues and Alternatives

Although cork oak is known as a fire-resistant and resilient species, wildfires can cause major economic and ecological impacts on cork oak ecosystems. A particular concern exists if trees are exploited for cork production, which is the situation in most cases.

Usually the first bark harvest occurs when tree d.b.h. reaches 19–22 cm (20–40 years old), with subsequent yields at 9–15 year intervals, meaning that a tree can be stripped about 12–20 times during its productive lifetime (150–200 years, although cork oak can live up to 500 years; Natividade 1950; Pereira 2007). The risk of fire damage in exploited trees is at its highest level just after bark harvesting and then it will decrease with time until cork reaches about 3–4 cm thick (see Sect. 9.3), which usually occur at the end of the stripping cycle. This means that most of the time trees face a considerable risk from wildfires, and managers should be aware of it.

### 9.4.1 Defining Management Objectives

After fire, it is important to define the management objectives and to plan the restoration actions accordingly. In general, the most common objective for burned cork oak stands is to restore cork production as soon as possible.

The post-fire management alternatives in cork oak forests will largely depend on fire severity, thus a multidisciplinary damage assessment should be performed first to identify the direct and indirect economic and ecological impacts and risks (see also Chaps. 1 and 5).

After a wildfire, a strong negative economic impact is expected, both because the charred bark looses its value and productivity decreases. The minimum time required to start extracting good quality cork again (i.e. cork that can be used for stoppers) will be about 40 years for trees that died and need to be replaced, 30 years for the surviving trees with stem mortality, and 10 years for trees with good crown regeneration. At the ecosystem level the more common ecological consequences of fire include factors such as: decrease of tree cover and vigor, decrease acorn production reducing the regeneration potential and food for live-stock and wildlife, decrease carbon, nutrients and water retention, and increase soil erosion risk. All these economic and ecological issues should be considered when defining the post-fire management objectives and evaluating the possible alternatives to achieve them.

After the evaluation of the fire impacts and associated risks, the burned area can be divided into units or blocks with homogeneous characteristics. Then, the prescriptions for each management unit should take into account the urgency, resource value, and success possibilities.

## 9.4.2 Current Post-Fire Management Practices

Management practices in burned cork oak forests can be quite variable from one region to another, depending on managers' objectives and perception of fire impacts, and on available funds. Here we briefly present some of the more common practices.

Usually the decision to cut or not cork oak trees after fire is mostly dependent on field assessments of fire severity and on the cork age. Burned trees with younger (thinner) cork bark (i.e. < 4 years old) or having severe inner bark damages are not expected to recover the crown and are logged, while trees with thicker cork in most cases are left to regenerate.

When trees are not expected to show adequate post-fire crown recover, and in order to make use of their basal sprouting capacity, the official recommenda-tions (in Portugal; DGRF 2006) are that younger trees (less than 40 years, or perimeter at breast height less than 90 cm) should be cut as soon as possible, preferably before the next growing period (i.e. end of following winter) to increase resprouting vigor. Actions to manage basal sprouts include shoot selec-tion, clearing of shrub or herbaceous vegetation, and avoiding animal browsing. Older trees (over 60 years) are assumed not to originate economically interest-ing resprouts, and are often uprooted and replaced by new trees (seeding or planting). In both natural and artificial regeneration, thinning and shoot selec-tion are usually carried out.

The cut material is either removed from the site, or logs and branches are left on the ground. In some cases, groups of trees or individuals that are less damaged and that can contribute to post-fire regeneration are maintained.

The decision to plant or seed after fire is mainly based on the existence of financial incentives (and market value), and occurrence of scarce post-fire natural regeneration. Active seeding or planting are also both carried out to increase tree density, usually in the period of 1–3 years after fire.

In Portugal and Spain, there are several legal issues related to the post-fire management of burned cork oak stands. First of all, the species is protected by law, thus official permission is needed to cut trees and the land cover cannot be changed after fire. Secondly, the cork of trees with d.b.h. smaller than 19–22 cm cannot be extracted. Thirdly, although cork cannot be extracted before 9 years after the previous extraction, some exceptions are allowed, including the case of burned trees (see Sect. 9.4.6).

## 9.4.3   Tree Logging

Cork oak trees that died or suffered stem mortality as a consequence of fire can be logged (after getting a permit). In some cases trees showing poor crown regeneration, and particularly those with severe stem damages, can also be logged (Fig. 9.7; see also Sect. 9.5.1).

From a silvicultural point of view, the most interesting cuttings are those aimed to take advantage of the remarkable resprouting capability of cork oaks. Sprouts originating from dormant buds at or near the base of severely damaged trees can be used to regenerate forest stands (see Sect. 9.4.4). The snag reduces sprouting energy and provokes the leaning of sprouts (Barberies et al. 2003). Dormant buds from stumps near or under the soil surface have better chances to survive than buds located higher in a rotting trunk; therefore *liberation cuttings* should be done as soon as possible after fire and lower as possible in the trunk (Cardillo et al. 2007). Trunk cuttings should be made horizontally or slightly inclined, leaving a smooth surface (DGRF 2006).

Sometimes cuttings can also be done for sanitary reasons. Burned cork oaks are exposed to the attack of pests such as ambrosia beetles *Platypus cilindrus* and *Xyleborus monographus* (see Sect. 9.4.8). Rarely the presence of these wood borers is a threat to the nearest forest stands but if their populations increase to outbreak proportions, sanitary cuttings and burning are recommended (Sousa and Inácio 2005). Logging can also be needed for security reasons; trees with seriously damaged trunks located close to buildings and roads can be wind thrown, thus selective cutting should be allowed. In some cases, and depending on management objectives, dead trees can also be left standing or the wood can remain in the ground to increase biodiversity.

Usually the wood from coppiced cork oaks can be only used as firewood or good quality charcoal; thus a market for this wood exists, particularly in the forests near charcoal kilns. In this case salvage logging, with subsequent debarking, is possible. Otherwise cuttings are a net expense and only can be thought as a silvicultural treatment.

**Fig. 9.7** Post-fire cork oak management: selective logging of most severely damaged trees (*left*), and shrub clearing 20 years after fire (*right*) avoiding soil ploughing (Extremadura, Spain) (Photos: F. Catry)

## 9.4.4 Assisting Natural Regeneration

In most cases, if trees were not recently debarked before the fire, burned cork oaks will show vegetative regeneration (i.e. resprouting; Fig. 9.8). When crown resprouts homogeneously, usually no interventions are required. Otherwise, if crown regeneration is absent or is very poor, basal sprouts are a viable way to regenerate cork oak stands, and this method is considerably faster, more effective and cheaper than seeding or planting (see Sect. 9.4.5). Stool sprouts and root sprouts are not frequent in cork oak but they have not silvicultural value since they originate from adventitious buds (Johnson et al. 2009).

A few years after cutting many sprouts have often crowded the stump and begin competing each other, thus thinning is highly recommended (see Chap. 8). One to three of the most vigorous sprouts per stump could be retained depending on stump diameter. Sufficiently spaced trunks (at least 40 cm) could be debarked easily in the future (Cardillo et al. 2007). Sprouts well inserted into the stump below soil surface are best joined to roots and should be preferred instead of those attached to higher parts of stump and exposed to rot. Early pruning is not recommended because sprout canopy helps to control excessive undesirable resprouting (Johnson et al. 2009).

Natural regeneration from seeds is much less common because acorns and flowers are destroyed by fire in most cases, and even if the crowns survive, trees will take at least 2 or 3 years to produce acorns again. The habitual year to year and tree to tree variations in acorn production, the activity of seed-dispersal agents and the action of herbivores and drought over seedlings, are the main factors affecting regeneration from seeds. In addition, seedlings usually need many years to establish and develop. The shelterwood method provides light, shelter and more recruitment while acorn and cork production are in part conserved.

Exclusion from livestock and other herbivores may be required to minimize the negative impacts on natural regeneration (see Sect. 9.4.7). Latter thinning, seeding or planting can also be needed in case of uneven spatial regeneration.

**Fig. 9.8** Cork oak forests post-fire natural regeneration: 16 months after fire (*above*; Algarve, south Portugal) and about 20 years after fire (*below*; Extremadura, west Spain) (Photos: F. Catry)

## 9.4.5   Seeding and Planting

Before planning reforestation actions in burned cork oak stands the presence of natural regeneration should be checked carefully. However, when natural regeneration is not enough to achieve the objectives (in terms of the desired tree density), reforestation by direct seeding or planting is an alternative. The main limitations when using these techniques are the availability of quality seeds, acorn or seedling predation, and summer drought.

Sometimes the number of acorns is not enough because of insufficient production or excessive predation. Mice are efficient in detecting and consuming acorns (although they can also act as short-distance dispersers especially in mast years; Pons and Pausas 2007). Sowing tests can be done in order to evaluate their presence and, if they are present, acorns can be protected with small tree shelters. Wild boars (*Sus scrofa*) are frequent in forested areas and they are able to consume large quantities of acorns. In this case shelters are not effective in protecting acorns against them, but well maintained electric fences can be very effective in relatively small areas. Large herbivores can also exert a negative impact on seedlings, thus protective measures should be taken when they are present (see Sect. 9.4.7).

Seeding season is also a very important issue. On one hand, early seeding in autumn will expose acorns to predation during winter, when food supply in the burned area is reduced, thus, higher success rates can be achieved if seeding is performed in the early spring, after recovery of grasses and shrubs. On the other hand, the summer drought is the main cause of seedling mortality in Mediterranean forests (Cortina et al. 2009), thus the earlier the seedlings reach the soil water table, the higher is the chance of survival. In summary, if the predation and frost risk are low, seeding should be done in autumn and winter; otherwise it is better to perform it in spring, as early as the temperatures begin to stimulate growth and frost risk is minimal. If the objective is to perform seeding in spring, acorns need to be preserved under controlled conditions because they germinate easily during winter. Acorn moisture must be reduced to 45–50% and the seeds stored in a dry and cold place. Moisture should be monitored because a decrease under 35–40% is lethal. Immediately before seeding the acorns should be rehydrated sinking them in water during 24 h. Those floating, light brown colored, with holes or wrinkles should be discarded.

Plantation is another option to reforest burned stands, although it is more expensive and disturbing than seeding. It should be initiated as soon as possible (first autumn or winter after fire) in order to avoid competition with the regenerating vegetation. Soil mobilization should be performed in a way to avoid erosion. Snags can be an obstacle to machinery movement and careless logging or site preparation can increase dramatically erosion rates in slopes. Site preparation should take into account the effects of mechanical operations over remaining root systems of sprouting species, the soil seed bank and the presence of hydrophobic soil layers. For example, subsoiling, a common method used in cork oak reforestation, is a very effective preparation work that improves water infiltration. However it should not be used if significant number of stumps can still sprout; in this case soil preparation in small spots is better. If a young plantation existed in the area before the fire, the shelters should be rapidly removed (and eventually replaced) since they usually melt, physically preventing emergence of seedlings sprouts, that are often vigorous.

As for seeding, one of the most critical issues in plantations (besides herbivory) is the low seedling survival during the summer drought period. A crucial step in restoration projects in the Mediterranean is thus achieving seedling survival during the first growing season. For example, some studies (Mousain et al. 2009) showed that ectomycorrhizal fungi improve water (and mineral) absorption when its availability is reduced, thus inoculation is one of the possible methods helping seedlings to survive long-term drought.

Restoration based on artificial regeneration is a long-term investment, thus different issues and alternatives should be carefully considered from the first stages. Several important aspects such as use of suitable genetic material, nursery cultivation regimes, sowing date, type of container, growing substrate, watering and fertilization, will largely determine the success of reforestation programs in the long-term (see Almeida et al. 2009 for more details). Additionally, several techniques can also be used in the field to improve cork oak seedlings establishment (see Cortina et al. 2009).

### 9.4.6  Cork Harvesting and Branch Pruning

Cork oaks with post-fire stem survival usually have energy reserves (mainly in the form of carbohydrates) to restore the crown foliage and to heal wounds. However certain stressing silvicultural practices such as cork harvesting and branch pruning, particularly when performed during the years immediately after the fire, will originate new energy demands, resulting in situations of great weakness. Additionally, pests and diseases may take advantage of this weakness and open wounds to attack trees causing more damages (see Sect. 9.4.8).

One of the most controversial issues in relation to cork oak trees affected by fire (when at least part of the crown survives) is related to the time at which the first post-fire cork harvesting should be performed. The charred cork is not useful to make stoppers with enological quality. This product is only useful as composition cork for insulation and it is sold at prices under harvest costs. Therefore managers are usually interested in debarking trees as soon as possible to initiate a new and clean cork production. However debarking too early after fire is not always suitable for tree health or owner economy. Some trees have burns under the cork cracks and need time to healing. Debarking can cause bigger wounds and slow the healing process. Moreover, charred bark offers less resistance to axe penetration, thus more wounds can occur. This causes early stripping to be more expensive than an ordinary debarking operation because workers have to progress slowly, suffering discomfort due to cinder and soot. Finally, and more important, less vigorous trees will produce less cork, representing lower incomes in the medium to long term.

In general, the factors determining the decision about the time to start debarking again should be the cork age (thickness) when the fire occurred, the fire severity, and the tree vigor (e.g. Cardillo et al. 2007). The existing legislation do not clearly defines what can (or cannot) be done. In Portugal, the world leading country in terms of cork production, cork harvesting is not usually allowed until cork is at least 9 years old, but there are a few exceptions (subjected to authorization) including the harvest of burned cork after verification of tree recovery. However the law has no reference to what is meant by recovery, thus the decision can be quite subjective. A recent publication (Portuguese Forest Services, DGRF 2006) recommends that cork stripping should only be performed on trees having at least 75% of the crown covered with foliage, but still, doubts may arise and in several cases this probably will not be enough to guarantee tree recovery. In Spain the IPROCOR (Instituto del Corcho, la Madera y el Carbón Vegetal) have more conservative, explicit and easy to follow guidelines, recommending that managers should wait a minimum of 2–3 years, until the crown has recovered 75% of its pre-fire volume and the cork is at least 2 cm thick.

When the cork is thinner than 2 cm, the odds of producing wounds on inner bark during the harvest increases significantly (Cardillo et al. 2007). Cork stripping should be done early in the season and conservatively, leaving the trees where cork does not detach easily, or reducing the cork harvest height. Another option is to wait until the trees develop a complete layer of cork suitable to stopper production under the charred layer. In this case lower growth rates and lower prices can be obtained but this can be better than waiting less years but debarking without revenues. It is difficult to know what choice is economically the best, but a harvesting delay of a few years for trees slightly damaged and a full technological rotation period for trees with damages of medium severity could be recommended.

Concerning tree pruning (of live branches) there are no specific post-fire regulations, but this should also be avoided during the first years. In a study in Sardinia (Italy) on the post-fire recovery of exploited cork oaks (Barberis et al. 2003), the percentage of viable trees among those that were pruned a few months after fire ranged from 20% to 28% (older and younger trees, respectively), while in not pruned trees (control group) the percentage of viable plants was two to four times higher (about 62% and 82%, respectively).

In fact, both cork harvesting and pruning are known to be stressing activities for trees, thus the law (regardless of fire) establishes a minimum period of time between these two operations (3 years in Spain and 2 years in Portugal), in order to enable tree recovery (e.g. Cardillo et al. 2007). Given that fires often causes crown defoliation and wounds, at least as severe as those caused by pruning, it would be prudent to establish a minimum time interval between fire and subsequent cork harvest or pruning, which should be at least 2 or 3 years.

### 9.4.7   Protection Against Herbivory

The presence of large wild or domestic herbivores (such as deer, goat, sheep or cattle) may represent a serious factor hindering cork oak regeneration after fire (also regardless of fire). In adult stands where all trees have crown regeneration, the presence of these herbivores is not usually a problem since they will not be able to reach the crown. However, when cork oaks are top-killed, regenerating only from basal sprouts, or when the objective is either to preserve the natural seed regeneration, or to reforest by seeding or planting, the presence of large herbivores in the burned area will likely constitute a serious problem (unless their densities are very low) and

**Fig. 9.9** Herbivory can negatively affect cork oak regeneration regardless of fire occurrence, but the impacts are likely to be much stronger in a post-fire situation: deer feeding on cork oak crown foliage and acorns (*left*), and individual protection to prevent post-fire deer browsing (*right*). Photos: M. Bugalho (*left*) and F. Catry (*right*)

some protective measures should be taken (Catry et al. 2010b; Whelan 1995; when seeding, other animals such as wild boars and mice can also be a problem because of acorn predation, see Sect. 9.4.5). This can be done by reducing the number of animals during the first years after fire or, more often, by protecting the plants. The reduction of animal densities to levels that are compatible with plant regeneration could be a good solution; however this may not be feasible or compatible with the area management objectives, and in that case other solutions, such as the physical protection of plants, must be adopted. This may involve fencing of large areas or individual plant protection during time periods that allow the regeneration and re-establishment of vegetation (Fig. 9.9).

The protection of individual trees is adopted in many countries (regardless of fire occurrence), when animals have access to regeneration areas or plantations. Various types of protections of variable prices and efficiency are available. The most common approach is to protect each tree with a protective cylindrical-shaped wire mesh shelter. To adequately fulfill its objective the wire mesh must be sufficiently strong and inelastic, and in areas where red deer (*Cervus elaphus*) are present the protection must be at least 2 m tall (preferably 2.5 m; Catry et al. 2007). Another possible protection method involves the application of chemical repellant but in most cases its effectiveness is short-lived or is still unproved.

Fencing parts of the area to regenerate may be also a good option. Generally for larger areas and higher tree density, this technique is cheaper than the protection of individual plants. Possible disadvantages of this option are the limited access to the area and higher fuel accumulation that may increase fire danger. However, if the fenced area is not very extensive and the surrounding areas have low fuel accumulation, the fire danger is reduced and these areas may act as important refuges for many animal species. Temporary protection by electric fencing is also possible, but it is most suitable for the domestic species or open woodlands, being ill adapted to forest environments where dense vegetation is present (Bonnet and Klein 1966).

### 9.4.8 Pests and Diseases

Cork oaks can be affected by pests and diseases in various ways and at all stages of their lives. Several insect species and microbial pathogens can negatively affect cork oak, from seeds and seedlings to mature trees. At moderate to high levels of incidence, they may increase mortality and reduce tree vigor, threatening the sustainability of cork oak forests (Branco and Ramos 2009).

Wood-boring insects affect primarily trees that are weakened or decaying, thus their economic impact is usually minor. However in favorable circumstances some species may become major pests. Three main groups of bark- and wood-boring insects are associated with cork oak trees: ambrosia beetles (especially *Platypus cylindrus*), two buprestids of the genus *Coroebus*, and longhorn beetles (*Cerambix cerdo*, *C. welensii*, and *Prinobius* spp.). The longhorn beetles are xylophagous species whose immature stages develop inside the trunks of decaying trees, but despite being secondary pests, *Cerambix* spp. (particularly *C. cerdo*) are associated with oak decline and are able to induce tree death (Martín et al. 2005; Branco and Ramos 2009). Tree weakening caused by increasing aridity in Mediterranean areas benefits *C. cerdo* and several other xylophagous pests. Damages caused by inappropriate cork harvesting or pruning may be a prime cause of the increase in holes made by *C. cerdo* which acts as entryways for fungal infection by *Biscogniauxia mediterranea* (Martín et al. 2005).

Moths (namely *Lymantria dispar*, *Malacosoma neustria*, *Euproctis chrysorrhoea*, and *Tortrix viridiana*) are the most important cork oak defoliators throughout the Mediterranean (Luciano et al. 2005). Severe cork oak defoliations reduce acorn production, stem diameter growth, and cork growth. Cork quantity, quality and cork stripping are also affected in subsequent years. Like for bark- and wood-boring insects, the attacks of defoliators are likely to be more severe on weakened trees. For example Luciano and Roversi (2001) suggested that infestations by *L. dispar* can occur more frequently in declining cork oak stands, such those subjected to overgrazing in Sardinia.

Among the cork oak diseases, cork oak cancer (causal agent *Botryosphaeria stevensii*), charcoal disease (causal agent *B. mediterranea*), and root diseases caused by *Armillaria mellea* and *Phytophthora cinnamomi*, are the four main fungal diseases of cork oak stands (Robin et al. 2001; Branco and Ramos 2009). *P. cinnamomi* has been regarded as the principal cause of cork oak mortality in Portugal and southern Spain (Brasier 1996; Moreira and Martins 2005). Stress and trunk wounds are the main predisposing factors for both cancer and charcoal diseases; therefore, the best control for these diseases lies in proper management practices to improve tree vigor and prevent trunk injuries (Branco and Ramos 2009).

Although the effects of wildfires on insect and diseases dynamics in cork oak forests is poorly known, the existing information suggests that the weakened status of burned trees will predispose them to suffer more severe attacks. On the other hand, fire may drastically impact herbivore arthropod populations directly by altering habitat, abundance, and species composition, or indirectly via cascading effects caused by alterations in food quality and availability (Rieske et al. 2002). Indirect effects of fire on herbivory may manifest themselves through plant growth and changes in foliar chemistry by increasing nutrient concentrations in the soil (Roth et al. 1994). Defensive phenolic compounds may also be affected by the increase in soil nutrients (Hunter and Schultz 1995) or increased sunlight (Dudt and Shure 1994). This is particularly relevant to defoliators such as *L. dispar*, which is responsive to enhanced nutritional substrate and alterations in defensive phenolic compounds (Roth et al. 1994).

### 9.4.9   Climate Change

Cork oak is adapted to highly variable climatic conditions (both between and within years). However, since the 1970s the frequency of droughts in the Mediterranean has increased significantly, and a long-term process of aridification seems to be under way as a part of the generalized trend of global warming (Pereira et al. 2009). Climate change scenarios suggest an aggravation of environmental conditions for cork oak in the Mediterranean, namely through increasing temperatures and decreasing precipitation (Giannakopoulos et al. 2009; Pausas 2004; Pereira et al. 2009). In general these factors are likely to increase the severity of plant water stress and increase the rate of nutrient losses from the soil (Pereira et al. 2009). More frequent and longer term droughts may negatively affect cork oak ecosystems in the future, by decreasing tree health and increasing the conditions conductive to the spread of some pests and diseases. For example one of the main cork oak diseases (*Phytophthora cinnamomi*) can be widely extended in the next decades due to climate change (Bergot et al. 2004).

Additionally, climate change is also expected to affect the current fire regimes in many regions, including the Mediterranean, by extending the fire season and increasing fire danger (Flannigan et al. 2009; Pausas 2004; Westerling et al. 2006). Wildfires are already a serious concern in the Mediterranean Basin burning nearly half million hectares every year, and most of this area (~87%) concerns the western Mediterranean countries where cork oak occurs (FAO 2006). In Portugal, the world leading country in terms of cork oak area and cork production, wildfires affected 15–20% of the cork oak area since 1990.

Thus, improving fire prevention and restoration techniques, throughout clearly defining the objectives, promoting natural regeneration to allow genetic variability and the possible selection of drought-tolerant genotypes, could increase the ability of cork oak forests to cope with climate change (Pereira et al. 2009).

### 9.4.10   Preventive Actions to Reduce Fire Damage

Several alternative or complementary actions can be taken in order to reduce the risk of fire damage in cork oak stands. Surface fuel reduction below and around the trees (just before debarking, i.e. every 9–15 years) could be an effective preventive action to avoid severe fires (this should be performed without soil ploughing or just with superficial tillage, in order to prevent tree root damages; Fig. 9.7).

On the other hand, the management of the cork harvesting activities could also decrease the risk of fire damages. Striping wounds are common in exploited cork oaks and they significantly reduce tree vigor and its ability to resist wildfires (Costa et al. 2004; Catry et al. unpublished). Reducing wounding (by employing skilled workers or by using automatic equipment for harvesting) could significantly increase tree resistance to fire. Other measures could include debarking trees of a given stand in different years (reducing the overall risk), increasing the debarking cycle (not necessarily meaning lower economic incomes; Natividade 1950) or decreasing the stripped surface.

Since cork is still the main economical income from these forests, stopping bark exploitation is not a realistic possibility. However, in fire-prone areas where conservation is the main objective, this would probably be the more effective option to increase ecosystem resilience (Fig. 9.10). The valorization of many other services provided by cork oak ecosystems (Bugalho et al. 2011) could create the economic incentives necessary to maintain these systems less dependent on bark exploitation.

## 9.5   Case Studies

### 9.5.1   Predicting Post-Fire Crown Recovery of Exploited Cork Oak Trees in Serra do Caldeirão (Algarve, Southern Portugal)

#### 9.5.1.1   The Wildfire

The study area is located in "Serra do Caldeirão", a mountain area in the Algarve region, southern Portugal. The climate is Mediterranean, with mean annual temperature of 16.6°C and mean annual rainfall of 900 mm. Altitude ranges from 150 to 570 m and soil type consists mainly of shallow schist lithosols. The landscape is characterized by cork oak forests with varying tree cover, with an understory of *Arbutus unedo*, *Cistus* spp., *Ulex* spp., and *Erica* spp. Other vegetation types include shrublands dominated by *Cistus ladanifer*, and scattered stands of maritime pine (*Pinus pinaster*) and eucalyptus (*Eucalyptus globulus*), sometimes mixed with cork oak stands. In July 2004 a large wildfire burned about 25,000 ha in this region.

**Fig. 9.10** Cork oak showing very good crown recovery only 3 years after wildfire and following complete defoliation (Mafra, Portugal). Trees in this area were not debarked for about 30 years and had a thick bark (Photo: F. Catry)

### 9.5.1.2 Objectives

One major decision that managers face after wildfires is whether the burned cork oak trees should be coppiced or not and when. Several authors mentioned that trunk coppicing is a good option when trees have serious stem damages that compromise future cork production, and when the crown regeneration is predicted to be nil or very weak (Pampiro et al. 1992; Cardillo and Bernal 2003, Barberies et al. 2003; DGRF 2006). One possible advantage of early coppicing is that it can promote the regeneration from basal sprouts, along with reducing mortality and speeding up the recovery on much damaged trees (Barberis et al. 2003). But, on the other hand, by cutting soon after fire, there is the risk of cutting trees that could show good crown recovery in the future, and cutting a tree implies waiting at least 30 years to start debarking good quality cork again.

The aim of this scientific study was to evaluate whether it is possible to identify, immediately after fire, trees that will likely show good or poor crown recover in the future. For this purpose, models were developed aiming to constitute decision-support tools helping managers to identify trees that will likely recover well, and trees that will likely die or show poor crown regeneration (and thus, potential candidates for trunk coppicing).

### 9.5.1.3  Methods

One and a half years after the fire, 858 trees being exploited for cork production (these trees represent the more common tree type in cork oak stands and constitute the main concern of managers because of their economic value) were sampled in a total of 40 plots spread across the burned area. Each tree was classified as having poor crown regeneration if regeneration appeared in <50% of the main branches or if it was much localized (also including trees which only resprouted from basal buds or dead trees). On the other hand trees were classified as having good crown regeneration if more than 75% of the main branches in the crown showed a homogeneously distributed regeneration. Trees with an intermediate regeneration state were not assigned to any of the previous groups.

Along with regeneration status, each tree was classified as a function of topographic variables (aspect and slope) of the plot where it was located, the amount of shrub and tree cover at the time of fire (based on aerial photos and burned remains in the field), tree size (height and d.b.h.), bark thickness and bark age (since last stripping), and minimum char height (an indicator of fire damage) expressed as proportion of tree height.

Logistic regression was used to explore which variables had a significant influence on good or poor post-fire crown regeneration in exploited cork oak trees. Different models were built, using original variables and simpler variables that can be easily assessed by forest managers.

### 9.5.1.4  Results

One-and-a-half years after fire occurrence, 31% of all trees presented a nil or poor crown regeneration (i.e. low probability of maintaining an economic interest in the near future), and 37% presented good crown regeneration, while the remaining 32% presented an intermediate state. The trees which were considered with poor crown regeneration included dead trees (18% of the total).

Bark thickness (and, therefore, cork age) was the most important variable affecting crown regeneration (better regeneration for increasing bark thickness). Char height and aspect (lower probability of good regeneration in drier southern slopes) were also significant variables influencing crown regeneration. Finally, larger trees were more likely to show poor crown regeneration (Fig. 9.11).

The probabilities obtained from the application of the two management models (to predict poor and good crown regeneration) to a given tree are negatively correlated as expected, meaning a decreasing probability of poor crown regeneration as the probability of good crown regeneration increased ($r = -0.874$; $P < 0.001$).

The obtained management models provide an easy way of getting an estimate of crown regeneration probability from only four variables that can be easily measured in the field immediately after a wildfire. More details of this study can be seen on Catry et al. (2009).

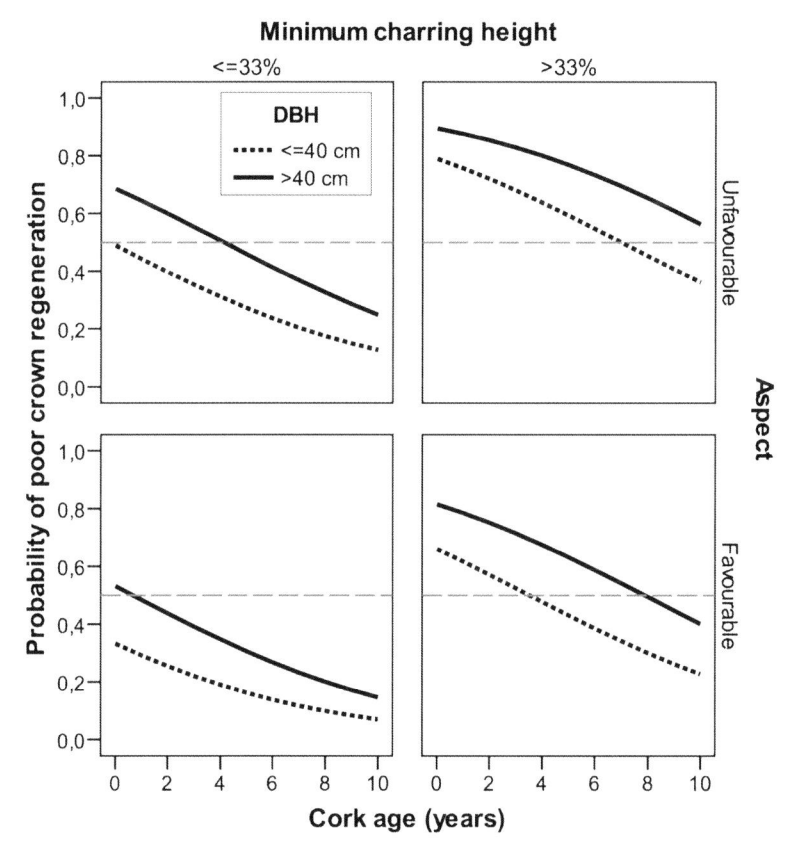

**Fig. 9.11** Logistic model prediction of poor crown regeneration in cork oak, 1.5 years after a wildfire. Different combinations of minimum charring height (larger or smaller than one-third of tree height), aspect (favorable vs. unfavorable), cork age when the fire occurred (in years), and d.b.h. (larger or smaller than 40 cm) are shown. The 50% probability line is also shown for each graph. (Reproduced from Catry et al. 2009)

## 9.5.2 Post-Fire Management of Cork Oak Woodlands in Sierra de San Pedro (Extremadura, West Spain)

### 9.5.2.1 The Wildfire

In early August 2003 during a dry thunderstorm dozens of lightning discharged upon the cork oak forests in the southern foothills of Sierra de San Pedro in the centre of Extremadura in western Spain, causing a large wildfire (more than 1,000 ha). *Coto de Santa Eulalia* is a private forest farm situated in a southern slope of these hills, with an uneven aged forest of cork oaks dedicated to cork production and hunting.

Before the fire, there were three different landscapes in the farm: in the sierra slopes a dense shrubland with scattered trees called locally *mancha*; in the foothill cork oak savanna-like woodland *dehesa*; and connecting both a narrow forest ecosystem of hardwoods and an orchard of fruit-trees associated to a small seasonal stream. At time of fire, fuels were very dry because they suffered a heat wave during the previous week with temperatures over 30°C. Therefore a very intense fire (with flames more than 20 m high) developed in the *mancha* stands. In the *dehesa* and stream areas the tall grass led to a medium intensity fire moving very fast. Nine months after the fire a diagnostic and restoration plan was carried out by the farm owner and a local forest research centre (IPROCOR).

### 9.5.2.2 Objectives of the Management Applied

The main objectives of restoration plan were: (1) Avoid soil erosion and water quality degradation, (2) Reach the normal level of cork production as soon as possible, and (3) Maintain hunting activities where possible.

To achieve the objectives the following activities were carried out:

1. In the sierra slopes, 1 year old cork oaks seedlings were planted in a furrow opened with a winged subsoiler by contour level. Sprouting vegetation was not disturbed between rows and in the area near of root systems of resprouting oaks;
2. Fences in the sierra slopes stands were repaired and strengthened to avoid game browsing over sprouts and seedlings;
3. In rolling or flat areas, acorns were seeded in small furrows (just to remove soil impervious layer and grass seeds in a narrow band). Tube shelters were used to protect seeds from mice predation (detected in previous seeding tests);
4. Dead stems were logged to improve growth and avoid leaning of stump sprouts. Two year later sprouts clumps were thinned to leave one or two vigorous stems per stump. Logs were used as erosion barriers in specific sites;
5. Deciduous broadleaved trees (*Fraxinus angustifolia* and *Celtis australis*) were planted along the stream. Small trees and bushes of the less fire resistant species were planted in a few small plots in order to help their future recovery in the farm.

### 9.5.2.3 Results

All mature cork oaks in the farm were debarked 1 year before the fire. All trees lost their crowns (stem mortality), and saplings, bushes and grasses disappeared. The soil became impermeable and was covered with a thick layer of ashes. Despite the fact that soils of stepper slopes were subsoiled (with help of local government funds), the ashes and fine soil particles begun to be drawn by the first winter showers and to accumulate in ponds and water lines.

The main conclusions of the post-fire diagnostic (9 months after fire) were:

1. All mature cork oak trees were top-killed, but trees with a diameter less than 50 cm could resprout from stumps vigorously (larger trees died). Tree and shrub species more adapted to wet conditions (those located along the stream) were eliminated;
2. Cork production was totally lost and this lack will last during next 20 years. After that some cork might be harvested from stump sprouts, but pre-fire production level will not be reached before 40 years;
3. Grasses were recovering successfully in gently slopes and bushes were sprouting or germinating from seed bank and covering the soil again moderately (~40%);
4. Game that escaped to nearest forests after fire come back to browse over plants regeneration;
5. The water ponds were filled with ashes and soil particles. Evident signs of erosion could be seen in the stepper slopes.

Today (2011; nearly 8 years after fire) most of the burned area has the same shrub cover existing before the fire. Shrub canopies have reached 2 m height and no new erosion signs are visible. Some log barriers have 5 cm of soil accumulation upslope but wood is rot and very decomposed. Reforestation was more successful at low areas than in the sierra but an average of 400 trees/ha are growing today. Nevertheless no more than 20 trees/ha are obtained from stump sprout (pre-fire mean density was about 45 trees/ha).

## 9.6   Key Messages

- Cork oak forests and woodlands constitute very important ecosystems providing a large number of socio-economic and ecological services. Thanks to its insulating bark (cork), cork oak trees have a remarkable fire-resistance and resilience, being one of the few tree species in Europe with the ability of crown resprouting after severe fires. This extraordinary resprouting capacity makes the cork oak a very good candidate for reforestation programs in fire-prone areas;
- In spite of remarkable cork oak ability to cope with fire, the periodical bark harvesting activity makes exploited trees much vulnerable to wildfires. Thus, fire risk should be taken into account by managers;
- Several alternative or complementary actions can be taken in order to reduce potential damage from wildfires in cork oak stands. Surface fuel reduction and the management of the cork harvesting activities (not debarking all trees in the same year, reducing wounding, increasing the debarking cycle) could significantly decrease the risk of fire damages;
- Management actions such as cork harvesting or pruning are not advisable at least during the first 2 or 3 years after the wildfire. Depending on factors such as fire severity, crown recovery and bark thickness, managers can decide the time to act, but in general we recommend that cork should be at least 2 cm thick and more than 75% of the pre-fire crown volume should be recovered;

- Restoration of burned areas using artificial regeneration (direct seeding or planting) is usually more expensive, slower and less successful than using natural regeneration of vegetative origin (sprouts);
- Domestic and wild animals (herbivores or omnivores such as goats, sheep, deer, wild boar) can compromise the restoration success of burned cork oak forests, by consuming acorns, seedlings, and resprouts, thus protective measures usually need be adopted when they are present in the areas to recover (unless their densities are very low).

**Acknowledgments**  To all people who contributed in some way to the contents of this chapter, with special thanks to Francisco Rego, Paulo Fernandes and Thomas Curt. To Fundação para a Ciência e a Tecnologia (PhD grant SFRH/BD/65991/2009), FFP (project Recuperação de Áreas Ardidas), EU (COST FP0701 and TRANZFOR), and the Spanish government (project VARQUS, CGL2004-04325/BOS).

# References

Almeida MH et al (2009) Germplasm selection and nursery techniques. In: Aronson J, Pereira JS, Pausas JG (eds) Cork oak woodlands on the edge: conservation, adaptive management, and restoration. Island Press, Washington, DC

APCOR (2009) APCOR yearbook 2009. Associação Portuguesa de Cortiça, http://www.realcork.org/userfiles/File/Publicacoes/AnuarioAPCOR2009.pdf. Accessed 10 April 2011

Barberis A, Dettori S, Filigheddu MR (2003) Management problems in Mediterranean cork oak forests: post-fire recovery. J Arid Environ 54:565–569

Bergot M, Cloppet E, Pérarnaud V, Déqué M, Desprez-Loustau ML (2004) Simulation of potential range expansion of oak disease caused by Phytophthora cinnamomi under climate change. Glob Change Biol 10:1539–1552

Bernal C (1999) Guia de las plantas del alcornocal. Dpto. Recursos Naturales Renovables, Instituto CMC, Junta de Extremadura. Artes Gráficas Boysu, s.l., Mérida

Bonnet G, Klein F (1966) Le cerf. Société Géologique de France, Paris

Branco M, Ramos P (2009) Coping with pests and diseases. In: Aronson J, Pereira JS, Pausas JG (eds) Cork oak woodlands on the edge: conservation, adaptive management, and restoration. Island Press, Washington, DC

Brasier CM (1996) Phytophthora cinnamomi and oak decline in southern Europe. Environmental constraints including climate change. Ann Sci For 53(2/3):347–358

Bugalho MN, Caldeira MC, Pereira JS, Aronson J, Pausas JG (2011) Human-shaped cork oak savannas require human use to sustain biodiversity and ecosystem services. Front Ecol Envir 9:278–286

Cardillo E, Bernal C (2003) Recomendaciones selvícolas para alcornocales afectados por el Fuego. Cuadernos Forestales 1/2003. Instituto del Corcho, la Madera y el Carbón. Junta de Extremadura, Mérida

Cardillo E, Bernal C, Encinas M (2007) El alcornocal y el fuego. IPROCOR, Instituto del Corcho, la Madera y el Carbón Vegetal, Mérida

Catry F, Bugalho M, Silva J (2007) Recuperação da floresta após o fogo. O caso da Tapada Nacional de Mafra. Centro de Ecologia Aplicada Prof. Baeta Neves – Instituto Superior de Agronomia, Lisboa

Catry FX, Moreira F, Duarte I, Acácio V (2009) Factors affecting post-fire crown regeneration of cork oak (Quercus suber) trees. Eur J For Res 128:231–240

Catry FX, Rego F, Moreira F, Fernandes PM, Pausas JG (2010a) Post-fire tree mortality in mixed forests of central Portugal. For Ecol Manag 206:1184–1192

Catry FX, Bugalho M, Silva JS, Fernandes P (2010b) Gestão da vegetação pós-fogo. In: Moreira F, Catry FX, Silva JS, Rego F (eds) Ecologia do fogo e gestão de áreas ardidas. ISA Press, Lisboa, pp 289–327

Correia OA, Oliveira G, Martins-Loução MA, Catarino FM (1992) Effects of bark-stripping on the water relations of Quercus suber L. Sci Gerund 18:195–204

Cortina J, Pérez-Devesa M, Vilagrosa A, Abourouh M, Messaaoudène M, Berrahmouni N, Silva LN, Almeida MH, Khaldi A (2009) Field techniques to improve cork oak establishment. In: Aronson J, Pereira JS, Pausas JG (eds) Cork oak woodlands on the edge: conservation, adaptive management, and restoration. Island Press, Washington, DC

Costa A, Pereira H, Oliveira A (2004) The effect of cork-stripping damage on diameter growth of Quercus suber L. Forestry 77:1–8

EEC (1992) Council Directive 92/43/EEC of 21 May 1992 on the conservation of natural habitats and of wild fauna and flora. Off J Eur Commun L206 (22.7.1992):7–50

Cramp S (ed) (1994) Handbook of the birds of Europe, the middle East, and North Africa. Oxford University Press, Oxford

Curt T, Bertrand R, Borgniet L, Ferrieux T, Marini E (2010) The impact of fire recurrence on populations of Quercus suber in southeastern France. In: Viegas DX (ed) Abstracts of the VI international conference on forest fire research. ADAI/CEIF, University of Coimbra, Portugal, CD Rom 9 pp

DeBano LF, Neary DG, Ffolliott PF (1998) Fire's effects on ecosystems. Wiley, New York

DGRF (2006) Boas práticas de gestão em sobreiro e azinheira. Direcção-Geral dos Recursos Florestais, Lisboa

Du Merle P, Attié M (1992) Coroebus undatus (Coleoptera: Buprestidae) sur chêne liège dans le Sud-Est de la France: estimation des dégâts, relations entre ceux-ci et certains facteurs du milieu. Ann For Sci 49:571–588

Dudt JF, Shure DJ (1994) The influence of light and nutrients on foliar phenolics and insect herbivory. Ecology 75:86–98

EEA (2007) European forest types. Categories and types for sustainable forest management reporting and policy. European Environment Agency (EEA) technical report no 9/2006, 2nd edn, Copenhagen

FAO (2006) Global forest resources assessment 2005 – Report on fires in the Mediterranean Region. Food and Agriculture Organization of the United Nations, Rome

Flannigan MD, Krawchuk MA, Groot WJ, Wotton BM, Gowman LM (2009) Implications of changing climate for global wildland fire. Int J Wildland Fire 18:483–507

Giannakopoulos C, Le Sager P, Bindi M, Moriondo M, Kostopoulou E, Goodess CM (2009) Climatic changes and associated impacts in the Mediterranean resulting from a 2 degrees C global warming. Glob Planet Chang 68:209–224

Hunter MD, Schultz JC (1995) Fertilization mitigates chemical induction and herbivore responses within damaged oak trees. Ecology 76:1226–1232

IUCN (2010) IUCN Red list of threatened species. Version 2010.4. www.iucnredlist.org. Accessed 5 March 2011

Johnson PS, Shifley SR, Rogers R (2009) The ecology and silviculture of oaks, 2nd edn. CABI Publishing International, Oxford

Kutiel P, Lavee H (1999) Effect of slope aspect on soil and vegetation properties along an aridity transect. Isr J Plant Sci 47:169–178

Luciano P, Roversi PF (2001) Oak defoliators in Italy. Industria Grafica Poddighe s.r.l, Sassari

Luciano P, Lentini A, Cao OV (2005) La lutte aux lépidoptères défoliateurs des subéraies dans la Province de Sassari. Industria Grafica Poddighe s.r.l, Sassari

Marques MA, Mora E (2003) The influence of aspect on runoff and soil loss in a mediterranean burnt forest (Spain). Catena 19:333–344

Martín J, Cabezas J, Buyolo T, Patón D (2005) The relationship between Cerambyx spp. damage and subsequent Biscogniauxia mediterranum infection on Quercus suber forests. For Ecol Manag 216:166–174

Mendes AMSC, Graça JAR (2009) Cork bottle stoppers and other cork products. In: Aronson J, Pereira JS, Pausas JG (eds) Cork oak woodlands on the edge: conservation, adaptive management, and restoration. Island Press, Washington, DC

Miller M (2000) Fire autecology. In: Brown JK, Smith JK (eds) Wildland fire in ecosystems: effects of fire on flora. General Technical Report. RMRS-GTR-42-vol 2, U.S. Department of Agriculture, Forest Service, Rocky Mountain Research Station, Ogden

Moreira AC, Martins JMS (2005) Influence of site factors on the impact of Phytophthora cinnamomi in cork oak stands in Portugal. For Pathol 35:145–162

Moreira F, Duarte I, Catry F, Acácio V (2007) Cork extraction as a key factor determining post-fire cork oak survival in a mountain region of southern Portugal. For Ecol Manag 253:30–37

Moreira F, Catry F, Duarte I, Acácio V, Silva J (2009) A conceptual model of sprouting responses in relation to fire damage: an example with cork oak (Quercus suber L.) trees in Southern Portugal. Plant Ecol 201:77–85

Mousain D, Boukcim H, Richard F (2009) Mycorrhizal symbiosis and its role in seedling response to drought. In: Aronson J, Pereira JS, Pausas JG (eds) Cork oak woodlands on the edge: conservation, adaptive management, and restoration. Island Press, Washington, DC

Natividade JV (1950) Subericultura. Ministério da Economia, Direcção Geral dos Serviços Florestais e Aquícolas, Lisboa

Oliveira G, Correia O, Martins Loução M, Catarino FM (1992) Water relations of cork oak (Quercus suber L.) under natural conditions. Vegetatio 100:199–208

Pampiro F, Pintus A, Ruiu PA (1992) Interventi di recupero di una giovane sughereta percorsa da incêndio. In: Instituto de Promoción del Corcho (ed) Simpósio Mediterrâneo sobre Regeneración del Monte Alcornocal, Mérida

Pausas J (1997) Resprouting of Quercus suber in NE Spain after fire. J Veg Sci 8:703–706

Pausas JG (2004) Changes in fire and climate in the eastern Iberian Peninsula (Mediterranean basin). Clim Chang 63:337–350

Pausas JG, Pereira JS, Aronson J (2009) The tree. In: Aronson J, Pereira JS, Pausas JG (eds) Cork oak woodlands on the edge: conservation, adaptive management, and restoration. Island Press, Washington, DC

Pereira H (2007) Cork: biology, production and uses. Elsevier Publishing, Amsterdam

Pereira P, Fonseca M (2003) Nature vs. nurture: the making of the montado ecosystem. Conserv Ecol 7(3):7

Pereira JS, Correia AV, Joffre R (2009) Facing climate change. In: Aronson J, Pereira JS, Pausas JG (eds) Cork oak woodlands on the edge: conservation, adaptive management, and restoration. Island Press, Washington, DC

Pons J, Pausas JG (2007) Acorn dispersal estimated by radio-tracking. Oecologia 153:903–911

Rieske LK, Housman HH, Arthur MA (2002) Effects of prescribed fire on canopy foliar chemistry and suitability for an insect herbivore. For Ecol Manag 160:177–187

Robin C, Capron G, Desprez-Loustau ML (2001) Root infection by *Phytophthora cinnamomi* in seedlings of three oak species. Plant Pathol 50:708–716

Roth SK, Lindroth RL, Montgomery ME (1994) Effects of foliar phenolics and ascorbic acid on performance of the gypsy moth (Lymantria dispar). Biochem Syst Ecol 22:341–351

Rundel PW (1973) The relationship between basal fire scars and crown damage in Giant Sequoia. Ecology 54:210–213

Silva JS, Catry F (2006) Forest fires in cork oak (Quercus suber) stands in Portugal. Int J Environ Stud 63:235–257

Sousa E, Inácio ML (2005) New aspects of platypus cylindrus fab. (Coleoptera: Platypodidae) Life history on cork oak stands in Portugal. In: Lieutier F, Ghaioule D (eds) Entomological research in Mediterranean forest ecosystems. INRA Editions, Paris

Sternberg M, Shoshany M (2001) Influence of slope aspect on Mediterranean woody formations: comparison of a semiarid and an arid site in Israel. Ecol Res 16:335–345

Wargo M (1996) Consequences of environmental stress on oak: predisposition to pathogens. Ann For Sci 53:359–368

Westerling AL, Hidalgo HG, Cayan DR, Swetnam TW (2006) Warming and earlier Spring increase western US forest wildfire activity. Science 313:940–943

Whelan RJ (1995) The ecology of fire. Cambridge University Press, New York

WWF (2007) Beyond cork – a wealth of resources for people and nature. World Wide Fund for Nature, Madrid

# Chapter 10
# Post-Fire Management of Exotic Forests

**Joaquim S. Silva and Hélia Marchante**

## 10.1 Introduction

The expression "exotic forests" may be associated to a wide range of forest types. One first differentiation should be made between the plantations of exotic trees established with a specific economic purpose, and the self-regenerated stands of naturalized species. According to the report on European Forest Types (EEA 2007) this wide range of forest formations is classified as Forest Type 14.2 – Plantations of not-site-native species and self-sown exotic forest. This forest type is included in the broader Forest Category 14 – Plantations and self-sown exotic forest.

Plantations of not-site-native species include the forest stands aimed at providing raw material for industrial purposes. In many cases these are managed using intensive short-rotation silviculture, falling into what is frequently known as industrial forestry. Between the species most commonly used in commercial plantations in Europe we may find: *Eucalyptus* spp., *Populus* spp., *Picea sitchensis, Pinus radiata, P. contorta, Pseudotsuga menziesii* and *Tsuga heterophylla*. The second sub-type, self-sown exotic forest, includes species which have been naturalized in many European countries, like: *Robinia pseudoacacia, Ailanthus altissima, Prunus serotina* and different species of *Acacia*. However, the assignment of a species to one of

J.S. Silva (✉)
Centre of Applied Ecology, Institute of Agronomy, Technical University of Lisbon, Lisbon, Portugal

H. Marchante
College of Agriculture, Polytechnic Institute of Coimbra, Coimbra, Portugal

Department of Life Sciences, Centre for Functional Ecology, University of Coimbra, Coimbra, Portugal

F. Moreira et al. (eds.), *Post-Fire Management and Restoration of Southern European Forests*, Managing Forest Ecosystems 24, DOI 10.1007/978-94-007-2208-8_10, © Springer Science+Business Media B.V. 2012

these two sub-types of forest is not straightforward, as some of them may be represented in both groups. The cases of *Robinia pseudoacacia* and *Eucalyptus* spp. are good examples since both are used in plantations and both are referred as invasive species, as explained later in this chapter. Invasive species are able to regenerate and spread naturally, competing successfully with local species. In many situations the natural regeneration of these species intermingles with native species, originating mixed stands which are increasingly altering the forest composition of natural communities (Richardson 1998; Silva et al. 2011). In ecological terms, forest plantations are characterized by a highly simplified forest structure and composition and low richness of associated fauna, when compared to natural and semi-natural forests (EEA 2007), whereas self-sown exotic forests may present different composition and structure, depending on the characteristics of the species involved.

The post-fire management of exotic woody species poses a series of different questions which basically depend: on fire severity, on the stand characteristics (plantation, self-sown forest), on the species composition previous to fire and on the objectives pretended for the burned area (which are very much associated with the stand characteristics). The post-fire management of a burned plantation is frequently directed to replant the same tree species, whereas the post-fire management of burned stands of self-regenerated exotic trees, either it is absent or, if existing, will most likely be directed to the conversion to a different forest type. This aspect is crucial because in many cases, self-sown exotic forests are the result of poor or absent management (Silva et al. 2011). Therefore, it is expectable that abandoned land colonized by invasive species, will stay abandoned after a wildfire, which will in many cases aggravate the invasion problem (Keeley et al. 2011). For either case we may have species which are relatively well adapted to fire occurrence, such as the case of many *Eucalyptus* spp. (Gill 1977, 1997; Gouveia et al. 2010; Marques et al. 2011) and others which have no specific traits known to be related with fire occurrence, like some North American conifers introduced in Europe. Considering this diversity of situations, it becomes clear that there is no typical procedure to be adopted in the post-fire management of exotic forests, contrarily to what may be the case in other forest types presented in this book. In fact, given the different variables to consider, there are different measures that can be taken. Because of the scarcity of information on the post-fire conversion of burned exotic plantations, we assume that in most situations, land managers will basically decide on replanting the burned stand or simply converting it to another productive type of plantation. However, these direct replacements of one plantation by another do not always follow good management practices, frequently presenting negative consequences at different levels, particularly in the conservation of soil and water regime (e.g. Shakesby et al. 2002; Smith et al. 2011).

On the other hand there is a big challenge in managing the self-regenerated stands of exotic species because of the associated difficulties. These stands may even result from former or from existing burned plantations. In fact, as we will see further in the present chapter, forest plantations of exotic species can be one of the pathways leading to plant invasions, and fire may have a positive effect in this context. Therefore the paradigm of ecosystem restoration, as it is presented in this book for most forest types, is difficult to apply to the specific case of exotic forests, even if there are common aspects to consider such as the case of soil and water conservation, for example.

This chapter aims at presenting a broad perspective on the post-fire management of exotic forests taking into account a European context and, given its increasing importance, clearly emphasizing the problem of invasive tree species. It is divided in two main sections. The first includes a characterization of the specific issue of exotic forests and fire in the European context. The second presents post-fire management alternatives, and includes detailed information for common types of exotic forests, taking into account their ecological and economic importance in southern Europe and their relevance in terms of post-fire management. An additional section presents two case studies concerning two important exotic species and their invasive behavior.

## 10.2   Exotic Forests and Fire

### 10.2.1   Basic Definitions and Concepts

Exotic species can be classified in terms of their regeneration level and introduction pathway, among other criteria. According to the definitions by Lambdon et al. (2008) we apply the term *exotic* (the same as *alien, introduced, non-native*, or *non-indigenous*) to those species in a given area whose presence there is due to intentional or unintentional human involvement, or which have arrived there without the help of people from an area in which they are alien. From these we should distinguish different regeneration levels. *Casual* species are those that may reproduce occasionally outside cultivation in an area, but that eventually die out because they do not form self-replacing populations, and rely on repeated introductions for their persistence. *Naturalized* species are those that sustain self-replacing populations for a period of time long enough to experience extreme climatic events in the area, and reproduce without direct intervention by people (or in spite of human intervention) by recruitment from seed or vegetative parts capable of independent growth. *Invasive* species are a subset of naturalized plants that produce reproductive offspring, often in very large numbers, at considerable distances from the parent plants and thus have the potential to spread over a large area. The original definition from Richardson et al. (2000) presents additional criteria in order to quantify "considerable distances from the parent plants": >100 m over <50 years for species spreading by seeds and other propagules; >6 m over 3 years for species spreading by roots, rhizomes, stolons or creeping stems. Exotic species can also be classified according to their introduction pathways (Lambdon et al. 2008). *Released* species have been released deliberately into the wild (e.g., for the enrichment of the native flora, landscaping, etc.), whereas *escaped* species have escaped into the wild from cultivation. Escapes can be associated with different activities, including forestry (e.g. *Eucalyptus* spp.), ornamental purposes (e.g. *Ailanthus altissima*) or soil stabilization (e.g. *Acacia* spp.) for example.

## 10.2.2 Distribution and Importance of Exotic Forests in Europe

There is a clear correlation between the establishment of plantations and the area of forests dominated by exotic species (MCPFE 2007). In total, about 8.1 million ha, (5.2%) of the total forest area of Europe, are dominated by exotic tree species in 32 reporting MCPFE countries (Ministerial Conference for the Protection of Forests in Europe, excluding the Russian Federation), including 10% dominated by invasive species (MCPFE 2007). According to EEA (2007) European countries have a total of 7% of forests classified in forest category 14 – Plantations and self-sown exotic forest and there are 5 countries with more than 50% of forest area assigned to this forest category (Ireland, Belgium, The Netherlands, United Kingdom and Denmark, in decreasing order) (Table 10.1). Combining the information from EEA (2007) and from MCPFE (2007), France and Spain are the countries presenting the largest surface occupied by plantations and self-sown exotic woody species with more than two million ha each, followed by the United Kingdom and Germany, with more than 1.5 million ha each. If we retain only the countries from southern Europe, corresponding to higher fire proneness (JRC 2010), the country presenting the highest percentage of forest Category 14 is Portugal (18%), followed by France (15%) and Spain (12%). These figures may be underestimated at least for Portugal where, according to the last National Forest Inventory, there are 23% of *Eucalyptus globulus* stands (AFN 2010), all of them obviously corresponding to forest category 14. All other southern European countries are referred as having less than 10% of plantations and self-sown exotics. The MCPFE report (MCPFE 2007) presents an incomplete list of countries with data on the forest surface dominated by exotic species. The countries presenting the highest surfaces are the United Kingdom (1.4 million ha), France (1.1 million ha), Hungary (0.8 million ha) Sweden (0.6 million ha) and Italy (0.4 million ha).

When looking at the forest surface of exotics which are considered invasive, Hungary leads the list (426,000 ha), followed by Italy (282,000 ha), the Russian Federation (55,000 ha), Slovakia (26,000 ha) and Austria (22,000 ha). However, this information is far from being complete and excludes some southern countries which are prone to invasive species, like Portugal and Spain. On the other hand, the criteria to consider a species as invasive may result in misleading information because it was not, up to our knowledge, based on quantitative studies. For example *Eucalyptus globulus* is referred by some authors as having invasive behavior in some areas in Portugal (Marchante et al. 2008a) and Spain (Dana et al. 2004) but was not apparently considered as so in the MCPFE report (MCPFE 2007). This calls for common procedures and common objective criteria among European countries in order to have a reliable assessment on the real situation in terms of exotic forests and invasive woody species (Hulme et al. 2009).

According to estimations by Köble and Seufert (2001) the main tree species exotic to Europe in terms of occupied surface are (in decreasing order): *Eucalyptus* spp., *Picea sitchensis, Robinia pseudoacacia, Pseudotsuga menziesii* and *Pinus contorta*. According to the European database on exotic species (DAISIE 2008) there are 10 different species of *Eucalyptus* essentially distributed in the southern western Europe.

In particular, *Eucalyptus* globulus is a very important species, occupying around one million ha in the Iberian Peninsula (Schelhaas et al. 2006). *Robinia pseudoacacia* is the most widely distributed exotic tree species in Europe being naturalized in 32 regions (Pyšek et al. 2009). *Robinia pseudoacacia* has been cultivated in several countries due to its excellent regeneration capacity and strong competitiveness, but now it accounts for half of the area dominated by invasive tree species in Europe. Within the southern European countries we should also consider *Pinus radiata* which has been widely planted in northern Spain particularly in the Basque Country, where it occupies around 150,000 ha (Mena-Petite et al. 2004; EEA 2008).

Besides the referred species, we should also add some problematic invasive tree species even if they still do not occupy the top rank in terms of forest surface. The team of the European project DAISIE proposed a list of 100 of the worst invasive species in Europe (DAISIE 2009). This list is composed of 18 plant species including 4 tree species: *Acacia dealbata*, *Ailanthus altissima*, *Prunus serotina* and *Robinia pseudoacacia*.

## 10.2.3   When Exotics Become Invasive

### 10.2.3.1   Social and Ecological Aspects

The problem of invasive species is becoming more evident as it is quantified in terms of economic costs. In Europe, most expenses generated by invaders are in the form of management costs (eradication, control, monitoring, and environmental education programs) in natural areas (Vilà et al. 2010). The crudest estimate of total known monetary impact of alien species (including animals and plants) in Europe is close to 10 billion Euros annually (Hulme et al. 2009). Andreu et al. (2009) have evaluated the economic costs of management initiatives associated with invasive species only in natural protected areas of Spain as more than 50 million Euros in 10 years. From this amount, around 60% was allocated to the management of *Eucalyptus* spp. Nonetheless and despite the enormous amounts of funds which are allocated to solve or at least to mitigate the problem of plant invasions, these are perceived by stakeholders as not so important, in comparison to other environmental concerns, according to a survey performed in Spain by Andreu et al. (2009). At the European level only 2% of the public feel biological invasions as important threats to biodiversity (Hulme et al. 2009). Therefore, in what concerns the particular issue of post-fire management, it is not surprising the lack of concern even amongst specialists. Within the European COST FP0701 Project (Post-fire forest management in southern Europe) a survey was distributed among country representatives about the importance of invasive exotic species in burned areas. The results of the questionnaire showed that among the 14 respondents only 4 (Israel, Lithuania, Portugal and Switzerland) considered invasive species in burned areas a relevant problem. The tree species referred by the respondents were: *Acacia* spp. (Israel, Portugal), *Pinus halepensis* (Israel), *Robinia pseudoacacia* (Lithuania, Switzerland) and *Ailanthus*

**Table 10.1** Estimated surface occupied by plantations and self-sown exotic forests (Category 14) according to EEA (2007); estimated surface occupied by introduced forest species (Type 14.2) according to MCPFE (2007); and estimated surface occupied by invasive tree species according to MCPFE (2007)

| Country | Forest surface 1000 ha | Category 14 1000 ha | % | Type 14.2 1000 ha | % | Invasive 1000 ha | % |
|---|---|---|---|---|---|---|---|
| Andorra | 16 | 0 | 0 | – | – | – | – |
| Austria | 3,862 | – | – | 53 | 1 | 22 | <1 |
| Belarus | 8,436 | – | – | 1 | <1 | 0 | 0 |
| Belgium | 672 | 538 | 80 | 259 | 39 | <1 | <1 |
| Bulgaria | 3,651 | 292 | 8 | 173 | 5 | 0 | 0 |
| Croatia | 2,135 | 43 | 2 | – | – | – | – |
| Cyprus | 174 | – | – | 1 | 1 | 0 | 0 |
| Czech Republic | 2,647 | 238 | 9 | 11 | <1 | 0 | 0 |
| Denmark | 500 | – | – | 314 | 63 | 0 | 0 |
| Estonia | 2,264 | – | – | 1 | <1 | 0 | 0 |
| Finland | 22,130 | – | – | 26 | <1 | 0 | 0 |
| France | 15,554 | 2,333 | 15 | 1,051 | 7 | – | – |
| Germany | 11,076 | 1,551 | 14 | – | – | – | – |
| Greece | 3,752 | 38 | 1 | – | – | – | – |
| Hungary | 1,948 | – | – | 820 | 42 | 426 | 22 |
| Ireland | 669 | 669 | 100 | – | – | – | – |
| Italy | 9,979 | 599 | 6 | 406 | 4 | 282 | 3 |
| Latvia | 3,035 | – | – | 1 | <1 | 0 | 0 |
| Lithuania | 2,121 | – | – | 4 | <1 | 0 | 0 |
| Luxembourg | 87 | – | – | 26 | 30 | 0 | 0 |
| Netherlands | 365 | 234 | 64 | 91 | 25 | 0 | 0 |
| Norway | 9,387 | – | – | 262 | 3 | 0 | 0 |
| Poland | 9,200 | 92 | 1 | – | – | – | – |
| Portugal | 3,783 | 681 | 18 | – | – | – | – |
| Romania | 6,391 | 383 | 6 | – | – | – | – |
| Russia | 808,790 | – | – | 71 | <1 | 55 | 0 |
| Serbia | 1,813 | 218 | 12 | 2 | <1 | 0 | 0 |
| Slovakia | 1,932 | 77 | 4 | 41 | 2 | 26 | 1 |
| Slovenia | 1,264 | 89 | 7 | 16 | 1 | 11 | 1 |
| Spain | 17,915 | 2,150 | 12 | – | – | – | – |
| Sweden | 27,871 | – | – | 636 | 2 | 0 | 0 |
| Switzerland | 1,220 | 24 | 2 | 4 | <1 | 1 | <1 |
| United Kingdom | 2,845 | 1,736 | 61 | 1,420 | 50 | 0 | 0 |

Inconsistent data between the two sources (higher surface in type 14.2 than in category 14) are not shown for Category 14 (–). Absent data in columns corresponding to Type 14.2 and to Invasive species, are also represented by symbol "–"

*altissima* (Swizerland). Therefore it seems that the concerns posed by plant invaders in burned areas are also not similar within the southern European countries. It is likely that moister coastal temperate regions like the north and west Iberian Peninsula may be more prone to plant invasions after fire, than other more dry regions of the Mediterranean (Chytry et al. 2009). For further details see also Chap. 5 of this book.

Nonetheless, it is consensual that the increasing number and intensity of introduction pathways associated with the non-stopping process of economical and social globalization, leads to an increased concern in respect to the problem of plant invaders. Other factors have also been associated with this increasing trend, namely: climate change (Walther et al. 2007), the expansion of urban areas (Chytry et al. 2009) and land abandonment (Silva et al. 2011).

The process of invasion does not normally occur immediately after introduction. It encompasses different phases and it is valid for different organisms, including trees (Mack et al. 2000; Murphy et al. 2005). In fact, few introductions result in the establishment (naturalization) of organisms and from these only a small part becomes invasive. The probability that a certain introduced species will pass the next stages depends on the ecological resistance, composed of demographic, biotic and environmental factors (Lodge 1993). Demographic resistance varies inversely with the number of propagules. Biotic resistance refers to features shaping the invaded community, like a high species richness which may work as a barrier to invasions (Elton 1958). Environmental resistance may involve abiotic resource availability (suitable climatic factors or light intensities, for example) and may preclude a species altogether, or may induce a significant time lag between arrival and establishment of a species (Murphy et al. 2005). In order to overcome the described resistance factors, plant species need to include a series of traits allowing passing the different invasion process stages. Some of the most successful invaders present a set of characteristics which become competitive advantages when the alien plants share the same space with native plants. Thus the most aggressive invasive species have the ability to reproduce sexually and grow clonally, are fast growing species, present phenotypic plasticity and have a high tolerance of environmental heterogeneity (Murphy et al. 2005). The most aggressive invasive tree species in Europe, present these characteristics all together.

### 10.2.3.2 The Role of Fire

Fire both influences and is influenced by plant community composition and structure, resulting in a complex relationship with the invasion by exotic plants (Mandle et al. 2011). There is clear evidence that many invasive species can be promoted by altered fire regimes. Reciprocally plant invaders can alter significantly the fire regimes, leading to what we may call a fire cycle. In fact, the rapid regeneration of exotic species in recently burned areas often creates a positive feedback as it may favor the spread of high intensity wildfires, which in turn may increase the regeneration of more invasive plants (D'Antonio and Vitousek 1992; D'Antonio 2000; Pauchard et al. 2008). Depending on the species concerned and on the fire regime, this fire cycle may be extremely difficult to break and it is probably the most challenging problem related with post-fire management. In many cases, the available technical solutions are either too expensive to be feasible or have not proved to be fully effective in the resolution of the problem.

One of the most referred mechanisms influencing the promotion of invasive species by fire, is the thermal effect which acts as a cue for triggering the germination of seeds (Whelan 1995). Seeds of different *Acacia* spp. (see the study case presented at the end of this chapter) are strongly stimulated to germinate after wildfires, as it happens with many other Fabaceae species (Hill 1982; Auld and O'connell 1991; Marchante and Marchante 2007; Pauchard et al. 2008; Hanley 2009). *Acacia dealbata* presents hard coated seeds that may stay viable for at least 200 years in the Tasmanian forests where the species is native (Gilbert 1959; Hunt et al. 1999). Another competitive advantage of aggressive invaders like *Acacia* spp., *Robinia pseudoacacia* or *Ailanthus altissima* is the vigorous resprouting after burning or cutting when compared to native species (Burch and Zedaker 2003; Lee et al. 2004; Zengjuan et al. 2006). Invaders can simply benefit from human disturbance leading to the creation of open spaces which are afterwards occupied by these opportunistic species such as the case of *Robinia pseudoacacia* in Hungary (Krízsik and Körmöczi 2000), elsewhere in Europe (Lambdon et al. 2008) and in other parts of the world (Lee et al. 2004; Mandle et al. 2011).

The second aspect to be considered is the fact that invasive species are known to affect fire regimes (Keeley et al. 2011; Mandle et al. 2011; Vilà et al. 2001). Fire regimes can be changed in any of its components: frequency, intensity, extent, type and seasonality (Brooks et al. 2004). This is due to the potential complete change in fuel properties caused by the replacement of existing plant communities by invasive species. Fuel properties can be altered in terms of flammability, load, vertical and horizontal continuity and packing ratio (Brooks et al. 2004). However, not always the changes in fire regime will lead to increased fire hazard. In fact there are many references in the literature reporting for example a decrease of flammability (Brooks et al. 2004; Mandle et al. 2011). It is worth to mention the case of *Robinia pseudoacacia* which is referred to be less flammable than native species in the northwest United States where it is exotic (Richburg et al. 2004). Among the many opposite examples it is worth referring the case of *Acacia* spp. woodlands in Portugal, considered as one of the most fire hazardous forest types (Fernandes 2009).

In conclusion, the relationships between invasive plants and fire regime have to take into account both the direct effects of fire and the different interactions between fire and the different ecosystem components including the native plant communities and other ecosystems properties. Brooks et al. (2004) presented a very comprehensive model which illustrates the invasive plant-fire regime cycle. After the introduction phase, the subsequent stages of the invasion process may have consequences in terms of the fuel properties, the native plants and the other ecosystem properties. In turn, these can have feedbacks which affect directly or indirectly the plant invaders. This cycle proceeds according to 4 phases. In phase 1, the propagules arrive to a new region; in phase 2, the plant species is naturalized or invasive but has not yet caused significant ecological impact; in phase 3, the plant species has had significant ecological impact but has not yet changed the fire regime; in phase 4, the plant species has changed the fire regime establishing an invasive plant-regime fire cycle.

## 10.3  Managing Exotic Forests After Fire

### 10.3.1  General Aspects

The post-fire management of exotic forests deals with aspects which are obviously common to other forest types. However, there are specificities which are a consequence of the exotic nature of the species involved and a consequence of the specific purpose of exotic forests, in particular in the case of plantations. As to the first aspect, it is consensual that exotic tree species normally have detrimental effects in terms of the conservation value of the occupied landscapes, when compared with local species (EEA 2007). This has obvious implications in terms of the difficulty to include ecosystem conservation objectives in the post-fire management of burned exotic plantations, as it is the case of other forest types. As to the second aspect, the specific purpose of exotic plantations is normally the production of raw material to supply the forest industries. This frequently leads to intensive silvicultural practices which may have an increased impact in terms of soil conservation (e.g. Ferreira et al. 1997). Therefore, it is important to define which are the major differences and specificities when dealing with burned stands of exotic species. With this purpose we can refer to Chap. 1 of this book, particularly to the definition of a restoration approach (Fig. 1.2) and to the framework to planning that restoration (Fig. 1.4). Again, we should clearly distinguish the case of plantations from the case of self-sown exotic stands.

The restoration approach of exotic plantations can be quite straightforward. If no other constraints are present, it can simply consist in felling the burned trees, and replanting the burned area. If we consider the definitions from Chap. 1, then the term restoration has little meaning in this case, and the term replacement should be used in most cases. However, that is not always the case, depending on the approach to be adopted which in turn will obviously depend on the fire severity, the tree species and the management objectives. The first two aspects are obviously connected, since different species provide different fuel characteristics, which influence fire intensity and fire severity. The cases of *Robinia pseudoacacia* and *Eucalyptus globulus* are well representative of species with different flammability and fire resistance, which may lead to different fire severity. Despite its high flammability (Dickinson and Kirkpatrick 1985), eucalypt stands can be also very resilient to fire (Gouveia et al. 2010), which may simply determine that the best solution for a burned stand is to coppice the trees and wait for the new resprouts. In fact in many cases the management objectives are just to maintain the previous productivity potential. However, other objectives may be present even considering the main production role of the regenerated stand. Soil and water conservation should always be a concern, leading to best practices in order to minimize soil losses or altered hydrological regimes. Similarly, ecosystem conservation should also be considered when replacing a burned exotic plantation. In fact, a forest fire can be a window of opportunity in order to implement measures of environmental mitigation, such as the establishment of ecological corridors within the plantation area (e.g. Silva et al. 2007).

In this case the term rehabilitation (sensu Chap. 1) may also be appropriate. If the cultivated tree has invasive potential, rehabilitation measures should consider this aspect, because the eventual escape from plantations may lead to the unwanted colonization of neighbour areas.

As to the planning of post-fire management in burned plantations, the basic steps proposed in Chap. 1 are essentially valid, with a few remarks. The mapping of vulnerable areas may have little utility in the case of plantations, since this type of forests have a rather simple and homogeneous structure and composition (EEA 2007). Therefore, despite differences induced by relief or by the maintenance of ecological corridors, each plantation has a previously known risk level (sensu Bachmann and Allgöwer 2001) which should be roughly similar for the whole area. Nonetheless, some pre-assessment may be needed, in particular to identify areas with potential risk of erosion. All remaining four steps are equally valid for plantations of exotic species. In the particular case of invasive species, a critical attention should be given to monitoring (step 5).

The management of burned self-sown exotic forests may involve difficult challenges, particularly for those species which are well adapted to fire. Contrarily to the case of exotic plantations, self-sown exotic forests may present heterogeneous structure and composition because, in many cases, they may include also naturally regenerated native vegetation (see the case studies in this chapter). Therefore, the ecosystem type, the fire severity and consequently the ecosystem response may vary in a wide range. On the contrary, restoration objectives should normally aim to help the regeneration of native species. Besides this critical objective, another important aspect to consider is fire prevention through the reduction of fire hazard.

The planning of post-fire management in burned non-native self-sown stands should give a particular emphasis to steps 4 (Long term planning) and 5 (Monitoring). As we will further describe, the post-fire management of invasive vegetation can be a long lasting task, including different approaches and subjected to different stages. Therefore the restoration of burned areas of invasive vegetation can be highly demanding both economically and technically, which makes very important the need to plan and to follow the implemented measures.

## 10.3.2 Soil Protection

The management of exotic plantations can be highly intensive. Most species have high productivity and fast growth which allow the use of short rotations leading to frequent soil disturbance. Additionally, the native understorey, which helps preventing the risks of soil erosion, is frequently removed, in order to reduce fire hazard and to reduce competition with the planted trees. The replacement or regeneration of burned stands frequently makes use of heavy machinery for clear cutting and soil preparation. Often the result is a lower level of soil protection and a higher risk of erosion, when compared with other less intensive forest systems. Therefore, in this type of forests, the problem of soil protection after fire is of paramount importance.

One of the important issues in this context is salvage logging. In intensively managed forest systems there is frequently little concern about the impact of this post-fire management practice. Direct impacts on soil may result in altered soil properties, lower nutrient levels, increased erosion and modified hydrological regimes. Besides the negative direct impacts on soil, salvage logging may also prevent the recovery of natural vegetation, facilitate the colonization of invasive species, and alter patterns of landscape heterogeneity (McIver and Starr 2000; Lindenmayer and Noss 2006). The consequences of intensive practices of salvage logging on soil result from the heavy machinery frequently used for cutting and transporting the wood, but also from the exposure of the soil when all material is removed (Merino et al. 2005).

There are different alternatives which can be used in order to minimize the effects of intensive logging. Shakesby et al. (1996) found that the litter produced by logging, if remaining in the soil, could reduce soil losses by up to 95%. Besides soil losses, the use of intensive harvesting including the mechanical removal of logging residues in industrial plantations can drastically reduce soil nutrients (Merino et al. 2005). This negative impact can be mitigated if felling and debarking is performed on site next after fire. Nutrient exports in plantations can also be diminished by adopting less intensive silvicultural models, which may include the reduction of tree density and the increase of the rotation length (Merino et al. 2005).

The other aspect to consider has to do with the intensive soil laboring frequently used in the replacement of the burned plantation. Shakesby et al. (2002) studied the impact of commonly used practices of rip-ploughing. These authors found that soil erosion resulting from a wildfire-rip-ploughing cycle was estimated to be up to 174 t/ha, potentially leading to ultimate physical degradation for typically thin soils within 50–100 years.

One of the most studied aspects in burned plantations is the formation of a repellent soil layer after fire. Hydrophobicity has been referred by many authors to be responsible for increased erosion rates after fire occurrence (e.g. Dekker et al. 2003).This phenomenon has been found to be particularly important in burned eucalypt plantations, contributing to considerable soil losses through erosion (Shakesby et al. 1993; Doerr et al. 1998; Ferreira et al. 2000). On the other hand it has been shown that even in the absence of fire, eucalypt plantations present a naturally high level of hydrophobic compounds leading to the formation of highly hydrophobic soils (Abelho and Graça 1996; Doerr et al. 1998; Zavala et al. 2009). Soil hydrophobicity can be enhanced after fire due to the volatilization of organic compounds which condense downwards, following the temperature gradient in the soil. This process leads to the formation of a hydrophobic layer underneath the soil surface which may contribute to an increased overland flow and increased soil erosion (Neary et al. 1999; Varela et al. 2005). Different practices have been proposed to break the repellent hydrophobic layer, including the use of surface tillage. However, this is also a risky alternative as soil labour may contribute to higher soil erosion rates in the denuded soils which result from wildfires in burned plantations. Further details on post-fire management measures and alternatives to prevent and mitigate soil erosion can be found in Chap. 5, of the present book.

## 10.3.3   The Restoration of Natural Vegetation
## and the Control of Invasive Species

The restoration of natural vegetation in burned exotic forests may present different aspects, mainly depending on the type of exotic forest. In the case of plantations, the demands of the international forest markets are leading forest industries to certify their products, according to sustainable criteria developed by different certification schemes (Georgiadis and Cooper 2007; Pettenella 2000). Therefore, besides the more common objective of intensive wood production, managers are being directed towards more sustainable and diverse management models. These models aiming to enhance ecological services such as biodiversity, involve tradeoffs in order to achieve a certain balance between goods and ecological services (Carnus et al. 2006). This is leading for example to the necessity of keeping patches of natural vegetation within the managed area, taking advantage of fire occurrence as a window of opportunity to convert areas of lower productivity into conservation areas (Silva et al. 2007). However there are few reported experiences of conversion of burned exotic forests into natural vegetation, at least for southern Europe. Depending on the species concerned, the conversion of burned areas of exotic forests into natural vegetation may involve a considerable effort both in technical as in economical terms. Particularly the species better adapted to fire occurrence may be difficult to replace or to control. These adaptations may result from a strong resprouting capacity, as it happens with common exotic broadleaved species planted in Europe like *Eucalyptus* spp. or *Robinia pseudoacacia*. Adding to this problem, the existence of seminal regeneration leading to plant invasions, may involve even larger efforts in order to prevent the expansion to neighbour areas (Catry et al. 2010a).

Given the importance of this latter aspect, we should distinguish different common situations involving the removal of exotic species in burned exotic forests. First, we should consider the rehabilitation of burned plantations in strategic areas. The rehabilitation may involve just felling, or in the case of resprouters it may involve additional operations of stump removal or stump killing on site. If the species regenerates from seeds, it may also involve the control of seminal regeneration. Second, we should consider the control of other arriving invasive species to a burned plantation. Given the opportunistic nature of most invasive species, a burned exotic plantation may allow the establishment of other exotic species which were not present before or which were present in a much smaller proportion. The more intensive management of plantations leads to a higher degree of disturbance, which may favour the establishment of invasive species (e.g. Fernández-Lugo et al. 2009). This will eventually lead to a conversion to mixed stands, if followed by abandonment (Silva et al. 2011) or on the contrary may assure the control of invaders, if there is a complete understory removal. Other aspects may have to be taken into account, when assessing the risk of plant invasions. Namely the proximity to roads, urban areas and human disturbance in general, has been proved to have a strong influence in the increase of invasive plants (Lee et al. 2004; Fernández-Lugo et al. 2009).

Finally, we should consider the post-fire regeneration and expansion of exotic self-sown woodlands. In this case the pre-fire vegetation is already the result of a plant invasion, which may become worst after fire occurrence.

Given its importance we should take a closer look at the problem of post-fire management of invasive species. In general, the management of invasive species should integrate a sequence of "key steps", including: (1) prevention; (2) early detection and rapid response; (3) eradication; (4) containment; (5) control; (6) restoration and mitigation; (7) monitoring and evaluation (Wittenberg and Cock 2005; Hulme 2006). Several of these steps are sequential, but not necessarily all, as some of them may be simultaneously applied.

Before embarking in expensive management options, it is essential to prioritize the management actions, concerning the species and the areas to be treated, taking into consideration factors such as the level of impacts, probabilities of success, value of ecosystem after recovery and available resources (Pyšek and Richardson 2010). To achieve suitable levels of success, persistence, scientific-based decisions and consideration of species at distinct stages of invasion (i.e. need for distinct management operations) are essential.

Preventing the introduction of species with high risk of becoming invasive is one of the most cost-effective management strategies (Pyšek and Richardson 2010). Several actions are considered in prevention, such as risk assessment, border interception, pathway and vector management, legislative frameworks and public awareness. Considering post-fire management in particular, and bearing in mind that fire works as a trigger to invasion by several plant species, prevention would also target the prevention of fire itself.

When prevention fails, the most effective way of minimizing impacts is to detect new introductions, small invasions or spreading propagules of species already spread, as early as possible, when populations are still localized, and quickly put in place procedures to eradicate, contain or control them. Early detection and rapid response (EDRR) is highly cost-effective, justifying major research and management efforts to improve protocols and techniques (Jarnevich et al. 2010). EDRR in the context of post-fire management can be somewhat easier as the management unit to monitor for early detection is well defined, i.e. corresponds to the burned area, instead of a complete region or country. Therefore, managers should focus on the early monitoring of the burned area in order to detect the occurrence of plant invasions. Nevertheless, as in other situations, it is essential that the infrastructure is in place, including control strategies, monitoring schemes, funding and human resources.

Eradication is the elimination of all the individuals of a species (including the seed-bank) within the management unit (Parkes and Panetta 2009). It is still a cost-effective option because it precludes negative impacts before they appear/aggravate. Nevertheless, it should only be attempted if it is considered to be feasible to avoid failure and associated wasted effort and financial costs. This includes the following conditions: (1) increase of the population estimated to be lower than planned removal rate; (2) there is not a source of new propagules; and (3) benefits of removing the species outweigh adverse effects of the removal. In post-fire situations eradication can sometimes be immediately precluded, e.g., if the burned area is surrounded by stands of the target invasive plant, which will work as a permanent source of new propagules. Commonly, costs of invasive plants eradication increase exponentially as the area of invasion expands, reason why it is more frequently considered for species with distributions still limited (Parkes and Panetta 2009).

Once eradication is no longer feasible, containment (i.e. limit the spread of an invasive species in the periphery of its range, in order to prevent its range from reaching its full potential (Kriticos et al. 2006) and control (reduce the impact and the abundance of an invasive species to an acceptable level in the long-term) are the most obvious options. Both should be planned and developed as long-term strategies, with special care given to guaranteeing resources and continued participation of the various stakeholders (Grice 2009). In general, the earlier the operations commence the higher is the likelihood of success and the more cost-effective they become. When fire promotes en masse germination of invasive species, as frequently happens with *Acacia* spp., it is crucial that control operations commence before the saplings/young trees reach maturity and start producing seeds that will replenish the seed bank.

To achieve a satisfactory level of success, control should include: (1) initial control, aiming to achieve a drastic reduction of the invasive species, which is usually very costly; (2) follow-up control, aiming to reduce any reinvasion after the initial control, which can include elimination of seedlings, root suckers and stump resprouts; and (3) maintenance control to sustain the invader at reduced levels, usually at lower costs and in the long term (Campbell 1993).

For eradication, control and containment, methods applied include mechanical, chemical, and biological control, habitat management and integrated management. All these methods are dependent on the species to manage.

Mitigation focuses on the affected native species rather than on the invasive species (Wittenberg and Cock 2001). The interventions of restoration and mitigation range from the simple removal of an invasive species to a variety of options that aims to favour natives, or even more complex options that involve engineering, native species reintroduction or translocation of a viable population of the species at risk to a "healthy", non-invaded ecosystem. In fact, both processes may be connected, as restoration processes may contribute simultaneously to eradication and to mitigation. One good example is the promotion of high densities of native plants in the recently burned areas in order to provide a closed canopy which may prevent the spread and growth of exotic plants (Keeley 2001).

When resources are scarce or no effective control measures are available, or when the degradation is already so profound that chances of successful recovery are very low or even nonexistent, doing nothing can be the best option, at least until some of the above scenarios change. This approach will avoid wasting resources that can be most effectively used into other areas or management actions.

Feedback from results of management actions is needed in order to modify, or even abandon, ineffective strategies, to allow other managers/stakeholders to learn from experience and even to validate the management programme. These issues can only be achieved through monitoring and evaluation of the actions taking place. Without these important steps, the management programme is not complete.

## 10.3.4  Post-Fire Management Practices and Alternatives for Common Types of Exotic Forests

### 10.3.4.1  Plantations of *Robinia pseudoacacia*

*Robinia pseudoacacia* (black locust) was introduced for the first time in France in the seventeenth century (Sykora 1990) being planted since then for ornamental and forestry purposes. It is now a very widespread species being naturalized in 32 European countries/regions (Lambdon et al. 2008). Although it is a prolific seed producer, germination rate is low and only on sites free of competition the seedlings are able to rapidly grow. The vegetative reproduction largely contributes to its success as invasive, forming vigorous root suckers and sprouting from stumps (Krízsik and Körmöczi 2000). It is less flammable than many other species, reducing fire frequency or intensity, particularly when invading pine ecosystems (Richburg et al. 2000). Because of these characteristics there are very few references to burned forests of *R. pseudoacacia* and, up to our knowledge, no references at all to the post-fire management of *R. pseudoacacia* plantations. It is not particularly associated with fire, but it can benefit from the absence of competition typical of burned areas. Krízsik and Körmöczi (2000) present a study describing the post-fire regeneration of a burned stand of *R. pseudoacacia* in Hungary, where the species has been widely planted. This study describes that the species was more aggressive and successful than the native competitor *Populus alba* in the occupation by resprouting of newly opened sites as a consequence of fire. A similar description is made by de Groot and Bordjan (2007) in Slovenia where the species is referred to invade burned areas.

As with other invasive tree species, seedlings and young saplings can be pulled or dug up, but roots must be extracted. Adult trees can be controlled mechanically, but follow up control (repeated more than once) is always needed due to its vigorous vegetative reproduction, including the formation of stump resprouts and root suckers. Repeated fires, can be used to control resprouting (Richburg et al. 2000), being at least temporarily efficient to reduce root suckers (Wieseler 2005).

Application of herbicides to cut stumps (immediately after cutting), or to stems, or using injection methods in the tree trunk, usually produce better results but still demand follow up control (often more than once). Spring is the best season to control, when new leaves are already developed. Special attention has to be given to monitoring plants after initial control, as plants apparently dead can resprout several years after treatment with herbicide (Wieseler 2005).

### 10.3.4.2  Plantations of *Eucalyptus* spp.

One important fire-prone group of exotic forests of southern Europe fire-prone (Moreira et al. 2009; Silva et al. 2009) requiring a particular attention in terms of post-fire management is constituted by the large area of *Eucalyptus globulus* plantations. *Eucalyptus globulus* (Tasmanian blue gum) is native to Australia and was introduced in Europe more than 150 years ago (Goes 1977). Portugal holds one of the largest surfaces of *E. gloublus* plantations in the world. According to the National Forest Inventory it is now the second most widespread tree species of the country (AFN 2010). *Eucalyptus globulus* is cultivated through a coppice system using rotations of 10–12 years. *E. globulus* plantations represent an important income, since they represent practically the only source of raw material to supply the paper pulp industry. Given the demands of the global paper market, the pulp industry and the forest managers are becoming more and more concerned about the sustainable management of *E. globulus* plantations.

*Eucalyptus globulus* is not a prolific seeder, but in turn it has strong resprouting capacity, in part due to the presence of a lignotuber which supplies the necessary carbohydrates to the regenerated tissues (Whittock et al. 2003). However, some mortality occurs in burned plantations. Silva et al. (2007) present mortality data (only graphical information) from burned plantations in Portugal after the intense fires of 2003. In stands up to 5 years of age there was 100% mortality (% wood volume), whereas older stand-age classes, presented mortality rates roughly between 50 and 60%. An important difference was also registered for the different coppice rotations, with mortality rates around 20–30% in the first rotation, reaching a maximum of around 80% in the third (last) rotation. However, there are inconsistent references, as in the case reported by Catry et al. (2010b) who found 100% survival in a sample of 60 burned trees monitored along 3 years. Marques et al. (2011) used data from 1648 trees to model mortality at the stand and tree levels. These models show that mortality increases with stand basal area and that it is higher for irregular stands, but trees with higher diameters are less prone to die when a wildfire occurs. According to these results the authors propose that silvicultural models should be directed to lower densities and higher tree diameters.

Given the specific economical role of forest plantations, very few references report the concern with ecosystem restoration in burned eucalyptus forests. Nonetheless, Silva et al. (2007) present a description of post-fire management problems and solutions encountered by a pulp company in Portugal in a burned area resulting from the 2003 fire season. Post-fire interventions in this area were decided according to three different management models: (a) clear cutting, stump removal and reforestation with the same species in productive areas; (b) clear cutting and selection of stump resprouts along one more rotation in areas with marginal productivity, for posterior conversion to natural vegetation; and (c) clear cutting, stump removal and conversion to native forest in ecological corridors, including stream lines. Other options included the postponing of salvage logging, for example. Postponing the removal of wood had the purpose of allowing charcoal wash by rainfall, in order to prevent negative effects in the quality of the paper pulp. The authors recognize that from a soil protection point of view it would be preferable to cut the trees earlier, since the resulting slash can help preventing soil erosion. Therefore, in order to provide some protection, trees are debarked on site up to 5 m high, which helps producing some mulching to cover the soil. The remaining bark is used to produce energy in the pulp mills. The authors refer the need of preserving snags from species other than *E. globulus*, resulting from fire. These should be left on site in order to provide habitat for animals. After logging, site preparation included stump destruction using heavy machinery, and posterior soil ripping and ploughing, following the contour lines. In the case of stream lines, interventions were in the sense of preserving the native vegetation, while reducing the fuel load, for fire hazard reduction. Here the problem encountered was the seminal regeneration of *E. globulus* which was preventing the restoration of these ecological corridors. Other cases of conversion to native vegetation included areas of lower productivity and steep slopes. Since the dominant native tree species in the area was *Quercus suber*, the intervention was also in favour of this species, by preserving and protecting the existing saplings.

One of the problems encountered in this post-fire management example, was the removal of *E. globulus* plants. Given its characteristics (fast growth, vigorous resprouting, seed dispersal and fire resistance), *E. globulus* can be very difficult to control. We should distinguish the control of adult trees from the control of seedlings and saplings. *E. globulus* adult trees can be successfully controlled by felling the tree cutting as low as possible (less than 10–15 cm), followed by immediately brushing/spraying the stump with a solution of chemical herbicides (glyphosate or triclopyr) (Bossard et al. 2000). The stump should be cleaned from sawdust in order to obtain maximal absorption. In resprouted stumps where cut and brush is not effective, it is advised to make incisions at 45° on the tree trunk with active vascular cambium and apply the herbicide in the incisions. If no chemicals are to be used, the continuous removal of resprouts will exhaust the tree resources and can be an alternative in a limited number of stumps. Alternatively stumps may be grinded or extirpated using heavy machinery down to 60 cm bellow soil level (Bossard et al. 2000). This is a costly operation which has been used in Portugal by pulp companies at the end of the last rotation, to reinstall a new plantation. Another alternative consists of covering the tree stumps and the surrounding ground 1 m out from the base of the trunk with landscape fabric and leave for 6–12 months (Cal-IPC 2006).

A different problem is the control of seedlings and saplings. In a technical report, Catry (2000) found that the seminal regeneration of *E. globulus* in a partially burned protected area, increased 16 times in 23 years (from only 4 ha to 64 ha). The invasive behavior of *E. globulus* in other regions of the world has been reported by different references, particularly in California (Cal-IPC 2006), Chile (Becerra and Bustamante 2008) and Western Australia (Virtue and Melland 2003), where the species is exotic. The control of *E. globulus* at the seedling/sapling stage can use both mechanical and chemical methods. If seedlings/saplings are small, they can just be easily pulled out, when the soil is wet (Cal-IPC 2006). An alternative is to spray with herbicide before saplings reach 3 m in the years following germination, although results are not always satisfactory (Bossard et al. 2000). This latter alternative has been used in Portugal along highways.

### 10.3.4.3  Plantations of *Pinus radiata*

*Pinus radiata* (Monterrey pine) is native in California and it is one of the most widely cultivated pine species in the world. It is also a very important tree species in southern Europe, particularly in Spain where it is cultivated for timber because of its fast growth. In terms of fire ecology it is considered a fire sensitive species, showing higher rates of mortality than other common pine species in Europe (Fernandes et al. 2008). However, the use of prescribed burning for fuel management in plantations is referred by several authors (see Fernandes et al. 2008). It is considered an invasive species in different parts of the world including South Africa, South America, New Zealand and Australia (Allen et al. 1995; Richardson 1998; Williams and Wardle 2005; Becerra and Bustamante 2008). Although it has established in Spain, Portugal (mainland and Madeira island) and France (DAISIE 2008), it is absent from the list of invasive species in Spain (Dana et al. 2004) where it has been widely planted, and it is referred as a non-regenerating species in the Canary islands (Fernández-Lugo et al. 2009). Therefore, the problem of invasive behavior after fire occurrence is something which has not yet gained importance in Europe, at least in the available literature. However, there are references mentioning post-fire germination in northern Spain, which may be an indication that the naturalization process is still ongoing (Verdú 2003). In terms of fire ecology *P. radiata* has been considered a fire evader, sensu Rowe (1983), as seeds are more easily released from the serotinous cones after heating by fire (Reyes and Casal 2002; Fernandes et al. 2008). Nonetheless seeds are not stimulated to germinate by fire nor by ashes (Reyes and Casal 2004).

Also few references are available reporting the post-fire management of *P. radiata* plantations in Europe, although different references can be found for other parts of the world. Lindenmayer and Noss (2006) report the post-fire management practices in a burned plantation of *P. radiata* in Australia, arguing about the correctness of the salvage logging practices which overlooked the possibility of conversion to natural vegetation. Also Smith et al. (2011) report the combined effect of fire and salvage logging in a *P. radiata* catchment in Southeastern Australia, which caused a substantial increase in runoff compared to nearby burned native forest catchments and to pre-fire conditions. The burned stand showed strong water repellency and the highest runoff velocities occurred in log drag-lines formed by cable harvesting. According to these findings, the authors present several management recommendations which are extendable to other plantation types, for mitigating salvage logging impacts. Post-fire salvage operations using cable harvest techniques should adopt measures to avoid the formation of drag-lines, particularly where the drag-lines are likely to occur in radial patterns converging downslope. Other proposed measures include the distribution of harvest slash, the application of seeding and mulching and the use of contour-felled logs as barriers to trap sediments on hillslopes.

Burned plantations of *P. radiata* have been associated with large increases in storm-flows and soil losses, caused in part by reduced infiltration resulting from water repellency in the soils of the burned stands (Scott 1993; Garcia-Chevesich et al. 2010). Garcia-Chevesich et al. (2010) suggest the use of different techniques to prevent post-fire water repellency. These practices may include the mechanical disaggregation by scarification of the first few centimeters of the mineral soil surface or deeper ripping, increasing soil pH, the use of wetting agents and avoiding the accumulation of large amounts of organic matter previous to fire.

As with other types of industrial plantations, poor management practices associated with site preparation for reforestation purposes may have high negative impacts in nutrient and soil losses. Olarieta et al. (1999) report losses between 30% and 100% of nutrients and 70% of organic matter, after practices of scalping and down-slope ripping after site preparation for *P. radiata* reforestation in the Basque Country in Spain.

#### 10.3.4.4   *Acacia* spp. Woodlands

The exotic genus *Acacia* (wattles), in particular the Australian acacias, have been widely planted around the world and many species have revealed invasive behaviour in several southern European countries (e.g. Portugal, Spain, France and Italy). According to data produced by the 1995–96 National Forest Inventory *Acacia*-dominated woodlands were estimated to cover 18,500 ha in Portugal (Godinho-Ferreira et al. 2005). The invasiveness of the genus is partially attributable to the prolific production of seeds, with high longevity, which accumulate in soil seed banks. Seeds are stimulated by fire which makes the management of this genus particularly difficult in post-fire situations. In fact, the high frequency of fires in some southern European countries is impossible to dissociate from the expansion of several *Acacia* species (e.g. *Acacia dealbata, A. melanoxylon, A. longifolia*).

Besides being prolific seed producers, some (but not all) species of the genus *Acacia* resprout vigorously from stump or root suckers which also contributes significantly to their invasiveness, being especially problematic in post-control situations. For example, *A. dealbata, A. melanoxylon and A. mearnsii* are vigorous resprouters (i.e., follow up control has to remove seedlings, stump resprouts and root suckers). When deciding the best methods to contain/eradicate/control *Acacia* species, the correct identification (as with other species) is thus crucial, as misidentifications can lead to adoption of unsuitable (or at least not the best) methods, jeopardizing levels of success and/or implying waste of resources.

In post-fire situations (but also after other disturbances that create large gaps) the en masse germination is very frequent, originating massive quantities of seedlings. In general, seedlings and saplings of *Acacia* species can be hand-pulled or dug up. The success of this method depends on the type of soil (saplings in sand dunes are much easier to pull than in more aggregated soils) and on the season (saplings in moist soils are easily pulled). In order to improve the sustainability of this method, two factors should be considered: (1) many seedlings die in the early stages of growth and removing them very early would be a waste of resources; (2) seedlings should not be mistaken with root suckers as those would demand a different method to be eliminated. In post-fire situations, *Acacia* seedlings are generally more frequent than root suckers.

Moderate-intensity burning in moist conditions is another option to eliminate saplings, with the additional benefit of contributing further to deplete the invader seed banks (Holmes and Cowling 1997; Richardson and Kluge 2008). In general, moderate-intensity burning destroys fewer *Acacia* seeds but is less detrimental for native seed banks and is therefore probably the best option.

Several methods are available to treat older saplings and adult trees, including debarking, cut stump treatments with chemical application, or fell and burn treatments.

In species that grow upright forming a trunk (e.g. *A. dealbata or A. melanoxylon and A. mearnsii* with a few years), and as long as the stem is round and the bark is (still) smooth, debarking can be one of the best options. The bark should be stripped, all around the stem circumference, from waist height (ca. 70–100 cm) down to the soil surface. Ring debarking frequently fails with wattles, particularly in species that are vigorous resprouters, because it does not prevent basal resprouting. Debarking should only be applied when the vascular cambium is active (Campbell 1993). All trees must be treated, since untreated trees may help treated trees to survive. This technique is particularly appropriated for conservation or other sensitive areas, where chemicals are not an option, or when the infestations are not too extensive. Despite hard-laborious, and consequently expensive, it has the great advantage of frequently being able to prevent resprouting (from both stumps and root suckers).

Another control option is cutting followed by immediate application of herbicide (glifosate has given good results) in the stumps. Resprout from stumps is easily prevented with the right herbicide concentration but root suckers may occur, demanding follow up control. Follow up control to remove seedlings, root suckers and stump resprouts may include hand pulling (more successful for seedlings), and cutting with application of herbicide or spraying when regrowth is up to a 1 m height in dense infestations. Spraying generally gives better results with species that have compound leaves (e.g. *A. dealbata, A. mearnsii*) than with phyllodinous species (e.g. *A. longifolia, A. melanoxylon*).

Parallel to control operations, silvicultural approaches can provide more sustainable results. These include the promotion of high densities of native plants in the recently burned areas in order to provide a closed canopy.

Monitoring, at least annually, is needed to determine the best time interval for follow up control (which frequently needs to be repeated several times) and later for maintenance control. Without follow up and maintenance control, reinvasion of the area will be the most probable outcome, with total waste of the resources initially allocated. Several *Acacia* species can reach reproductive maturity after 1–5 years. This stage is the time limit to apply control measures in order to prevent the seed rain from replenishing the seed bank.

Biological control with exotic mono-specific agents, namely gall formers from the genus *Trichilogaster* (Hymenoptera) and *Melanterius* weevils (Curculionidae) which target seed production, is successfully used in South Africa to control Australian *Acacia* species. This is considered by several authors as the most cost effective, sustainable and reliable option for long term management of established invasive Australian wattles (Dennill et al. 1999; Wilson et al. 2011). This metho-dology is not yet an option in Europe despite is being considered in Portugal (Marchante et al. 2011a).

### 10.3.4.5  *Ailanthus altissima* Woodlands

*Ailanthus altissima* (tree of heaven) is a widespread invasive species in Europe, including both northern and southern countries (DAISIE 2008). It is a dioiceous species which establishes in disturbed habitats, such as urban environments, surrounding agricultural fields or transport networks (Burch and Zedaker 2003; Constán-Nava et al. 2010). Although it is not particularly associated with fire, it may benefit from the absence of competition provided by a disturbance regime typical of burned areas. It is a prolific seed producer, with very efficient vegetative reproduction (root suckers and stump resprouts), which enables it to grow rapidly and form impenetrable thickets. Vigorous and repeated resprouting after control, make management plans particularly long-lasting (Constán-Nava et al. 2010).

Seedlings or young saplings may be hand-pulled or dug up, which is easier when soil is moist. The entire plant should be removed, including all roots and fragments, which may otherwise resprout. Root suckers should not be confused with seedlings, as they demand a different approach to be removed.

Adult trees and large saplings can be controlled with herbicides, which may be applied using different treatments: foliar spray, directly on the bark 30 cm above soil level, on the stump, or injection in wounds made on the tree stem. Despite being relatively easy to kill the above ground portion of the trees, the root system has also to be killed or seriously damaged to prevent or limit stump sprouting and root suckering. This can be better achieved with standing-tree methods, as intact trees will consume some of the carbohydrate root reserves. The injection or hack and squirt method is one of the most effective, as it minimizes sprouting and suckering when applied during the summer. The cuts should be spaced leaving uncut living tissue between them. Cutting alone is usually counter-productive as it induces large numbers of stump sprouts and root suckers (Cal-IPC 2006; Swearingen and Pannill 2009).

Follow up control is essential in order to treat any new suckers or seedlings (cut, sprayed or pulled) as soon as possible, especially before they are able to rebuild root reserves. Establishing a thick tree or grass cover, should provide enough shade to prevent the establishment of seedlings. Targeting large female trees for control will help reducing the seed rain.

## 10.4   Case Studies

### 10.4.1   Fire Effects on Acacia longifolia Woodlands in Nature Reserve of São Jacinto Dunes – Possible Management Options

*Acacia longifolia* (long-leafed wattle) is a leguminous tree (Fabaceae) native to Australia whose seeds are stimulated by fire. It was introduced in many locations around the world, including the Portuguese coastal dunes. This was the case of the Nature Reserve of São Jacinto Dunes (hereafter São Jacinto), where *A. longifolia* became invasive forming pure and mixed stands covering around 350 ha (Fig. 10.1). São Jacinto is located in the central-northern coast of Portugal, on a strip of sand

**Fig. 10.1** Distinct views of the São Jacinto dune ecosystem. Some of the native habitats still present (*left*) *vs.* dunes invaded by *A. longifolia* after fire (*right*), and view of the understory, where *A. longifolia* excluded all plant species (*center*). Photos H. Marchante

dunes bordered by the Atlantic Ocean and the Vouga river estuary. *Acacia longifolia* was introduced in São Jacinto along the first 4 decades of the 20th century to curb movement of sand. After that it spread forming extensive, dense stands, particularly promoted by fire events. The most recent fire in São Jacinto occurred in the summer of 1995, burning ca. 200 ha and eliminating all the aboveground vegetation in the affected area. After the fire there was a drastic transformation of the dune landscape into arboreal, nearly monospecific stands of *A. longifolia* (Marchante et al. 2003). Previous to fire there was a mosaic of sparse native vegetation including herbs, few shrubs and trees and pine plantations. The understory of these plantations included a few scattered individuals of *A. longifolia*.

A study was conducted aiming to evaluate the impacts of *A. longifolia* invasion and the resilience of the ecosystem in burned coastal dunes. The results of the study were fundamental to support the establishment of post-fire management alternatives. Seven years after the fire, experimental plots were established including: (1) invaded plots where *A. longifolia* was left intact; (2) plots cleared of *A. longifolia* by cutting the trees with chainsaws at ground level; and (3) plots unburned/non-invaded. Parameters measured included: species richness, diversity and cover, native and invader seed bank and seed rain, soil C and nutrients, microbial processes as well as soil water content.

After the fire, the density of *A. longifolia* increased significantly in the burned areas reaching ca. 70% cover in 7 years (Marchante et al. 2011b).The establishment of *A. longifolia* resulted from the rapid fire-induced germination of seeds accumulated in the seed bank, although some input of seeds from neighboring areas may have also occurred. Although germination en masse after the fire may have contributed to deplete the seed bank, this was quickly counterbalanced because *A. longifolia* plants rapidly reached maturity due to the lack of management and started replenishing the seed bank vigorously. Annual seed rain has been quantified in São Jacinto as reaching an average of 12,000 seeds/m$^2$. Nonetheless, a considerable percentage of these seeds are lost before and during the formation of the soil seed bank. In 2004 (9 years after the fire), the accumulated seed bank in the burned areas was quantified as 500 seeds/m$^2$, particularly beneath the tree canopies (Marchante et al. 2010).

As plant density increased, additional changes included the accumulation of a thick litter layer (Marchante et al. 2008c), mainly composed of *A. longifolia* phyllodes, and the significant decrease of light intensity reaching the soil. As a consequence of this quick establishment of *A. longifolia* and resulting changes, the native plant species were outcompeted and both plant species richness and native plant cover decreased significantly. Similarity coefficient and species traits analyses revealed that ecological effects were aggravated, because several lost species were specific of the dune ecosystem. While species characteristic from dunes were far more numerous in non-invaded areas, the opening of large gaps that occurred as a result of fire, favored the entrance of propagules from generalist species (e.g. *Conyza* spp., *Sonchus* spp.,) into the seed bank of burned areas. These species became the most abundant in the burned areas, despite this was only revealed after experimental clearing of *A. longifolia*. In non-invaded areas (i.e. native dunes not burned in 1995) the spread of *A. longifolia* was not observed, despite the presence of A. longifolia seeds in the seed bank (Marchante et al. 2011c).

The native seed bank was also affected by the fire. It decreased both due to the seed germination of fire-adapted species and to the seed destruction of fire-sensitive species. In these nutrient-poor sand dunes, soil C and nutrients, especially total N, progressively increased after invasion, but these increments were not significant (only became evident in areas invaded for several decades). Nonetheless microbial processes (catabolic diversity) and mineral N (higher nitrification) were significantly affected (Marchante et al. 2008c, b).

Results observed in cleared plots, showed that these areas, despite the drastic changes observed, still maintained some resilience even if autogenic recovery was slow and partially threatened by reinvasion by *A. longifolia* and other exotic species (Marchante et al. 2009, 2011b). Native species (although many were generalists and not dune specific) noticeably dominated all cleared areas suggesting that at least some degradation thresholds had not yet been crossed. After removal of the invader, soil microbiology and chemistry (Marchante et al. 2009), light and litter layer conditions, native seed-bank (Marchante et al. 2011c) and several components of plant communities gradually recovered and increasingly began to resemble the situation in analogous non-invaded/unburned areas.

These data indicate that future management of areas invaded after fire (namely areas with high conservation status), should include active restoration practices after initial control in order to help the recovery of native vegetation. Additionally, follow up control to manage *A. longifolia* and the other invasive plants that germinate, should not be neglected. Moderate-intensity burning destroys fewer *A. longifolia* seeds than higher fire intensities, but it is less damaging for native seed banks and is therefore probably the best management option at this stage. Eradication of *A. longifolia* is an unrealistic goal in this system and containment should be a priority as new invasion foci (still manageable) occur several metres from the main thickets (Marchante et al. 2010; Rejmánek and Pitcairn 2003). In the case of *A. longifolia*, 2 years is the time limit after each operation, because plants begin producing seeds by that time. Additional measures may include introduction (sown or transplanted) of desirable species to accelerate recovery. This has been successfully achieved in some restoration projects (Hartman and McCarthy 2004). Transplanting of saplings will probably be most successful because these will have a height advantage over the invasive seedlings (Galatowitsch and Richardson 2005).

## 10.4.2 The Controversial Case of Eucalyptus globulus in Portugal

*Eucalyptus globulus* is a very important species in Portugal and Spain, because of the large area of plantations. Besides other post-fire management problems, the seminal regeneration of *E. globulus* in burned areas has gained importance in recent years. However, the literature about the regeneration processes is surprisingly scarce. It is well established that *E. globulus* starts producing seeds at 3–5 years of age, but closing canopies in plantations (not applicable to stand edge and isolated trees) tend to suppress flowering (Bossard et al. 2000; Potts et al. 2008). Virtue and Melland (2003) mention that seedlings normally appear close to the edge of the tree canopy (5–15 m) which is consistent with the references for other Eucalyptus species (Potts 1990). There is information from Bean and Russo (1986) indicating maximum distances between 30 and 90 m. The only published information for Portugal reports the existence of seedlings at 150 m from the stand edge (Catry et al. 2010c). Seedling colonization may be a consequence of different fire-related means of seed dispersal including the transportation by convection columns in crown fires or by surface or rill flow after heavy rain events in burned areas (Kirkpatrick 1977; Potts 1990; Virtue and Melland 2003). This aspect may be coupled with the existence of heat-resistant seed capsules from which seeds may be more easily released after fire occurrence (Gill 1977; Ashton 1981). Nonetheless, there seems to exist a negative effect of heat on the germination rates of *E. globulus* seeds and a similar relationship was applicable to ash treatments (Reyes and Casal 1998). In terms of soil conditions, Virtue and Melland (2003) report high germination rates in disturbed sites, including firebreak grading and roadsides. In general, the available literature refers that *E. globulus* does not regenerate profusely, contrarily to other alien species, outside its natural range (Virtue and Melland 2003; Cal-IPC 2006). Nonetheless Virtue and Melland (2003) mention observed densities in Kangaroo Island (outside *E. globulus* natural range) of up to 1–2 naturally-regenerated trees/m$^2$, which is

**Fig. 10.2** Images illustrating the post-fire establishment of *E. globulus* in different areas of Central Portugal. Relatively sparse recruitment at later (left) and earlier (right) stages. Dense recruitment (center). Photos above by J.S. Silva; photo below F.X. Catry

consistent with unpublished field observations in Portugal. In what concerns seedling demography (recruitment and mortality), a study from Chile presented a survival rate of only 15%, after 6 months (Becerra and Bustamante 2008).

A preliminary study was carried out in Central Portugal aiming to characterize *E. globulus* seminal regeneration and accompanying vegetation in burned areas (Fig. 10.2). With this purpose a set of 39 sampling plots was established in 2010, using the regular grid of the National Forest Inventory (2005/2006). The plots were circular with a radius of 5 m. Associated with each plot there was also a set of 4 sub-plots, with 1.78 m radius, corresponding to the four main directions (North, South, West and East). The main plots were used to characterize the stand and the site, whereas the sub-plots were used to characterise the seminal regeneration and the accompanying vegetation. *E. globulus* saplings were classified according to height and diameter. From these plots, 31 were previously occupied by pure *E. globulus* stands and 8 were mixed stands of *E. globulus* and *Pinus pinaster*. Most of the plots were set in areas which burned in 2005 in a large forest fire which has surrounded the city of Coimbra, reflecting a post-fire history of 5 years. In most of the plots there were post-fire management practices aimed at harvesting the burned wood. In 93% of pure stands and 80% of mixed stands, there were clear cuts or selective cuts on resprouted stumps. Presumably all plots corresponded to small private properties.

The preliminary results (Estevens 2011; Lameiras 2011) showed that 29 out of 39 plots (74%) presented seminal *E. globulus* regeneration. Saplings between 50 cm and 130 cm height presented a density of $0.36 \pm 0.35$ plants/m$^2$ (mean $\pm$ SD); saplings > 130 cm height and < 5 cm diameter presented a density of $0.62 \pm 0.69$ plants/m$^2$; saplings > 130 cm height and diameter between 5 cm and 7.5 cm presented a density of $0.06 \pm 0.02$ plants/m$^2$. In the overall (all subplots and all size classes) there was a sapling density of $0.88 \pm 0.96$ plants/m$^2$. These values ranged between 0.03 plants/m$^2$ and 3.29 plants/m$^2$. In order to assess the dispersal distance from putative mother plants, the nearest adult trees were identified and the respective distance to plot was measured. The average distance was $18.33 \pm 24.79$ m and maximum distance recorded was 105 m. Although these latter results match the referred literature, some saplings could have also resulted from felled trees removed afterwards, or from other unknown source. Therefore, further results are still needed in order to estimate the maximum seed dispersal potential of *E. globulus*. Given the abandonment of most plots, there was considerable regeneration of other plant species. On average, there were $14.9 \pm 6.3$ species registered on each plot in the pure stands, (no distinction made between herbaceous species except ferns) and $16.4 \pm 5.7$ species on mixed stands. This corresponded to Shannon-Wiener index values of $1.51 \pm 0.49$ and $1.71 \pm 0.28$, respectively.

These preliminary results are probably representative of the fate of burned areas dominated by former eucalypt plantations followed by lack of management, in Central Portugal. The result is a mixture of native and exotic vegetation which may lead to high fuel loads and increased fire hazard, with reduced ecological and economic value. These preliminary findings are in line with the conclusions from Silva et al. (2011) in two other distinct areas of Central Portugal.

## 10.5 Key Messages

- Exotic forests can present different post-fire management requirements and approaches depending on the specific type of forest. In particular, it is important to distinguish exotic industrial plantations from self-sown exotic woodlands.
- In the first case there are aspects related with soil conservation which deserve a particular attention, given the very intensive management practices common in industrial forestry.
- In the second case the problem of invasive species is gaining importance, since some species are actively promoted by fire occurrence.
- Nonetheless, the problems caused by plant invaders after fire do not seem to have a similar importance for all southern European countries. In fact, there are regions like the moister parts of the Iberian Peninsula, where invasive species seem to have a much greater importance than what is reported for other Mediterranean regions and countries.
- The exotic tree species which are relevant in terms of caused impact in a southern European context, being consistently referred by different authors are: *Acacia*

spp., *Ailanthus altissima, Eucalyptus* spp., *Pinus radiata* and *Robinia pseudoacacia*. With the apparent exception of *Pinus radiata*, these species are considered invasive in different lists produced by different authors and European organizations.

- From these it is particularly important to emphasize the case of *Acacia* species, because of the close relationship with fire, the high noxious impact in the countries where they were introduced, and the remarkable difficulties associated with their eradication or control in post-fire management situations.
- Nonetheless, well planned post-fire management may be used to manage these invader species, namely by depleting the seed bank through prescribed burning and by monitoring the invaded area. In the likely event of a reinvasion after a new fire, management is still possible.
- It is also important to refer the case of *E. globulus* plantations, both because of the high fire incidence, the potential post-fire management impacts on soil conservation, and the growing concern with its invasive behavior.
- Land abandonment after fire seems to be one of the commonest ways of promoting plant invasions, at least in certain areas. Therefore a particular attention should be given to areas where invasive species were present previous to fire, in order to prevent increased problems in the near future.

# References

Abelho M, Graça MAS (1996) Effects of eucalyptus afforestation on leaf litter dynamics and macroinvertebrate community structure of streams in Central Portugal. Hydrobiologia 324:195–204. doi:10.1007/bf00016391

AFN (2010) Relatório Final do 5º Inventário Florestal Nacional. Lisbon

Allen R, Basher L, Comrie J (1995) The use of fire for conservation management in New Zealand. Science and Research Division, Department of Conservation, Wellington

Andreu J, Vilà M, Hulme PE (2009) An assessment of stakeholder perceptions and management of noxious alien plants in Spain. Environ Manag 43:1244–1255

Ashton D (1981) Fire in tall open-forests (wet sclerophyll forests). In: Gill AM, Groves RH, Noble IR (eds) Fire and the Australian biota. The Australian Academy of Science, Camberra, pp 339–366

Auld TD, O'connell MA (1991) Predicting patterns of post-fire germination in 35 Eastern Australian Fabaceae. Aust J Ecol 16:53–70

Bachmann A, Allgöwer B (2001) A consistent wildland fire risk terminology is needed! Fire Manag 61:28–33

Bean C, Russo MJ (1986) Elemental Stewardship abstract for *Eucalyptus globulus*. The Nature Conservancy. http://www.invasive.org/gist/esadocs/documnts/eucaglo.pdf

Becerra PI, Bustamante RO (2008) The effect of herbivory on seedling survival of the invasive exotic species *Pinus radiata* and *Eucalyptus globulus* in a Mediterranean ecosystem of Central Chile. For Ecol Manag 256:1573–1578

Bossard CC, Randall JM, Hoshovsky MC (eds) (2000) Invasive plants of California's wildlands. University of California Press, Berkeley

Brooks ML, D'Antonio CM, Richardson DM, Grace JB, Keeley JE, Di Tomaso JM, Hobbs RJ, Pellant M, Pyke D (2004) Effects of invasive alien plants on fire regimes. BioScience 54:677–688

Burch PL, Zedaker SM (2003) Removing the invasive tree *Ailanthus altissima* and restoring natural cover. J Arboric 29:18–24

Cal-IPC (2006) California invasive plant inventory. California Invasive Plant Council. www. cal-ipc.org. Accessed 10 June 2011. Cal-IPC Publication 2006-02

Campbell P (1993) Wattle control, handbook no 3. Plant Protection Research Institute Handbooks. Plant Protection Research Institute, Agriculture Research Council, Pretoria

Carnus JM, Parrotta J, Brockerhoff E, Arbez M, Jactel H, Kremer A, Lamb D, OHara K, Walters B (2006) Planted forests and biodiversity. J For 104:65–77

Catry F (2000) Projecto de elaboração de cartografia digital de ocupação do solo para a Tapada Nacional de Mafra e área envolvente. Relatório de Projecto. Estação Florestal Nacional, Lisbon

Catry F, Bugalho M, Silva JS, Fernandes P (2010a) A gestão da vegetação pós-fogo. In: Moreira F, Catry F, Silva JS, Rego F (eds) Ecologia do fogo e gestão de áreas queimadas. ISAPress, Lisbon, pp 289–327

Catry F, Rego F, Moreira F, Fernandes P, Pausas J (2010b) Post-fire tree mortality in mixed forests of central Portugal. For Ecol Manag 260:1184–1192

Catry F, Silva JS, Fernandes P (2010c) Efeitos do fogo na vegetação. In: Moreira F, Catry F, Silva JS, Rego F (eds) Ecologia do fogo e gestão de áreas queimadas. ISAPress, Lisbon, pp 49–86

Chytrý M, Pyšek P, Wild J, Pino J, Maskell LC, Vilà M (2009) European map of alien plant invasions based on the quantitative assessment across habitats. Divers Distrib 15:98–107

Constán-Nava S, Bonet A, Pastor E, Lledó MJ (2010) Long-term control of the invasive tree *Ailanthus altissima*: insights from Mediterranean protected forests. For Ecol Manag 260:1058–1064

D'Antonio CM (2000) Fire, plant invasions, and global changes. In: Mooney HA, Hobbs RJ (eds) Invasive species in a changing world. Island Press, Washington, pp 65–93

D'Antonio CM, Vitousek PM (1992) Biological invasions by exotic grasses, the grass/fire cycle, and global change. Annu Rev Ecol Syst 23:63–87

DAISIE (2008) European invasive alien species gateway. http://www.europe-aliens.org

DAISIE (2009) Handbook of alien species in Europe. Invading nature-Springer series in invasion ecology, vol 3. Springer, London

Dana ED, Sobrino E, Sanz-Elorza M, Bañares A, Blanca G, Güemes Heras J, Moreno Sainz JC, Ortiz S (2004) Plantas invasoras en España: un nuevo problema en las estrategias de conservación. In: Bañares Á, Blanca G, Güemes J, Moreno J, Ortiz S (eds) Atlas y libro rojo de la flora vascular amenazada de España. Dirección General de Conservación de la Naturaleza, Madrid, pp 1010–1029

de Groot M, Bordjan D (2007) Possibilities for fire as a management tool on Kras (SW Slovenia): a bird's perspective. Acrocephalus 28:3–15

Dekker L, DeBano L, Oostindie K, Elsen E, Ritsema C (2003) More than one thousand references related to soil water repellency. In: Ritsema CJ, Dekker LW (eds) Soil water repellency: occurrence, consequences, and amelioration. Elsevier, Amesterdam, pp 315–345

Dennill GB, Donnelly D, Stewart K, Impson FAC (1999) Insect agents used for the biological control of Australian *Acacia* species and *Paraserianthes lophanta* (Willd.) Nielsen (Fabaceae), Biological control of weeds in South Africa (1990–1998). In: Olckers T, Hill MP (eds.) South Africa. African Entomology Memoir, Pretoria, 1:45–54

Dickinson K, Kirkpatrick J (1985) The flammability and energy content of some important plant species and fuel components in the forests of southeastern Tasmania. J Biogeogr 12:121–134

Doerr SH, Shakesby RA, Walsh RPD (1998) Spatial variability of soil hydrophobicity in fire-prone eucalyptus and pine forests. Port Soil Sci 163:313

EEA (2007) European forest types. Categories and types for sustainable forest management reporting and policy. European Environment Agency Report, vol 9. European Environment Agency, Copenhagen

EEA (2008) European forests–ecosystem conditions and sustainable use. European Environment Agency report, vol 3. European Environment Agency, Copenhagen

Elton CS (1958) The ecology of invasions by animals and plants. University of Chicago Press, Chicago

Estevens A (2011) Amostragem da regeneração natural de *Eucalyptus globulus* e de *Pinus pinaster* em áreas ardidas. Polytechnic Institute of Coimbra, Coimbra

Fernandes PM (2009) Combining forest structure data and fuel modelling to classify fire hazard in Portugal. Ann For Sci 66:415

Fernandes PM, Vega JA, Jimenez E, Rigolot E (2008) Fire resistance of European pines. For Ecol Manag 256:246–255

Fernández-Lugo S, Arévalo J, Sierra A (2009) Gradient analysis of exotic species in *Pinus radiata* stands of Tenerife (Canary Islands). Open For Sci J 2:63–69

Ferreira AJD, Coelho COA, Shakesby RA, Walsh RPD (1997) Sediment and solute yield in forest ecosystems affected by fire and rip-ploughing techniques, central portugal: a plot and catchment analysis approach. Phys Chem Earth 22:309–314

Ferreira A, Coelho C, Walsh R, Shakesby R, Ceballos A, Doerr S (2000) Hydrological implications of soil water-repellency in *Eucalyptus globulus* forests, north-central Portugal. J Hydrol 231:165–177

Galatowitsch S, Richardson D (2005) Riparian scrub recovery after clearing of invasive alien trees in headwater streams of the Western Cape, South Africa. Biol Conserv 122:509–521

Garcia-Chevesich P, Pizarro R, Stropki C, Ramirez de Arellano P, Ffolliott P, DeBano L, Neary D, Slack D (2010) Formation of post-fire water repellent layers in Monterry pine (*Pinus radiata* D. Don) plantations in South-Central Chile. J Soil Sci Plant Nutr 10:399–406

Georgiadis NM, Cooper R (2007) Development of a forest certification standard compatible with PEFC and FSC's management requirements. A case study from Greece. Forestry 80:113–135

Gilbert J (1959) Forest succession in the Florentine valley. In: Papers and proceedings of the Royal Society of Tasmania, vol 93. pp 129–151

Gill AM (1977) Plant traits adaptive to fires in Mediterranean land ecosystems. In: Mooney H, Conrad CE (eds.) Symposium on the environmental consequences of fire and fuel management in Mediterranean ecosystems. General technical report WO-3. U.S. Department of Agriculture, Forest Service, Palo Alto, pp 17–26

Gill AM (1997) Eucalypts and fires: interdependent or independent? In: Williams JE, Woinarsk JCZ (eds) Eucalypt ecology: individuals to ecosystems. Cambridge University Press, Cambridge, pp 151–167

Godinho-Ferreira P, Azevedo A, Rego F (2005) Carta da Tipologia Florestal de Portugal Continental. Silva Lusitana 13:1–34

Goes E (1977) Os eucaliptos - ecologia, cultura produção e rentabilidade. Portucel, Lisbon

Gouveia C, Camara C, Trigo R (2010) Post-fire vegetation recovery in Portugal based on spot/vegetation data. Nat Hazard Earth Syst 10:673–684

Grice T (2009) Principles of containment and control of invasive species. In: Williams PA, Clout MN (eds) Invasive species management: a handbook of principles and techniques. Orford University Press, Oxford, pp 61–76

Hanley ME (2009) Thermal shock and germination in North-West European Genisteae: implications for heathland management and invasive weed control using fire. Appl Veg Sci 12:385–390. doi:10.1111/j.1654-109X.2009.01038.x

Hartman KM, McCarthy BC (2004) Restoration of a forest understory after the removal of an invasive shrub, Amur honeysuckle (*Lonicera maacktii*). Restor Ecol 12:154–165

Hill RS (1982) Rainforest fire in Western Tasmania. Aust J Bot 30:583–589

Holmes PM, Cowling RM (1997) The effects of invasion by *Acacia saligna* on the guild structure and regeneration capabilities of South African fynbos shrublands. J Appl Ecol 34:317–332

Hulme PE (2006) Beyond control: wider implications for the management of biological invasions. J Appl Ecol 43:835–847

Hulme PE, Pyšek P, Nentwig W, Vilà M (2009) Will threat of biological invasions unite the European Union. Science 324:40–41

Hunt MA, Unwin GL, Beadle CL (1999) Effects of naturally regenerated *Acacia dealbata* on the productivity of a *Eucalyptus nitens* plantation in Tasmania, Australia. For Ecol Manag 117:75–85

Jarnevich CS, Holcombe TR, Barnett DT, Stohlgren TJ, Kartesz JT (2010) Forecasting weed distributions using climate data: a GIS early warning tool. Invasive Plant Sci Manag 3:365–375. doi:10.1614/IPSM-08-073.1

JRC (2010) Forest fires in Europe 2009. Joint Research Centre, Ispra

Keeley JE (2001) Fire and invasive species in Mediterranean-climate ecosystems of California. In: Galley KEM, Wilson TP (eds.) Proceedings of the invasive species workshop: the role of fire in the control and spread of invasive species. Fire conference 2000, vol 11. Tall Timbers Research Station, Tallahassee, pp 81–94

Keeley JE, Franklin J, D'Antonio C (2011) Fire and invasive plants on California landscapes. In: McKenzie D, Miller C, Falk DA (eds) The landscape ecology of fire, vol 213. Ecological studies. Springer Netherlands, Amsterdam, pp 193–221. doi:10.1007/978-94-007-0301-8_8

Kirkpatrick JB (1977) Eucalypt invasion in southern California. Aust Geogr 13:387–393

Köble R, Seufert G (2001) Novel maps for forest tree species in Europe. In: Proceedings of the 8th European symposium on the physico-chemical behaviour of air pollutants: "a changing atmosphere!". Torino

Kriticos DJ, Alexander NS, Kolometz SM (2006) Predicting the potential geographic distribution of weeds in 2080. In: Preston C, Watts JH, Crossman ND (eds.) 15th Australian weeds conference: managing weeds in a changing climate. Weed Management Society of South Australia, Adelaide, pp 27–34

Krízsik V, Körmöczi L (2000) Spatial spreading of *Robinia pseudo-acacia* and *Populus alba* clones in sandy habitats. Tiscia 32:3–8

Lambdon PW, Pyšek P, Basnou C, Hejda M, Arianoutsou M, Essl F, Jarošík V, Pergl J, Winter M, Anastasiu P (2008) Alien flora of Europe: species diversity, temporal trends, geographical patterns and research needs. Preslia 80:101–149

Lameiras T (2011) Amostragem da regeneração natural de *Eucalyptus globulus* e de *Pinus pinaster* em áreas ardidas e diversidade vegetal. Polytechnic Institute of Coimbra, Coimbra

Lee CS, Cho HJ, Yi H (2004) Stand dynamics of introduced black locust (*Robinia pseudoacacia* L.) plantation under different disturbance regimes in Korea. For Ecol Manag 189:281–293

Lindenmayer DB, Noss RF (2006) Salvage logging, ecosystem processes, and biodiversity conservation. Conserv Biol 20:949–958. doi:10.1111/j.1523-1739.2006.00497.x

Lodge DM (1993) Biological invasions: lessons for ecology. Trends Ecol Evol 8:133–137

Mack RN, Simberloff D, Mark Lonsdale W, Evans H, Clout M, Bazzaz FA (2000) Biotic invasions: causes, epidemiology, global consequences, and control. Ecol Appl 10:689–710

Mandle L, Bufford J, Schmidt I, Daehler C (2011) Woody exotic plant invasions and fire: reciprocal impacts and consequences for native ecosystems. Biol Invasions 13:1815–1827. doi:10.1007/s10530-011-0001-3

Marchante E, Marchante H (2007) As exóticas invasoras. In: Silva JS (ed) Do castanheiro ao teixo, série Árvores e Florestas de Portugal (vol V), vol V, Árvores e Florestas de Portugal. Público/Fundação Luso-Americana para o Desenvolvimento/Liga para a Protecção da Natureza, Lisboa, pp 179–198

Marchante H, Marchante E, Freitas H (2003) Invasion of the Portuguese dune ecosystems by the exotic species *Acacia longifolia* (Andrews) Willd.: effects at the community level. In: Child LE, Brock JH, Brundu G et al (eds) Plant invasion: ecological threats and management solutions. Backhuys Publishers, Leiden, pp 75–85

Marchante E, Freitas H, Marchante H (2008a) Guia Prático para a Identificação de Plantas Invasoras de Portugal Continental. Imprensa da Univ. de Coimbra, Coimbra

Marchante E, Kjøller A, Struwe S, Freitas H (2008b) Invasive *Acacia longifolia* induce changes in the microbial catabolic diversity of sand dunes. Soil Biol Biochem 40:2563–2568. doi:10.1016/j.soilbio.2008.06.017

Marchante E, Kjøller A, Struwe S, Freitas H (2008c) Short and long-term impacts of *Acacia longifolia* invasion on the belowground processes of a Mediterranean coastal dune ecosystem. Appl Soil Ecol 40:210–217. doi:10.1016/j.apsoil.2008.04.004

Marchante E, Kjøller A, Struwe S, Freitas H (2009) Soil recovery after removal of the $N_2$-fixing invasive *Acacia longifolia*: consequences for ecosystem restoration. Biol Invasions 11:813–823. doi:10.1007/s10530-008-9295-1

Marchante H, Freitas H, Hoffmann JH (2010) Seed ecology of an invasive alien species, *Acacia longifolia* (Fabaceae), in Portuguese dune ecosystems. Am J Bot 97:1–11

Marchante H, Freitas H, Hoffmann JH (2011a) Assessing the suitability and safety of a well-known bud-galling wasp, *Trichilogaster acaciaelongifoliae*, for biological control of *Acacia longifolia* in Portugal. Biological Control 56:193–201

Marchante H, Freitas H, Hoffmann JH (2011b) Post-clearing recovery of coastal dunes invaded by *Acacia longifolia*: Is duration of invasion relevant for management success? J Appl Ecol. doi:10.1111/j.1365-2664.2011.02020.x

Marchante H, Freitas H, Hoffmann JH (2011c) The potential role of seed banks in the recovery of dune ecosystems after removal of invasive plant species. Appl Veg Sci 14:107–119

Marques S, Garcia-Gonzalo J, Borges J, Botequim B, Manuela M, Oliveira JT, Tomé M (2011) Developing post-fire *Eucalyptus globulus* stand damage and tree mortality models for enhanced forest planning in Portugal. Silva Fennica 45:69–83

McIver JD, Starr L (2000) Environmental effects of post-fire logging: literature review and annotated bibliography. General technical report PNW-GTR-486. U.S. Department of Agriculture, Forest Service, Pacific Northwest Research Station, Portland, OR

MCPFE (2007) State of Europe's forests 2007-the MCPFE report on sustainable forest management in Europe. Ministerial conference on the protection of forests in Europe, UNECE, FAO Warsaw

Mena-Petite A, Estavillo JM, Duñabeitia M, González-Moro B, Muñoz-Rueda A, Lacuesta M (2004) Effect of storage conditions on post planting water status and performance of *Pinus radiata* D. Don stock-types. Ann For Sci 61:695–704

Merino A, Balboa MA, Rodríguez Soalleiro R, González JGÁ (2005) Nutrient exports under different harvesting regimes in fast-growing forest plantations in southern Europe. For Ecol Manag 207:325–339

Moreira F, Vaz P, Catry F, Silva JS (2009) Regional variations in wildfire preference for land cover types in Portugal: implications for landscape management to minimize fire hazard. Int J Wildland Fire 18:563–574

Murphy H, VanDerWal J, Lovett-Doust L, Lowtt-Doust J (2005) Invasiveness in exotic plants: immigration and naturalization in an ecological continuum. In: Cadotte MW, McMahon SM, Fukami T (eds) Conceptual ecology and invasion biology. Springer, Amsterdam, pp 65–105

Neary DG, Klopatek CC, DeBano LF, Ffolliott PF (1999) Fire effects on belowground sustainability: a review and synthesis. For Ecol Manag 122:51–71

Olarieta JR, Besga G, Rodríguez R, Usón A, Pinto M, Virgel S (1999) Sediment enrichment ratios after mechanical site preparation for *Pinus radiata* plantation in the Basque Country. Geoderma 93:255–267

Parkes JP, Panetta FD (2009) Eradication of invasive species: progress and emerging issues in the 21st century. In: Clout MN, Williams PA (eds) Invasive species management: a handbook of principles and techniques. Techniques in ecology and conservation. Oxford University Press, Oxford, pp 47–60

Pauchard A, Garcia RA, Pena E, González C, Cavieres LA, Bustamante RO (2008) Positive feedbacks between plant invasions and fire regimes: *Teline monspessulana* (L.) K. Koch (Fabaceae) in central Chile. Biol Invasions 10:547–553

Pettenella D (2000) Certification of forest activities: the state of the art in Italy. Sherwood-Foreste ed Alberi Oggi 6:23–29

Potts B (1990) The response of eucalypt populations to a changing environment. Tasforests 2:179–193

Potts B, McGowen M, Suitor S, Jones T, Gore P, Vaillancourt R (2008) Advances in reproductive biology and seed production systems of *Eucalyptus*: the case of *Eucalyptus globulus*. South For 70:145–154

Pyšek P, Richardson DM (2010) Invasive species, environmental change and management, and health. Annu Rev Environ Res 35:25–55

Pyšek P, Lambdon PW, Arianoutsou M, Kühn I, Pino J, Winter M (2009) Alien vascular plants of Europe. In: DAISIE (ed) Handbook of alien species in Europe. Springer, New York, pp 43–61

Rejmánek M, Pitcairn M (2003) When is eradication of exotic pest plants a realistic goal? In: Veitch C, Clout M (eds) Turning the tide: the eradication of invasive species. IUCN-The World Conservation Union, Invasive Species Specialist Group, Gland/Cambridge, pp 249–253

Reyes O, Casal M (1998) Germination of *Pinus pinaster. P. radiata* and *Eucalyptus globulus* in relation to the amount of ash produced in forest fires. Ann For Sci 55:837–845

Reyes O, Casal M (2002) Role of fire on seed dissemination and germination of *Pinus pinaster* and *P. radiata.* Nova Acta Cient Comp 12:111–118

Reyes O, Casal M (2004) Effects of forest fire ash on germination and early growth of four *Pinus* species. Plant Ecol 175:81–89. doi:10.1023/B:VEGE.0000048089.25497.0c

Richardson DM (1998) Forestry trees as invasive aliens. Conserv Biol 12:18–26

Richardson DM, Kluge RL (2008) Seed banks of invasive Australian *Acacia* species in South Africa: role in invasiveness and options for management. Perspect Plant Ecol Evol Syst 10:161–177

Richardson DM, Pysek P, Rejmanek M, Barbour MG, Panetta FD, West CJ (2000) Naturalization and invasion of alien plants: concepts and definitions. Divers Distrib 6:93–107

Richburg J, Dibble A, Patterson W (2000) Woody invasive species and their role in altering fire regimes of the North-east and Mid-Atlantic states. In: Galley KEM, Wilson TP (eds) Invasive species workshop: the role of fire in the control and spread of invasive species. Tall Timbers Research Station, Tallahassee, pp 104–111

Richburg J, Patterson W III, Ohman M (2004) Fire management options for controlling woody invasive plants in the Northeastern and Mid-Atlantic US. Department of Natural Resources Conservation, University of Massachusetts, Amherst

Rowe JS (1983) Concepts of fire effects on plant individuals and species. In: Wein R, MacLean D (eds) The role of fire in Northern Circumpolar ecosystems. Wiley, Chichester, pp 135–154

Schelhaas MJ, Varis S, Schuck A, Nabuurs GJ (2006) EFISCEN Inventory Database. European Forest Institute. http://www.efi.int/portal/virtual_library/databases/efiscen/

Scott D (1993) The hydrological effects of fire in South African mountain catchments. J Hydrol 150:409–432

Shakesby R, Coelho C, Ferreira A, Terry J, Walsh R (1993) Wildfire impacts on soil-erosion and hydrology in wet Mediterranean forest, Portugal. Int J Wildland Fire 3:95–110. doi:10.1071/WF9930095

Shakesby RA, David J, Richard A (1996) Limiting the soil degradational impacts of wildfire in pine and eucalyptus forests in Portugal: a comparison of alternative post-fire management practices. Appl Geogr 16:337–355

Shakesby RA, Coelho COA, Ferreira AJD, Walsh RPD (2002) Ground-level changes after wildfire and ploughing in eucalyptus and pine forests, Portugal: implications for soil microtopographical development and soil longevity. Land Degrad Dev 13:111–127. doi:10.1002/ldr.487

Silva JN, Feith H, Pereira JC (2007) Exploração e silvicultura pós-fogo em eucaliptais. In: Alves AM, Pereira JS, Silva JN (eds) O eucaliptal em Portugal – Impactes ambientais e investigação científica. ISAPress, Lisbon, pp 285–312

Silva JS, Moreira F, Vaz P, Catry F, Godinho-Ferreira P (2009) Assessing the relative fire proneness of different fores types in Portugal. Plant Biosyst 143:597–608

Silva JS, Vaz P, Moreira F, Catry F, Rego FC (2011) Wildfires as a major driver of landscape dynamics in three fire-prone areas of Portugal. Landsc Urban Plan 101:349–358

Smith HG, Sheridan GJ, Lane PNJ, Bren LJ (2011) Wildfire and salvage harvesting effects on runoff generation and sediment exports from radiata pine and eucalypt forest catchments, south-eastern Australia. For Ecol Manag 261:570–581. doi:10.1016/j.foreco.2010.11.009

Swearingen JM, Pannill P (2009) Tree of heaven – *Ailanthus altissima* (Mill.)Swingle. http://www.nps.gov/plants/alien/fact/aial1.htm. Accessed 12 May 2011

Sykora K (1990) History of the impact of man on the distribution of plant species. In: Castri FD, Hansen AJ, Debussche M (eds) Biological invasions in Europe and the Mediterranean Basin, vol 65. Kluwer, Dordrecht, pp 37–50

Varela ME, Benito E, de Blas E (2005) Impact of wildfires on surface water repellency in soils of northwest Spain. Hydrol Process 19:3649–3657. doi:10.1002/hyp. 5850

Verdú B (2003) Regeneración de pino radiata después de un incendio en el monte de UP [Utilidad Pública] número 147 "Posadero". Montes 73:69–73

Vilà M, Lloret F, Ogheri E, Terradas J (2001) Positive fire-grass feedback in Mediterranean Basin woodlands. For Ecol Manag 147:3–14

Vilà M, Basnou C, Pyšek P, Josefsson M, Genovesi P, Gollasch S, Nentwig W, Olenin S, Roques A, Roy D and DAISIE partners (2010) How well do we understand the impacts of alien species on ecosystem services? A pan-European, cross-taxa assessment. Front Ecol Environ 8:135–144

Virtue JG, Melland RL (2003) The environmental weed risk of revegetation and forestry plants. Department of Water, Land and Biodiversity Conservation, Adelaide

Walther GR, Gritti ES, Berger S, Hickler T, Tang Z, Sykes MT (2007) Palms tracking climate change. Glob Ecol Biogeogr 16:801–809

Whelan RJ (1995) The ecology of fire. Cambridge studies in ecology. Cambridge University Press, Cambridge

Whittock S, Apiolaza L, Kelly C, Potts B (2003) Genetic control of coppice and lignotuber development in *Eucalyptus globulus*. Aust J Bot 51:57–68

Wieseler S (2005) Black locust. http://www.nps.gov/plants/alien/fact/rops1.htm. Accessed 12 May 2011

Williams MC, Wardle GM (2005) The invasion of two native Eucalypt forests by *Pinus radiata* in the Blue Mountains, New South Wales, Australia. Biol Conserv 125:55–64

Wilson JRU, Gairifo C, Gibson MR, Arianoutsou M, Bakar BB, Baret. S, Celesti-Grapow L, DiTomaso JM, Dufour-Dror J-M, Kueffer C, Kull CA, Hoffmann J, Impson FAC, Loope LL, Marchante E, Marchante H, Moore JL, Murphy DJ, Rinaudo A, Tassin J, Witt A, Zenni RD, Richardson DM (2011) Risk assessment, eradication, containment, and biological control: global efforts to manage Australian acacias before they become widespread invaders. Diver Distrib 17:1030–1046

Wittenberg R, Cock MJW (2001) Invasive alien species: a toolkit of best prevention and management practices. CAB International, Wallingford

Wittenberg R, Cock MJW (2005) Best practices for the prevention and management of invasive alien species. In: Mooney HA, Mack RN, McNeely JA, Neville LE, Schei PJ, Waage JK (eds) Invasive alien species. A new systhesis, vol 63. Island Press, Washington/Covelo/London, p 368

Zavala LM, González FA, Jordán A (2009) Intensity and persistence of water repellency in relation to vegetation types and soil parameters in Mediterranean SW Spain. Geoderma 152:361–374

Zengjuan F, Chuanhong Z, Yongqi Z (2006) Invasive potential of two introduced tree species: *Acacia mearnsii* and *Acacia dealbata*. Sci Silv Sin 42:48–53

# Chapter 11
# Management of Threatened, High Conservation Value, Forest Hotspots Under Changing Fire Regimes

**Margarita Arianoutsou, Vittorio Leone, Daniel Moya, Raffaella Lovreglio, Pinelopi Delipetrou, and Jorge de las Heras**

## 11.1  The Biodiversity Hotspots of the Earth

Biodiversity hotspots are geographic areas that have high levels of species diversity but significant habitat loss. The term was coined by Norman Myers to indicate areas of the globe which should be a conservation priority (Myers 1988).

A biodiversity hotspot can therefore be defined as a region with a high proportion of endemic species that has already lost a significant part of its geographic original extent. Each hotspot is a biogeographic unit and features specific biota or communities. The current tally includes 34 hotspots (Fig. 11.1) where over half of the plant species and 42% of terrestrial vertebrate species are endemic. Such hotspots account for more than 60% of the world's known plant, bird, mammal, reptile, and amphibian

M. Arianoutsou (✉)
Department of Ecology and Systematics, Faculty of Biology, School of Sciences,
National and Kapodistrian University of Athens, Athens, Greece
e-mail: marianou@biol.uoa.gr

V. Leone
Faculty of Agriculture, University of Basilicata, Potenza, Italy
e-mail: vittorio.leone@tiscali.it

D. Moya • J. de las Heras
ETSI Agronomos, University of Castilla-La Mancha, Albacete, Spain
e-mail: daniel.moya@uclm.es; Jorge.heras@uclm.es

R. Lovreglio
Faculty of Agriculture, University of Sassari, Sardinia, Italy
e-mail: rlovreglio@uniss.it

P. Delipetrou
Department of Botany, Faculty of Biology, School of Sciences,
National and Kapodistrian University of Athens, Athens, Greece
e-mail: pindel@biol.uoa.gr

F. Moreira et al. (eds.), *Post-Fire Management and Restoration of Southern European Forests*, Managing Forest Ecosystems 24, DOI 10.1007/978-94-007-2208-8_11,

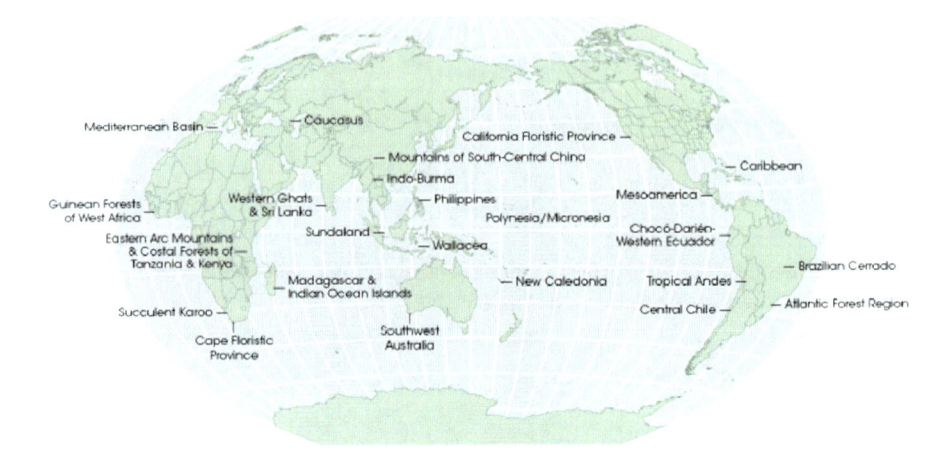

**Fig. 11.1** Global Biodiversity hotspots distribution. (Source: http://earthobservatory.nasa.gov/Features/Conservation)

species whose distribution area was originally restricted to 15.5% of the Earth's land surface and has now shrunk to 2.2%.

The principles of biodiversity hotspots have been the basis for conservation efforts by Conservation International and other leading environmental groups to distinguish a global set of high-priority terrestrial ecoregions for conservation.

## 11.1.1 The Biodiversity of the Mediterranean Basin Hotspot

The Mediterranean Basin is one of the 34 biodiversity hotspots included in the current list of Conservation International (2005). It encompasses North Africa (Morocco, Algeria, Tunisia, Libya, Egypt, Israel, Lebanon, Jordan and Syria), Turkey, Greece, Malta, Cyprus, all the Adriatic coastal countries (Montenegro, Albania, Croatia), almost all Italy and major islands, Mediterranean France (PACA Region and Corsica,) Spain and Portugal, including the islands in the Atlantic Ocean which pertain to the last two countries [the Macaronesian Islands of the Canaries, Madeira, the Selvagens, the Azores, and Cape Verde]. Of the 22,500 species of vascular plants in this hotspot, approximately 11,700 (52%) are found nowhere else in the world (Greuter 1991, 1994; Médail and Quézel 1999).

Having about 2,085,292 km$^2$ of surface, the Mediterranean Basin is located at the intersection of two major landmasses, Eurasia and Africa, which contributes to its high diversity and spectacular scenery, made of thousands of islands, abrupt coasts and sandy beaches, high elevation mountains up to 4,808 m (Monte Bianco, Italy), a rugged topography and a wide variety of climatic types. The broad climatic type prevailing over the Mediterranean Basin could be defined as a Dry Summer Temperate climate under the Köppen climate classification.

Although much of the Mediterranean Basin was thought to be once covered by evergreen and deciduous oak forests as well as conifer forests, 8,000 years of human settlement and habitat modification have distinctly altered the characteristic vegetation (Quézel 1985; Zohary 1973). Today, the most widespread vegetation type is hard-leafed or sclerophyllous shrublands called *maquis* or *matorral*, which include representatives from the plant genera *Juniperus, Myrtus, Olea, Phillyrea, Pistacia,* and *Quercus*. Some important components of Mediterranean vegetation (species of the genera *Arbutus, Calluna, Ceratonia, Chamaerops,* and *Laurus*) are tropical relicts from the ancient forests that dominated the Basin two million years ago. Frequent burning of maquis (mainly for grazing purposes) resulted in depauperate vegetation dominated by Kermes oak (*Quercus coccifera*), *Cistus* spp., *Sarcopoterium spinosum* or *Genista fasselata*, all of which regenerate rapidly after fire by resprouting or mass seed germination. Shrublands, including maquis and the aromatic, soft-leaved and seldom drought deciduous phrygana of *Rosmarinus, Salvia, Thymus* or *Coridothymus* persist in the semi-arid, lowland, and coastal regions.

The Mediterranean Region harbours a high degree of tree richness and endemism (290 indigenous tree species of which 201 endemics). A number of trees are important flagships, including the cedars (such as the famous cedar of Lebanon, *Cedrus libani*, which has been exploited since the rise of civilization in the Fertile Crescent); the argan tree (*Argania spinosa*), a species in the Souss region of southwest Morocco; oriental sweet gum (*Liquidambar orientalis*); and Cretan date palm (*Phoenix theophrastii*) in Crete (Greece) and western Turkey.

Many of the hotspots occur in regions with high levels of population density, a potential factor of threat for their conservation status and overall existence. Population trends in many biodiversity hotspots indicate a high risk of ongoing habitat degradation (Cincotta et al. 2000).

The greatest impacts of human civilization in the Mediterranean have been deforestation, intensive grazing and fires, and urbanization, especially along the coast. At present, Mediterranean forests cover about 9.4% of the region's total land area (Leone and Lovreglio 2004). The Mediterranean region is now facing a turbulent transformation of its political, human and settlement patterns. Dramatic changes in population density and distribution mark the area, which includes one of the most over-crowded coastlines in the world.

In terms of habitat deterioration, population density is, among others, strongly related with forest fires: a number of papers relate number of fires to population density with more or less robust relationships (Cardille et al. 2001; Leone et al. 2003; Syphard et al. 2007; Martínez et al. 2009). Among other factors threatening biodiversity hotspots fires represent one of the most detrimental.

## 11.2 Climate Change and Fire Regime Interaction Affect New Areas

Global warming represents perhaps the most pervasive of the various threats to the planet's biodiversity, given its potential to affect even areas far from human settlements. Recent reports outline the extensive biological changes that are ongoing because of global warming (Parmesan and Yohe 2003; Root et al. 2003). Malcolm et al. (2006) highlighted the potential impacts of global warming on biodiversity hotspots. Projected extinctions in hotspots under doubled $CO_2$ climates represented 39–43% of the biota, some 56,000 endemic plant species and 3,700 endemic vertebrate species. For specific hotspots, projections predicted extinctions of more than 3,000 plant species (Cape Floristic Region, Caribbean, Mediterranean Basin and Tropical Andes). For the Mediterranean Basin air temperature is foreseen to increase between 2% and 4% over the next century (Palutikof and Wigley 1996), while precipitation is predicted to decrease in autumn and increase in winter (Deque et al. 1998). Models predict changes in frequency, intensity and duration of extreme events, with more hot days, heat waves and heavy precipitation events, and fewer cold days (Lindner et al. 2008). The main expected impacts of climatic change on Mediterranean forests, namely on stability of forest ecosystems and natural disturbances (e.g. fire, pests, wind-storms) among others are strongly negative (Alcamo et al. 2007; Rosenzweig et al. 2007) and they are expected to affect plant life (e.g. growth, litter production), species recruitment, community composition and biodiversity, and regeneration processes after fire (Lavorel et al. 1998).

Climate warming is likely to have a rapid and profound impact on fire activity in several vegetation zones. For the Mediterranean Europe, Piñol et al. (1998) have studied a climatic series of 50 years from a locality in southern Spain and two fire hazard indices, and concluded that an effect of climate warming on wildfire occurrence is evident. Pausas (2004) analysed data from 350 meteorological stations in the eastern Iberian Peninsula covering a time period of 50 years (1950–2000) and fire records for the same area. He concluded that a clear pattern of increasing number of fires and size of area burned during the last century is observed, related to increasing mean annual as well as summer temperatures.

Climatic extremes observed recently (Founda and Giannakopoulos 2009; Tolika et al. 2009) have been clearly related to global warming. In particular, heat waves, when combined with droughts, can result in severe forest fires (Good et al. 2008; Moriondo et al. 2006; Rosenzweig et al. 2007). According to Dury et al. (2011) regions with more severe droughts might also be affected by an increase of wildfire frequency and intensity, which may have large impacts on vegetation density and distribution. For the Mediterranean Basin, the area burned can be expected to increase by a factor of 3–5 at the end of the twenty-first century, compared to present. Projections for 2070–2100 confirm a significant increase of fire potential for Europe, an enlargement of fire prone areas and a lengthening of fire season (Lavalle et al. 2009). Fire danger, length of the fire season, fire frequency and fire severity are very likely to increase in the Mediterranean, as effect of climatic change (Alcamo et al. 2007). Fire is a widespread process in the earth system and plays a key role in ecosystem composition and distribution (Bond et al. 2005). Fire has long ago been considered as a natural phenomenon, largely incorporated in Mediterranean climate systems evolution, having shaped their diversity (Cowling et al. 1996) and function (Rundel 1991). However, the Mediterranean Basin includes vegetation types that are either above the Thermo-Mediterranean climate zone or belong to non-fire prone ecosystems (such as sand dunes), which means that some types are not adapted to fire and cannot cope with it, since their biota had not been subject to its recurrent action through evolution. Because of differences in post-fire survival and regeneration strategies, a change in the future fire regimes will favour some Plant Functional Types (PFTs) such as resprouters and may cause a shift in forest composition (Weber and Flannigan 1997). The rapid response of fire regimes to changes in climate can potentially overshadow the direct effects of climate change on species distribution (Dale et al. 2001); actually climatic change leads to variations in the fitness of some species too, as a consequence of significant shift of timing in their phenological phases, with advance of spring phases and extension of the growing season (Menzel et al. 2006).

## 11.3 Case Studies

Forests that are already moisture limited (Mediterranean forests) or temperature limited (boreal forests) will adapt with more difficulty to climate change than other forests (Alcamo et al. 2007). For this reason, forest types which now grow in moderately fire prone areas could face unexpected difficulties, and gradually will become fire threatened, even though they apparently are fire adapted for some traits (e.g. good sprouting ability, serotiny).

Mountainous forests receiving high rainfall amount had not been subject to recurrent fires. However, a changing pattern is documented for some regions (e.g. Greece), where data indicate that both the numbers of fires occurring at the high altitude coniferous forests and the respective area burned during the last 20 years have increased (Arianoutsou et al. 2008, Fig. 11.2). A significant proportion of these areas have been assigned a protected area status, because of their biodiversity components. Yet, the dominant tree species in high altitude areas exhibit rarely, if ever, any specific adaptation mechanism to cope with fire, as fire has not played a selective role as a disturbance factor in their evolution (see case studies to follow).

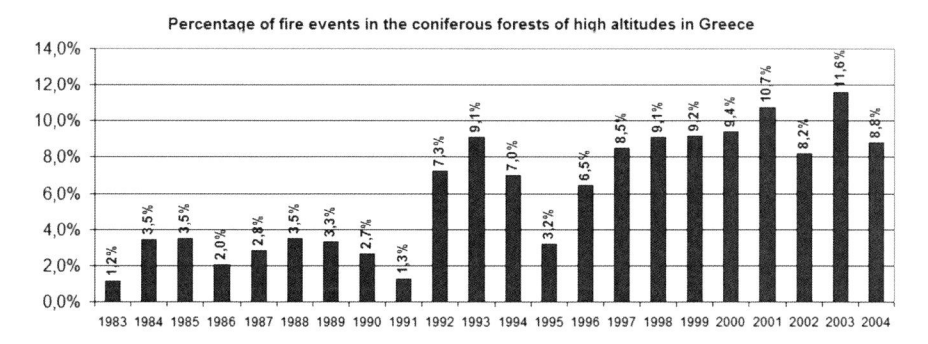

**Fig. 11.2** Number of fires occurred during the period 1983–2004 in the high altitudes coniferous forests of Greece. Note the increase observed since early 90s. (Source: adapted from Arianoutsou et al. 2008)

This situation provides both a challenge and an opportunity for the research and conservation agendas. As the biota face an increasing threat under global warming and subsequent climate change, the priority of these areas for conservation will increase.

In this chapter we will, therefore, deal with rare forest types which are expected to become more fire prone. Some of them exhibit adaptations to fire, which are however tuned to a certain fire regime different from the new, climate change-induced regime. Others do not even have any adaptation to fire. Among those, we have considered to include as study cases, forest types which are geographically restricted in the Mediterranean region, and may also consist of endemic tree species, namely *Abies cephalonica* and *Abies pinsapo, Juniperus macrocarpa, Quercus trojana, Tetraclinis articulata* and *Pinus leucodermis.* They are considered hotspots, in the commonly accepted sense of areas of specific high level of biodiversity, not in the sense of biodiversity hotspots at a planetary scale.

## *11.3.1 Abies Forests*

The biodiverse mountainous areas of the Mediterranean region have been a refuge for certain conifer taxa (species of *Abies, Cedrus, Cupressus, Juniperus* and *Pinus*) during the glacial periods (Bennett et al. 1991). Some of these taxa have been widely used as sources of wood and food (Farjon et al. 1993) and, as a result, many of these conifers have been overexploited and are now of considerable conservation concern. Most of the conifer forests in the southern Mediterranean are threatened as a result of historical or current deforestation and overgrazing (Barbero et al. 1990). *Abies* forests represent such a type of threatened ecosystems. Eight *Abies* taxa occur in the Mediterranean Basin (Farjon and Rushforth 1989), which is one of the distribution centers for the genus (Parducci 2000). *Abies nordmanniana* subsp. *equi-trojani* and *Abies borisii-regis* require management plans to guarantee their survival (Quézel and Barbero 1990). In northern Sicily *Abies nebrodensis* has been reduced to 29 individuals (Parducci et al. 2001) and the species is listed as Critically Endangered in the IUCN Red List (Farjon et al. 2006). Of the Mediterranean fir taxa, *Abies nebrodensis* (Lojac.) Mattei and *Abies cephalonica* Loudon are the only two with island populations too. *Abies nebrodensis* is an extremely rare species with very limited distribution in the Madonie range of Sicily (Morandini et al. 1994; Parducci 2000), while insular populations of the Greek endemic *A. cephalonica* (Greek fir) occur in two islands, Euboea in the Aegean Sea and Cephalonia in the Ionian Sea. *A. numidica* in Algeria and *A. pinsapo* var. *tazaotana* in Morocco are categorized as Vulnerable (Farjon and Page 1999). Although the other Mediterranean firs are categorized as Low Risk (i.e. they have been assessed and found not to be in danger of extinction), they still face the threats common to all Mediterranean mountain conifer forests, i.e. the combination of felling (often illegal), livestock raising, farming and, more recently, devastating fires.

Several researchers point out a comparatively low fire danger in *Abies* dominated forests (e.g. Wein and Moore 1977, 1979). The main reason for this is that they mostly occur in areas with moderate to high rainfall. The high relative humidity of the air, along with the rather dense canopy of trees which intercept radiation, results in relatively high moisture contents of the forest floor. This makes fire ignition and spread less probable. However, there is some evidence that, under certain conditions, *Abies* forests are susceptible to severe fires. A prolonged drought period may be a prerequisite for such fires in fir forests. In contrast to broad-leaved and pine forests, where self-pruning is common, little branch-pruning does occur in *Abies* stands. Thus, there is continuous fuel from the forest floor to the tree crown, which increases the probability of crowning should a fire occur (Furyaev et al. 1983).

### 11.3.1.1  *Abies cephalonica* in Mt. Parnitha National Park, Greece

Among the *Abies* taxa occurring in the Mediterranean Basin, the endemic to Greece *Abies cephalonica* Loudon (Greek fir) extends in central and southern mainland Greece and in Cephalonia, an island at the Ionian Sea. On the Greek mainland, a series of intermediate *Abies* forms occur, belonging to the putative hybrid species *Abies borisii-regis*. At the northern limit, the hybrid populations mostly resemble *Abies alba* and grow together with individuals of this species, while at the southern limit they mostly resemble *A. cephalonica* (Barbéro and Quézel 1976; Mitsopoulos and Panetsos 1987; Fady et al. 1991). *A. cephalonica* forests are included in category 6.10.6 'Mediterranean and Anatolian fir forest' in European Forest Types classification scheme (EEA 2007).

In Greece, fir forests of *A. cephalonica* and *A. borisii-regis* form pure stands managed for maximizing timber production and improving its quality in conjunction with optimizing their environmental benefits. Sylvicultural measures aim at converting their current structure of even–aged stands, partly due to the irrational use applied in the past (e.g. clearings, illegal cuttings and overgrazing) to uneven-aged stands, where all tree ages are present and continuous natural regeneration occurs (Dafis 1988). In several places, however, fir forests suffer from diseases that cause high tree mortality attributed to environmental changes during the last five decades (Raftoyannis and Radoglou 2001; Tsopelas et al. 2001).

The year 2007 will be regarded as a landmark in the environmental history of modern Greece, when more than 270,000 ha were affected by fire, with most of the area (70%) being burned by only 7 fire events (mega-fires) (Camia et al. 2008a). Both ecosystems of the thermo-mediterranean climate and high altitude forest ecosystems were burned. Among the most affected forest types were those of the Greek endemic *Abies cephalonica* Loudon (Greek fir) (Arianoutsou et al. 2009, 2010). Greek fir is vulnerable to fire as it does not produce serotinous cones and does not maintain a canopy seed bank when summer wildfires occur (Habrouk et al. 1999; Politi et al. 2011). Therefore, their natural post-fire recovery is limited, and strongly dependent on seed dispersal from neighboring unburned individuals or patches (Arianoutsou et al. 2009, 2010).

Mt. Parnitha, the highest (1,413 m) and most extended mountain of Attica, in central Greece, is a National Park since 1961 and it is included in Natura 2000 "network of sites for the conservation of species and habitats" (Directive 92/43/ EEC). The core zone of the National Park comprises the highest peaks of Parnitha, an area of ca. 3,800 ha, 90% of which had been covered with the endemic *Abies cephalonica* forest till 2007 (Fig. 11.3 – upper image). The altitudinal zone of the area is between 900 and 1,400 m and the climate is characterized by cool summers (usually air temperature does not exceed 18°C) and winter temperatures frequently near 0°C. The buffer zone of the Park is covered mostly by *Pinus halepensis* forest which stretches down to the foothills of the mountain (Amorgianiotis 1997). Snow is also frequent in the fir forest. One of the largest fires in the history of the Park took place in June 2007 and burned a great part (2,180 ha) of the strictly protected area and almost 50% of the *Abies cephalonica* forest. Fire did not burn the area in a homogeneous way, probably because of the dissected landscape physiography, the prevailing meteorological conditions and the tactics applied during the suppression phase. As a result, several unburned patches of various sizes have remained inside the burned area (Fig. 11.3 – lower image).

A field study was conducted during the 2nd post-fire year aiming at documenting the potential post-fire recovery in the burned fir community. No seedlings of *Abies cephalonica* were observed.

In *A. cephalonica* seeds mature during summer to early-autumn, dispersal usually begins in October, and dissemination is completed within 1 week (Politi et al. 2011). This means that a fire occurring in summer may easily consume the soft scales of the cones and the immature seeds they contain. Dispersed seeds may be buried under the organic soil or snow forming a transient soil seed bank during winter. Seedling emergence starts in late spring (Politi et al. 2009). Therefore, any seedling that might have appeared will be burned even by a surface fire.

The lack of active post-fire natural regeneration observed in the Greek fir may become worse over the long term given the masting behavior of *A. cephalonica* (Politi et al. 2009). The annual variability of cone production (masting) in populations of *A. cephalonica* has been reported following observations in northern Greece (Panetsos 1975; Dafis 1986) and it has been recently documented for *A. cephalonica* population at Mt. Ainos National Park (Cephalonia, Greece) by Politi et al. (2011). Although no *A. cephalonica* seedling was recorded in the burned sites, 2 years after fire a considerable amount of seeds was counted on the ground at distances varying up to 70 m from the adjacent unburned individuals (Fig. 11.4).

Nearly 90% of the seeds were recorded at distances ≤70 m from unburned trees, but some seeds were found up to 100 m from the unburned patch (Arianoutsou et al. 2009). The highest proportion of dispersed seeds (~90%) consisted of sound seeds. Density of seeds on the burned ground reached only 1/10 of that recorded on the floor of the unburned forest ($1.05 \pm 0.51$ versus $16.97 \pm 13.05$ per m$^2$).

*Juniperus oxycedrus,* the main shrub species of the understory also failed to resprout or germinate as it has been previously reported for other ecosystem types too (Kazanis and Arianoutsou 2004; Pausas et al. 2008). However, a considerable number of species regenerated in the burned forest community. Most of the species

**Fig. 11.3** Forest of *Abies cephalonica* Loudon in Mt. Parnitha National Park before (*upper image*) and after fire (*lower image*). Small unburned forest patches are shown in the upper right corner of the lower picture. (Source: Margarita Arianoutsou, University of Athens)

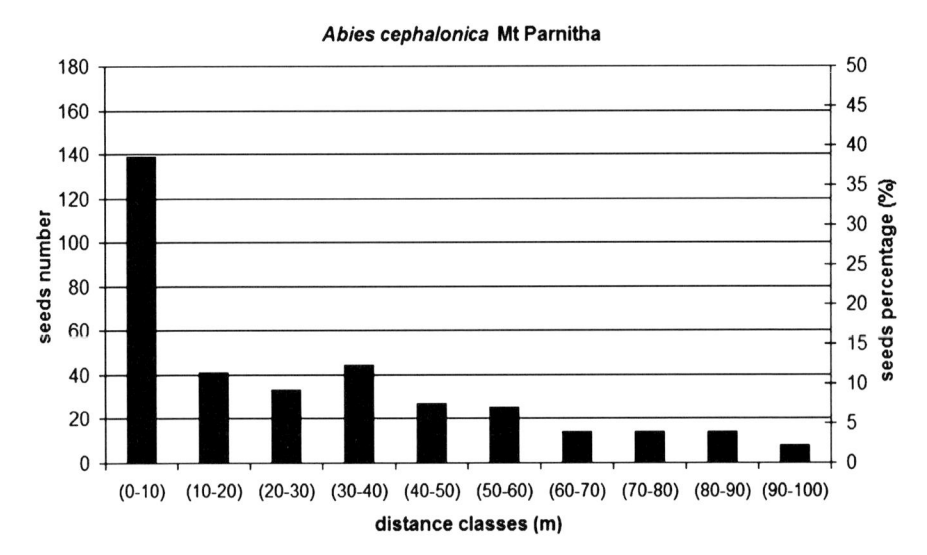

**Fig. 11.4** Number of seeds counted on the burned ground along an increasing distance from unburned *A. cephalonica* individuals in Mt. Parnitha National Park. (Source: adapted from Christopoulou et al. 2008)

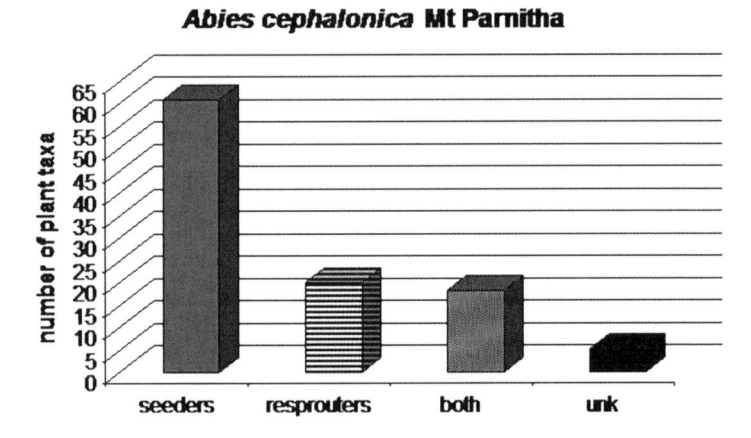

**Fig. 11.5** Regeneration mode of taxa appearing at the post-fire regenerating community of *A. cephalonica* forest in Mt. Parnitha National Park. (Source: adapted from Arianoutsou et al. 2010)

were annuals, which are absent from the understory layer of the unburned forests. The vast majority of the regenerating species recorded were seeders, many of which (32 taxa) seem to colonize burned areas through seed dispersal (Fig. 11.5). All colonizers are either annual or perennial herbs belonging to the families of Asteraceae and Poaceae.

Post-fire management was designed and performed by Park Authorities. Two main measures were applied: salvage logging and construction of wooden dams,

**Fig. 11.6** Extended log dams have been constructed following the contours after salvage logging (*upper image*). Planting of *Abies* saplings has been performed among the dams created by the logs or branches. Quite often cloth shelters have been used to protect the saplings from desiccation due to direct sunlight exposure (*lower image*). (Source: Margarita Arianoutsou, University of Athens)

and tree plantings (Fig. 11.6). Most standing burned trees were harvested and their trunks were used to construct series of log dams at a density of ~6.4 m/ha of dams made of the tree trunk. Log dams were placed even in flat areas. In general, the efficiency of this technique decreased with the formation of gaps between their placements and the ground surface and the total distance from the point where runoff started (Reppa, Detsis, Efthimiou, pers. comm.). Reforestation management consisted of plantings of *Abies* saplings, together with *Pinus nigra* and *Quercus pubescens* saplings. Forty five thousand (45,000) fir saplings, 8,000 Black pine and 2,000 oaks were used, while 2,000 Plane trees (*Platanus orientalis*) were planted along the streams in spring 2008. The evaluation of these plantations was made in autumn of 2009 in terms of individuals' growth and survival. Survival of the *Abies* plantings ranged between 45% and 84% and it varied as a function of aspect (Theodoropoulou, Detsis, Efthimiou, pers. comm.). Early results also suggest that when fir saplings were planted next to naturally resprouting individuals of *Arbutus andrachne* or *Quercus ilex*, survival was higher, while this effect was not observed when fir saplings were planted next to planted *Q. pubescens*. Certainly, these results are preliminary and do not cover an adequate time or space scale. More work is needed to formulate guidelines on post-fire management of these fire-vulnerable forests.

Time is certainly a critical parameter, as if the system was allowed to recover naturally, it would predictably have needed a 100 years. On the other hand, intervening in this process is not guaranteed to be a wise option. For example, extreme disturbance, like log removal, may indirectly cause soil degradation and encourage the invasion of alien species, e.g. *Robinia pseudoacacia*, which is usually planted along the roads for ornamental purposes. Constructing log dams is not necessary on flat areas and disturbance of the fragile burned soil may be another negative impact. Selection of planting material is also critical for ensuring the biological integrity of the species. If, for example, seeds used for producing the plantings are from another province, then genetic contamination becomes a potential risk.

### 11.3.1.2  *Abies pinsapo* in Spain

*Abies pinsapo* Boiss. is a conifer species of the Mediterranean Basin. It is considered a relict fir living in the south-western area of the Iberian Peninsula where it is endemic (Arista et al. 1997). According to the classification of the European Environmental Agency (2007) it is included in the category 6.10.6 '*Mediterranean and Anatolian fir forest*'. *Abies pinsapo* is found exclusively in the western area of the Spanish Betic Cordillera (Serrania de Ronda, Sierra Bermeja and Grazalema) frequently in shaded locations with northern exposure (Ceballos and Ruiz de la Torre 1979) in rocky places or not very deep stony soils and on limestone (European Environmental Agency 2007). It can be found over a broad elevation range, usually between 1,000 and 1,800 m, occupying 2,350 ha (Linares and Carreira 2006) of which 300 ha were included in UNESCO Biosphere Reserve in 1977. The *A. pinsapo* forests in Sierra de las Nieves were considered as relict by UNESCO Biosphere Reserve in 1995 and subsequently, they were included in Annex I of the Habitats' Directive (92/43/EEC) which protects the three main habitats in Spain. *A. pinsapo* occurs naturally in low aridity locations (Aussenac 2002) with average annual rainfall higher than 1,000 mm but nevertheless it has to cope with drought periods, as it needs more than 100 mm during summer (Ceballos and Ruiz de la Torre 1979). Usually, the Pinsapo fir forms monospecific forests although in the lower altitudinal range it can occur mixed with drought tolerant oaks (*Quercus faginea*, *Quercus ilex*

**Fig. 11.7** *Abies pinsapo* and *Quercus alpestris* mixed forests 4 years after fire. Naturally recovering vegetation of oak resprouts and companion shrubs. (Source: Jose Antonio Carreira, University of Jaen)

and *Quercus suber*) and other conifers such as *Pinus pinaster* or *Pinus halepensis* (Ceballos and Ruiz de la Torre 1979).

In the absence of fire or under very low fire recurrence, usually of low intensity, mixed uneven-aged *A. pinsapo* stands are established (Vega 1999). The low resilience of this species to fire has been confirmed in several studies and its presence has been used as a bio-indicator of non-fire-prone areas (Cabezudo et al. 1995). Several losses of *A. pinsapo* stands due to wildfires have been recorded in the last decades (Esteban et al. 2010), despite the fact that the species is not particularly flammable. Pinsapo fir shows no resprouting ability and has low regeneration success in burned and shaded areas. This has gradually induced the degradation and range contraction of Pinsapo fir forests (Rodriguez-Silva 1999), which are often replaced by more resilient species in areas with increasing fire recurrence.

The most important forest fire events in Pinsapo fir forests were recorded in the summers of 1966 and 1971 burning more than 200 ha of Pinsapo fir which has totally disappeared. In 1991, a large wildfire consumed 10,000 ha of maritime pine forest (*Pinus pinaster*) in Serrania de Ronda and burned the surrounding mixed forest including Pinsapo fir (in its lower elevation range) and also isolated Pinsapo fir woods developing in shaded and humid areas (approximately 30 ha in total) (Valladares 2009). Pinsapo fir stands showed some recovery, even if low, due to seedling recruitment in the areas of higher altitude originating from adjacent unburned Pinsapo fir stands (Arana et al. 1991).

In summer 2004, a fire burned over an old mixed stand with *Quercus alpestris* in Sierra Bermeja (Fig. 11.7). Since then no regeneration has been recorded mainly due to the lack of specific post-fire strategy of the species coupled with the sunny exposure and the poor soil quality of the site (Esteban and De Palacios 2007).

Climate change seems to induce smaller-sized trees and leads to a greater vulnerability to pests, as observed for Pinsapo fir forest in Spain and northern Morocco (Esteban et al. 2010). The increasing temperature and changes in rainfall dynamics in southern Spain (Parry et al. 2007) induce decline and dieback of the Pinsapo fir population (Linares et al. 2009b), decreasing the area covered by the species at the lower altitudes (Linares et al. 2010; Genova-Fuster 2007). Changes in fire regime, mainly the increase in number and recurrence of fire episodes, is the major problem for these forests (Vega 1999) which have been usually under-managed (or non-managed) and disappear in areas of low elevation due to the negligible recruitment (Arista 1994). These remaining forest stands became more closed and with lower canopy structural diversity, thus increasing their vulnerability to fire. The adaptation to climate change requires a shift to proactive management, directed towards the enhancement of canopy structural diversity at both stand and landscape levels (Carreira et al. 2008). The Second Programme for the Recovery of *A. pinsapo* (Andalusia Regional Government) implemented low-intensity thinning practices at low elevation sites (1,200 m), in order to achieve the structural objectives. This improved resistance to pests and promoted natural regeneration, including the establishment of *Quercus faginea, Q. rotundifolia* and *Pinus halepensis,* which increased plant diversity and resilience in the treated areas (Linares et al. 2009a; Carreira et al. 2008). The resistance to drought and pests has been increased by high intensity of thinning in the more humid areas of Spain (Lietor et al. 2002).

In other areas with no post-fire regeneration, restoration was carried out with reforestation using Pinsapo fir but it showed a very low success. Esteban et al. (2010) found out that at least 2 years of high rainfall and short, mild summers after seed dissemination or planting were needed for successful natural regeneration or reforestation. Grazing control and fire prevention to promote the natural regeneration have been suggested as proper management schemes (Arista et al. 1997). To promote seedling recruitment outside the closed forests, supplementary pollination management and artificial seed rain using vigorous seeds (including seedling protection from herbivores) have also been suggested (Arista 1995; Arista et al. 1997).

## 11.3.2   *Juniperus macrocarpa Woods in Italy*

*Juniperus macrocarpa* was formerly considered as one of the four subspecies of *J. oxycedrus* (Farjon 2005), named *J. oxycedrus* subsp. *macrocarpa* (Sibth. et Sm.) Ball 1878. Based on leaf essential oils and random polymorphic DNA amplifications analysis, Adams (2000) suggested that *J. oxycedrus* subsp. *macrocarpa* merits recognition at the specific level as *J. macrocarpa,* and recovered (Adams 2011) the original name of *Juniperus macrocarpa* Sm. which was originally published in

J. Sibthorp & J.E.Smith (1816 Fl. Graec. Prodr. 2: 263) and is now widely accepted (for instance in Germplasm Resources Information Network (GRIN).[1]

GRIN reports *J. macrocarpa* as native in Northern Africa (Algeria, Morocco Tunisia), Western Asia, Turkey, East Europe (Ukraine – Krym), South-Eastern Europe (Albania, Bosnia and Herzegovina, Bulgaria, Croatia, Former Yugoslavia, Greece, Italy, Malta), South-Western Europe (France, Spain), where the maritime juniper woodlands of *Juniperus macrocarpa* with *Juniperus phoenicea*, represent the mature ecosystem on outer dunes and cliffs (Adams 2011; Muñoz-Reinoso 2004).

Sand dune juniper woods are described as a priority habitat in Annex I of the Habitat Directive (92/43/EEC), under the code 2250* Coastal dunes with *Juniperus* spp[2]; they are included and in several classification systems such as CORINE (16.27 – Dune juniper thickets and woods) and European Forest Types (6.10.7 Juniper forest Woods dominated by *Juniperus* spp. of the Mediterranean and Anatolian mountains).

*Juniperus macrocarpa* lives exclusively in littoral zones, preferring sand dunes, where it can form single species stands (Fig. 11.8). Traditionally, *J. macrocarpa* was used in construction for its rotting and woodworms proof wood. Some use is documented for sculpture and as Christmas tree in Basilicata (province of Matera), and Apulia (Gargano promontory, province of Foggia) regions of Italy.

Oil (for veterinary medicine, soap, perfumes, detergents and disinfectants; FAO 1995) is extracted from the heartwood of *Juniperus macrocarpa* (Valentini et al. 2003; Velasco-Negueruela et al. 2005; Massei et al. 2006). The female cones (berries) are also sued in spirit (gin) production. Juniper woodlands have a high ecological value in relation to their sand retaining ability, their ability to withstand wind effects, drought, salt sprays and increased soil pH. They serve as habitat for many plant and animal species.

During the last century, juniper woodlands were almost destroyed or profoundly disturbed by logging, urban development, pine plantations, invasions by alien species and cultivation (Muñoz-Reinoso 2003, 2004; Picchi 2008). Air pollution and increase of atmospheric particulates are also considered to negatively interfere with pollination and consequently, with the reproduction and survival of these species (Pacini and Piotto 2004; Mugnaini et al. 2004).

---

[1] USDA, ARS, National Genetic Resources Program.*Germplasm Resources Information Network - (GRIN)* [Online Database].

[2] Juniper formations [*Juniperus turbinata* ssp. *turbinata* (=*J. lycia, J. phoenicea* ssp. *lycia*), *J. macrocarpa, J. navicularis* (=*J. transtagana, J. oxycedrus* ssp. *transtagana*), *J. communis*] of Mediterranean and thermo-Atlantic coastal dune slacks and slopes (*Juniperion lyciae*). *Juniperus communis* formations of calcareous dunes. This habitat type includes the communities of *J. communis* from the calcareous dunes of Jutland and the communities of *J. phoenicea* ssp. *lycia* in Rièges woods in the Camargue.

**Fig. 11.8** *Juniperus macrocarpa* stand on dunes in Apulian coast (*upper image*) and detail of an old *J. macrocarpa* on dunes with evident signs of trampling (*lower image*). (Source: *upper image*: Luigi Forte, University of Bari. *Lower image*: Vittorio Leone, University of Basilicata)

The species exhibits, in normal conditions and in absence of fire, a low regeneration potential with a total number of juvenile plants rarely reaching 10% of the adult individuals. Seed germination could be promoted by fire smoke which has been recently reported to act as a dormancy breaking cue in some Mediterranean species

(Crosti and Piotto 2006; Crosti et al. 2006). Production of seeds is low in J. macrocarpa and it is accompanied by high proportion of aborted female cones (Klimko et al. 2004; Juan et al. 2003). Furthermore, the number of sound, healthy seeds is very low (Juan et al. op. cit.). Seedling survival in the field is very low, even in the absence of fire, possibly due to abiotic stress (water, wind and salinity), intraspecific competition, grazing and/or trampling (Delipetrou 2010), (Fig. 11.8 lower image). On the other hand, *J. macrocarpa* propagates readily on sand dunes by adventitious root formation but this reaction and its possible relation to sand burial, as it has been shown for other sand dune trees (Dech and Maun, 2006) has not been studied. No particular silvicultural treatment of *J. macrocarpa* is recommended, as far as we know. Due to the small size of such woods, the harvest pattern, if any, was merely an occasional, uncontrolled tree logging, pinching here and there single trees of some economic interest, when needed, presumably the biggest ones, therefore males which grow faster than females (Massei et al. 2006).

*Juniperus* spp. forests in general, but mainly *J. oxycedrus* have little resistance to fire disturbance, therefore being associated with low incidence and recurrence of fire (Fernandéz-Gonzaléz et al. 2005). *Juniperus macrocarpa* is among the species for which fire is considered a completely destructive event (Piotto and Di Noi 2001). It has no adaptive trait to cope with fire, whereas, because of its dense evergreen foliage, full of aromatic substances (terpenes) it is particularly flammable.

Post-fire regeneration of *Juniperus* species (*J. oxycedrus* in western Central Spain, *J. phoenicia, J. communis, J. thurifera*) is reported as poor by Moreno (2010). The same is likely for *J. macrocarpa* for which no data are available. Post-fire establishment for *J. macrocarpa* does, in any case, depend on seed dispersal from unburned areas and the seeds have relatively poor germination rates, never exceeding 20% (Crosti and Piotto 2006; Picchi 2008; Thanos et al. 2010).

*J. macrocarpa* woodlands exhibit a rather reduced extent, all over the Mediterranean basin, as a consequence of human pressure, mainly on coastal areas: direct human activities, in which fire is a more and more predictable threat as related to accelerated climate change, literally squeeze the relict habitat. Possible guidelines for post-fire management could therefore be based upon the plans for the recovery of *J. macrocarpa* 2,250 priority habitat (Habitat Directive 92/43/EEC). The most relevant and complete is JUNICOAST, in the island of Crete (Delipetrou 2010; Kazakis et al. 2010; Thanos et al. 2010) and the one carried out in the region of Veneto, in Italy (Fiorentin 2006), both in the mainframe of LIFE Nature actions.

Juniper seeds collected from local plants on site should be planted in deep pots so as to permit sound root growth and to facilitate the placement of saplings in the dunes. The seedlings must be at least 30 cm high before being planted, the best results to get the plants established are achieved when about 1/3 of the aerial part of the plant is buried (Picchi 2008). Planting of first-year seedlings is not recommended as they seem to have low survival rate. A more efficient solution is planting of juvenile plants (1–2 years old), originating from seedlings and/or graft offshoots taken from the respective site and grown or rooted in a nursery for several months (Thanos et al. 2010). Young junipers should be planted in autumn, after the first rains. Small young plants should be planted in the most exposed areas since they are better able to survive (e.g. seafront of dunes), but they need to be protected with shelters. The biggest and oldest specimens (1 m/5 years) should be planted more landward, where protection from other species is greater. Wide spacing (not less than 2.5×4) is recommended (Fiorentin 2006). The ratio between young and mature individuals should be greater than 1/10 while female/male individuals' ratio in the population should be kept at 1:1 (Thanos et al. 2010).

### 11.3.3  *Tetraclinis articulata in Spain*

*Tetraclinis articulata* (Vahl) Master is an evergreen coniferous tree of the Cupressaceae family known as Mediterranean sandarac-cypress, Araar or Sictus tree. The species is native from north-western Africa, in the Atlas Mountains of Morocco, Algeria and Tunisia, its distribution range covering more than $10^6$ ha (Arman 1988). In Southern Europe small outlying natural populations can be found in Malta and near Cartagena in SE Spain, although it has been reintroduced in Seville and other provinces of the country (Mañez et al. 1997; Simón 1996). According to Baonza (2010), *T. articulata* was naturally distributed in several areas in southern Spain but was extinguished during the fifthteenth and sixteenth centuries due to overgrazing, high fire recurrence and clear cutting for timber. In the first half of the twentieth century, the *T. articulata* population in south-eastern Spain was reduced to less than 15 individuals due to habitat conversion to agricultural land, the high intensity of logging for timber, overgrazing and high recurrence of forest fires (López-Hernández 2000). However, protection by proper legislation contributed to a gradual population increase, although the limited distribution in Spain seems to be under threat given the predictions foreseen under the climate change scenario (Esteve 2009).

In Spain, the Araar tree currently covers 228 ha (Esteve 2009) with 12 different populations including a total of circa 3,000 individuals (Martín-Albertos and González-Martínez 2000). It is a relict species whose forests are considered rare or residual and are included in Annex I of the Habitats' Directive (92/43/EEC). It has been catalogued as a forest type of the class 6.10. '*Coniferous forest of the Mediterranean, Anatolian and Macaronesian regions*' according to EEA classification scheme (European Environmental Agency 2007).

It is one of only a small number of conifers able to coppice (re-grow by sprouting from stumps), an adaptation to survive wildfire and moderate levels of browsing by animals (Farjon 2005). It grows in a semiarid Mediterranean climate where annual precipitation varies from 300 to 375 mm per year. In the lowest precipitation range recruitment and growth are limited and the main companion species is *Periploca angustifolia* Labill. In areas with the highest precipitation range, *T. articulata* shows higher growth and recruitment. The main companion shrubs are *Pistacia lentiscus* L., *Chamaerops humilis* L., *Olea europaea* L. subsp. *sylvestris* (Mill.) Rouy ex Hegi & Berger, *Maytenus senegalensis* (Lam.) Exell subsp. *europaea* Boiss. and several *Rhamnus* species. It shows indifferent edaphic behaviour with preference for sunny spots although the distribution and tree density depend on the competition with *Pinus halepensis* Mill (López-Hernández 2000).

**Fig. 11.9** Araar tree resprouting after the forest fire which burned 55 ha of Calblanque Natural Park (Cartagena, Spain) in summer 1992. (Source: Juan José Martínez-Sánchez Universidad Politécnica Cartagena)

In summer 1992, a wildfire burned 55 ha in the Cartagena Mountains (Calblanque Peña del Aguila y Monte de las Cenizas Regional Park) located in south-eastern Spain (Fig. 11.9). This area is an ecotone of Mediterranean and subtropical arid climate where several iberoafricanic species are found (Nicolas et al. 2004). The area has been covered with open Araar and Aleppo pine forests. Two hundred twenty two (222) Araar trees were burned and they were subsequently monitored. In spring 1993 and 1994, the area was visited to record amount of resprouting and number of seedlings.

Fire intensity, which was visually estimated based on the amount of biomass burned and the colour of the remaining ash, varied from medium to very high. Resprouting ability was significantly related to fire intensity, although 99% of the 222 burned trees managed to resprout. The individuals affected by the higher intensity showed a lower number of resprouts produced only from lignotubers. In areas burned with a medium intensity, a higher number of resprouts emerging from both lignotubers and burned branches were recorded. Two years after fire, no Araar tree seedlings were found and the mortality rate of the regenerating sprouts was very low (López-Hernández et al. 1995). The main negative impact of wildfire was biomass loss. This effect leads to an investment of resources in growth (low productivity rate) with subsequent reduction of seed production. This in turn induces aging of the forest, which in addition to coppicing may lead to the change of forest to a more shrubby vegetation, reducing its resilience to fire and increasing the die out risk (López-Hernández 2000). In some areas, high fire recurrence has promoted *Calicotome intermedia* C. Presl. which shows a post-fire dual strategy, recovering by both resprouting and seedling (from soil seed bank) (López-Hernández et al. 1995). Furthermore, in shaded slots the Araar stands are dominated by Aleppo pines which implies a reduction of the Araar tree cover when fire recurrence is high (Nicolás et al. 2004).

To ensure the protection and improvement of the Spanish Araar tree populations, mainly after fire, the following management was proposed (Esteve 2009; Martín-Albertos and González-Martínez 2000):

- Population monitoring;
- Grazing control, mainly in regenerating stands;
- Creating a germplasm nursery and checking provenances;
- Increasing the occurrence area of the species, protecting and recovering potential habitat areas and reforesting areas with low natural regeneration using local provenances.

## 11.3.4 *Quercus trojana Woods in Apulia, Italy*

The Macedonian oak, *Quercus trojana* Webb (=*Quercus macedonica* A. DC.), is a semi-deciduous oak of eastern distribution, commonly present in Albania, Bosnia and Herzegovina, Bulgaria, Greece, Montenegro and Turkey. The western limit of its distribution area is exclusively in Italy, in the southeastern Murge hills territories, between the regions of Apulia and Basilicata, where *Quercus trojana* is considered a transadriatic endemic (Maselli 1940). It grows on "terra rossa" soils, frequently mixed with *Q. ilex*, *Q. pubescens*, *Carpinus orientalis* (Biondi et al. 2004), locally with *Ostrya carpinifolia* and *Q. frainetto*. Locally, a subspecies with big acorns (*Q. trojana* f. *macrobalana* Gavioli 1935) can occur.

*Q. trojana* (locally called *fragno*) occupies a well defined district in the sub-region of South-eastern Murge hills, with an extent of about 13,000 ha. Its presence there is made possible by the peculiar structure of the karstic environment, consisting of platform Cretaceous limestone and dolostone covered with thin layers of Pliocene-Quaternary rocks and soils. The carbonate rock is bedded, jointed, and subject to karstic phenomena (Polemio et al. 2009), which allow some water permeating the upper layers in summer (Macchia et al. 2000). Soil moisture is actually the main limiting factor for survival (Pignatti 1998).

The presence of Macedonian oak in Apulia can be interpreted as the result of a progressive withdrawal of the species during the last Ice Age, from a certainly wider range (Schirone and Spada 1995; Bozzano and Turok 2002).

*Quercus trojana* woods are included in Annex I of the Habitat Directive (92/43/EEC) as habitat type 9250[3] which corresponds to code 6.8.5 (Macedonian oak forest Supra- Mediterranean, and occasionally Meso-Mediterranean woods dominated by the semi-deciduous *Quercus trojana*) in the European Forest Types classification system. The distribution range is included in a trans-regional IPA (Important Plant Area) ITA 29 (Blasi et al. 2010) and in SIC (Site of Community Interest) IT 9120002 "*Murgia Dei Trulli*". Several shrubs appear as accompanying species.[4]

*Quercus trojana* usually occurs in a man shaped mosaic of scattered thickets or woods of small-medium size, rarely of considerable height (>15 m), usually appreciated for fodder production in farms with animal breeding (cows, horses and locally pigs, which eat the abundant acorn production). No example of stands of *Q. trojana* with natural structure (physical and temporal distribution of trees and other plants, sensu Oliver and Larson 1996) is known (Pignatti 1998). This means that regeneration, in the majority of cases, exclusively relies on the sprouting ability of stumps, which weakens with age.

Management, oriented to domestic fuel wood production and much less to timber, the latter limited to poles and small beams, is usually represented by pure or mixed coppices or low-forests [rarely simple coppice,[5] usually coppice with standards or reserves; more rarely and only in excellent, albeit rare, conditions of soil fertility compound coppice (Nyland 1996)]. In all cases structures are very simplified and rather far from the pristine multilayered mixed forests covering the area in the past. From such forests, harvesting of valuable wood of *Q. trojana* for shipbuilding by the Republic of Venice is reported but not adequately documented (Bellarosa et al. 2002).

Surface fires of low-medium intensity occasionally occur in such woodlands, where understory grazing, often with a very high stocking rate, reduces fire hazard. Consequently, grazing acts as an effective fire prevention practice, by elevating the crown base height which roughly corresponds to the browse line of animals and by decreasing surface fuel load, therefore strongly reducing the risk of crown fires.

---

[3] *Quercus trojana* woods Supra-Mediterranean, and occasionally meso-Mediterranean woods dominated by the semideciduous *Quercus trojana* or its allies (*Quercetum trojanae*) Sub-type 41.7/82 - Apulian Trojan oak woods]

[4] The main accompanying species are *Arbutus unedo* L., *Asparagus acutifolius* L., *Calicotome spinosa* (L.) Link, *Cistus creticus* L. subsp. *eriocephalus* (Viv.) Greuter & Burdet, *Cistus incanus* L., *Cistus monspeliensis* L., *Cistus salvifolius* L., *Crataegus monogyna* Jacq., *Evonymus europaeus* L., *Fraxinus ornus* L., *Juniperus macrocarpa* Sm., *Paliurus spina-cristi* Mill. *Phillyrea latifolia* L., *Pirus amygdaliformis* Vill., *Pistacia lentiscus* L., *P. terebinthus* L., *Prunus spinosa* L., *Rhamnus alaternus*, L *Rosa sempervirens* L., *Rubia peregrina* L, *Ruscus aculeatus* L., *Vitex agnus castus* L., (Macchia et al. 2000). Presence of *Allium subhirsutum* L: *Asphodelus ramosus* L. subsp. *ramosus*, *Brachypodium retusum* (Pers.) P. Beauv., *Bromus erectus* Huds., *Galium aparine* L:, *Geranium purpureum* Vill., *Geranium dissectum* L:, *Teucrium chamaedrys* L:, marks anthropic influence. (Misano and Di Pietro 2007).

[5] Coppicing is an ancient form of woodland management, mainly directed to fuelwood production, which involves repetitive felling on the same stump, near to ground level, and allowing the shoots arising from adventitious bud at the base of a woody plant to regrow from that main stump (coppice stool).

**Fig. 11.10** *Quercus trojana* stand a few months after coppicing (*coppice with reserves*): stools in rapid growth form the sparse understory under the reserves, in number of about 150 ha⁻¹ (*upper image*) the first spring after coppicing. Abundant grass understory may represent a fire hazard in this early phase of growth with low fire resistance (*lower image*). (Source: Vittorio Leone, University of Basilicata)

Situations of isolated wood fragments in a rural matrix, with high edge effects, increase fire risk.

Regeneration strategy of *Q. trojana* woods is mainly based on the high sprouting capacity, which is also the basis of coppicing. Epicormic sprouting is also present after fire (Fig. 11.10).

Seed germination is rather rare, being strongly dependent on climatic conditions and on abundant seed availability. Management of *Q. trojana* as coppice is a factor of progressive reduction of cover density, due to ageing of stumps and progressive loss of their sprouting capacity. Improper management and excessive grazing pressure can induce different stages of degradation of coppices too, which start as high forest and could in many cases tend toward a sparse stand of old trees ending up as barren carstic rock terrain.

Post-fire treatment is normally based on local Forestry Regulations which obliges owners of broadleaved forests to cut damaged stands as soon as possible, possibly in the first autumn after the fire event, for preventing dangerous fuel accumulation, pests and noxious insects' diffusion, but above all in the search of aesthetics. This treatment, in the case of coppice, corresponds to an ahead of time harvest.

Post-fire response, based on vigorous sprouting, is crucially dependent on grazing prohibition for the first 4–5 post-fire years and in the time interval between consecutive fires; a new fire in less than 5–6 years after the previous one means total destruction of such woodlands.

A possible improvement of post-fire treatment should consider the resilience and resistance to fire of *Q. trojana* at the level of single trees, avoiding current salvation felling, but applying a very selective salvage logging, harvesting surely compromised ones and avoiding elimination of individuals with good epicormic shoots from branches, which could survive. This treatment partially produces a shift from *coppice with reserves* (from 60 to 150 plants ha$^{-1}$) to a *compound coppice* in which some old standards are harvested and the remaining are left to grow for additional rotations in number decreasing with age. A longer rotation time should be also applied, at least 20–22 years.

A change of climate from wetter phases to drier ones in Southern Italy [warming in all seasons, marked precipitation decrease, long periods of drought and decrease in soil moisture (Castellari and Artale 2010)] could strongly interfere with the ecology of *Q. trojana*, representing a threat for its survival. Since a very significant increasing trend in fire danger is projected in Apulia (Camia et al. 2008b), an increment in forest fires is also expected in the distribution range of *Q. trojana*.

Increasing drought could amplify oak decline too. The oak stands of *Quercus pubescens* and *Q. trojana* in the region of Apulia show a dramatic dieback (Sicoli et al. 1998).

Such changes could cause increased *Q. trojana* mortality, reduce the occurrence of optimal conditions for germination, which are strongly related to moisture, and therefore progressively hamper species' presence. On the other hand, repeated fires and management as coppice could gradually reduce the vitality of coppices, depleting their reserves and inducing a progressive regression in the already restricted distribution range and abundance of such an interesting endemic.

### 11.3.5  *Pinus leucodermis Woods in the Pollino National Park, Italy*

Bosnian pine or palebark or white-bark pine (*Pinus heldreichii* Christ. (1863) syn. *Pinus leucodermis* Ant. (1864), or *Pinus leucodermis* (Antoine) Markgraf ex Fitschen (Richardson and Rundel 1998) is widespread in the western Balkan peninsula (Albania and Greece) at 1,000–2,500 m elevation, in the central western part of the Balkan peninsula (including Bosnia, Serbia, Montenegro and Bulgaria), and in Italy.

In the southern Italian Apennines *P. leucodermis*[6] occupies dry, sunny sites from 530 up to 2,240 m elevation, forming mixed stands with silver fir (*Abies alba* Mill.) and European beech (*Fagus sylvatica* L.) or pure stands on steep and dry rocky southern slopes (Avolio 1984, 1996). The already reduced range of *P. leucodermis* is split into five mountainous areas of Monte Pollino in the Pollino National Park (Gargano and Bernardo 2006). The scattered and isolated populations in Southern Italy of this Balkano-Italian restricted endemic are at the western limit of its geographical distribution range and represent a biogeographical island, interpreted as a relict of Tertiary flora of oro-Mediterranean forests which are genetically isolated from Greek populations (Bucci et al. 1997).

Populations are found at diverse ecological conditions, from the lower vegetation belt, where they are mixed with evergreen sclerophyllous vegetation, up to the alpine vegetation belt beyond the closed formations of *Fagus sylvatica* (Guerrieri et al. 2008). Stands at higher elevation are under the form of open woodland, characterized by large white snags and stumps mixed with scattered old monumental trees: such stands exhibit very low density (20–40 trees ha$^{-1}$) but relevant size (dbh up to 1.20 m) and age until 900 years and more (Guerrieri et al. op.cit., Corbetta and Pirone 1996). The soil is shallow, rendzina – like, with large outcropping rocks on bedrock of fissured greyish limestone (dolomite) from the Mesozoic.

Mean annual precipitation is 1,570 mm; mean annual temperature is 4°C; snow cover lasts from November to the end of May (Avolio 1996). According to the classification of Rivas-Martínez (2004) the climate varies from supra-Mediterranean up to subalpine/alpine (Gargano and Bernardo 2006).

*Pinus leucodermis* woods are included in Annex I of the Habitats Directive (92/43/EEC) as habitat type 95A0 which corresponds to 42.7[7] in the Palaearctic classification system and 6.10.5 (Alti Mediterranean pine forest) in the European Forest Types system. The distribution range in Basilicata is included in transregional IPA 30 "Pollino" (Important Plant Area) and in SICs IT 9210165, IT 9210245, IT 9210185.

White-bark pine's populations occur in different environmental conditions (Fascetti 2001) in dry and rocky mountain grasslands on the cliffs (Corbetta and Pirone 1981) and in screes (Corbetta and Pirone 1981), (Fig. 11.11 upper image). The main accompanying species are: *Daphne oleoides*, *Rhamnus pumilus*, *Sorbus graeca*, very short shrubs of *Juniperus hemisphaerica* (20–60% in cover) and *Gentiana lutea*, surrounded by patches of grassland of the *Caricetum– Seslerietum nitidae* association (Bonin 1978).

---

[6] We keep the name *Pinus leucodermis* as suggested in Conti et al. (2005).

[7] High oro-Mediterranean pine forests Subtype 42.71 White-barked pine forests Local tree line formations of *Pinus heldreichii* restricted to the southern Balkans, northern Greece and southern Italy, usually open and with an undergrowth formed by stripped grasslands on dry, often stony or rocky soils).

**Fig. 11.11** *Pinus leucodermis* typical community (old growths, matures, over matures, hard snags prevail on scarce saplings) (*upper image*) and sapling in a safe site between rocks with low nutrient availability near tree stumps or fallen logs (*lower image*) in the Pollino National Park, Italy. (Source: *upper image*: Pietro Civale, CODRA. *Lower image*: Antonio Romano, University of Basilicata)

The main value of Bosnian pine is currently as an ornamental and landscape tree rather than as a timber species. In the past it was very important in the local people's cultures (Fascetti 2001) and the wood was used for many handicrafts (timber, windows and doors frames, furniture, boxes and frames of musical instruments, oars, boats, wood chests etc.; resin also was tapped). Relevance in local culture is confirmed by its choice as symbol of the National Park of Pollino.

Since the 16th century *P. leucodermis* forests (approximately 3,000 ha in size) were used for grazing by goats, sheep, cows and mares arrived by transhumance from the Ionian Sea coast. Timber from *P. leucodermis* and *Fagus sylvatica* was the main resource for heating, lighting with torches, housing and dairying (Todaro et al. 2007). In the 1920s law enforcement favoured the increment of the resident population and their agricultural and pastoral activities. The number of browsing animals in the area increased till the beginning of 1970s. With the establishment of Pollino National Park in 1993, *P. leucodermis* became a protected species (Todaro et al. 2007) and grazing activities were somehow controlled.

Overgrazing has seriously affected population structure of the species with effects that are still evident. Only adult trees occur in the area and a marked lack of intermediate age classes is observed. A limited recruitment of new individuals at the tree-line ecotone is observed in many stands as a consequence of such long-lasting pressure (Morandini 1950; Longhi 1956). Bosnian pine is characterized as Low Risk in the IUCN Red List of threatened species. However, its natural distribution is currently rather limited and fragmented, mostly because of human pressure and forest fires (Vendramin et al. 2008).

Recent observations report that *P. leucodermis* forests are affected by recurrent fires in its lower elevation range but rather satisfactory post-fire regeneration occurs (Pierangeli pers. comm., Saracino pers. comm.) in more wet and favourable sites. Regeneration is ensured by seeds from adjacent areas or from surviving trees in the burned area; small or medium size trees generally do not survive fire, but old growth trees seem rather resistant to fires which are usually low-intensity ones. These observations confirm Gargano and Bernardo (2006) who uphold that in the SW part of the range and at low and intermediate elevation (900–1,300 m), fire appears to be a determining factor in establishing and maintaining young *P. leucodermis* populations (Fig. 11.11 lower image).

*P. leucodermis* exhibits great adaptability and great colonizing potential as a pioneer species. Its high level of self-fertilization, (probably due to the presence of a reduced number of recessive embryonic lethals (Morgante et al. 1994; Vendramin et al. 2008) as a consequence of genetic drift) represents a selective advantage for this highly competitive, pioneering and precocious species, (Gargano and Bernardo 2006).

To our knowledge, no specific protocol for post-fire treatment has been implemented for *P. leucodermis*. The general recommendations for post-fire management of pine forests could be also applied for *P. leucodermis* (see relevant chapters). An additional recommendation is to respect the natural patchiness of regeneration and the presence, if any, of "tree collective structure" or collectives, i.e. of a clump or cluster of seedlings which is functionally structured so that the outer trees act as protective of the inner, most fit ones. At favourable sites cluster planting can be used too, in order to minimise seedling mortality and is therefore a valuable strategy for afforestation in non-suitable sites at higher elevations (Schönenberger 2001; Souček and Špulák 2011), minimizing browsing damages. In high mountainous areas, tree collectives should be considered as resistant and stable stand elements. Because of the large gaps between the tree collectives, they radiate from very few individuals, gradually expanding around the edges by natural seeding (or layering for some species) with the result that uneven-aged cohorts are formed (Fillbrandt 1999). In the tree collectives humus development is faster. This is especially important in these habitats where the upper soil layers are subject to erosion (Ucler at al. 2007). In the higher elevations of Pollino National Park, where the number of trees is much reduced, surface fire frequency is expected to increase with climatic change. Fire could cause rapid and drastic reduction of the population size by killing surviving adult trees with consequent further losses in seedling genetic diversity and adaptive potential (Vendramin et al. 2008). In addition, the number of already scarce sapling could be further reduced. Seedling establishment can be compromised by the high post-fire competitive pressure from herbs. In such areas, if natural regeneration is absent, a possible management option could be seeding in the microsites where regeneration generally occurs (mainly bare lithosoils strips between rock outcrops). Given the high inbreeding and low outcrossing rates of this species, ideally seeds should be collected from a large number of trees (approx. 100) and from trees growing at least 100 m apart as suggested in Euforgen guidelines (Vendramin et al. 2008). Seeding is costly but in the strictly protected zone of the National Park (integral reserve) it seems the most natural treatment possible, since no alteration of habitat is induced.

## 11.4   Key Messages

- Significant climate changes are projected by the XXI century at a world scale, with temperature increase and decreased precipitation. Such changes could affect the frequency and severity of conditions suitable for the ignition and spread of fires;
- Any change in fire regime (frequency, severity, and seasonal timing of fire) may change the ecological consequences of such fires, leading to shifts in community structure and species composition;
- Species growing in the Mediterranean region exhibit adaptations and resistance to fire that lead to predictable responses to fire, in the current fire regime, but can prove unfit to cope with significant variations. As a consequence, species and communities can be vulnerable showing decreased resilience, along with possible shifts in their distribution and performance;
- Many forest species which share this vulnerability status are high altitude species, or narrow endemics with a restricted distribution range. These species may be unable to cope with increased fire frequency, or with the simple fire occurrence, in the case of forest types in which species have no adaptations to fire;
- We discussed the management of increasingly fire-prone forest with species of high conservation value (*Abies pinsapo*), restricted or very restricted endemics (*Abies cephalonica, Pinus leucodermis, Tetraclinis articulata*), species of narrow distribution range (*Quercus trojana*) or under the threat of range reduction and anthropic pressure (*Juniperus macrocarpa*). Some of these may lack in production value, but they all considered as emblematic of high biodiversity conditions;
- For their conservation, further scientific research and the development of specific protocols for post-fire restoration are needed to cope with the threat of increased fire hazard.

# References

Adams RP (2000) Systematics of Juniperus section Juniperus based on leaf essential oils and random amplified polymorphic DNAs (RAPDs). Biochem Syst Ecol 28:515–528

Adams RP (2011) Junipers of the world: the genus *Juniperus*, 3rd edn. TRAFFORD publishing Co, Vancouver

Alcamo J, Moreno JM, Nováky B, Bindi M, Corobov R, Devoy RJN, Giannakopoulos C, Martin E, Olesen JE, Shvidenko A (2007) Contribution of working group II to the fourth assessment report of the intergovernmental panel on climate change. In: Parry ML, Canziani OF, Palutikof JP, van der Linden JP, Hanson CE (eds) Europe. Climate change 2007: impacts, adaptation and vulnerability, Cambridge University Press, Cambridge

Amorgianiotis G (1997) Management plan of the National Park of Parnitha. vol A–H. Forest Service of Parnitha, Acharnes (in Greek)

Arana JJ, Molina J, Maldonado R, Osuna L, Ortiz FJ, Romero M (1991) Impactos causados por el incendio de 1991 sobre formaciones de pinsapos de la Serranía de Ronda. Jábega 74:81–85

Arianoutsou M, Kaoukis C, Kazanis D (2008) Fires in mountain coniferous forests of Greece: random events or a climate change symptom? In: Book of abstracts of the 4th conference of the Hellenic Ecological Society, Volos (in Greek)

Arianoutsou M, Christopoulou A, Ganou E, Kokkoris Y, Kazanis D (2009) Post-fire response of the Greek endemic Abies cephalonica forests in Greece: the example of a NATURA 2000 site in Mt Parnitha National Park. In: Miko L, Boitani L (eds) Proceedings of the 2nd European congress of conservation biology, Czech University of Life Sciences, Faculty of Environmental Sciences, Prague

Arianoutsou M, Christopoulou A, Kazanis D, Tountas Th, Ganou E, Bazos I (2010) Effects of fire on high altitude coniferous forests of Greece. In: DX Viegas (ed), 6th international conference of wildland fire, Coimbra

Arista M (1994) Supervivencia de las plántulas de *Abies pinsapo* Boiss. en su habitat natural. An Jard Bot Mad 51(2):193–198

Arista M (1995) The structure and dynamics of an *Abies pinsapo* forest in southern Spain. For Ecol Manag 4:81–89

Arista M, Herrera J, Talavera S (1997) *Abies pinsapo* Boiss.: a protected species in a protected area. Bocconea 7:427–436

Arman M (1988) La sabina mora. Especies singulares de la región de Murcia. Editora Regional de Murcia, España

Aussenac G (2002) Ecology and ecophysiology of circum-Mediterranean firs in the context of climate change. Ann For Sci 59:823–832

Avolio S (1984) Il pino loricato (*Pinus leucodermis* Ant.). Ann Ist Sp Selvicoltura 15:77–153

Avolio S (1996) Il Pino Loricato (*Pinus leucodermis* Ant.). Prometeo Edizioni, Castrovillari (CS)

Baonza J (2010) *Tetraclinis articulata* (Vahl) Mast. especie probablemente autóctona en Doñana. Ecología 23:139–150

Barbéro M, Quézel P (1976) Les groupements forestieres de Grèce centro-meridionale. Ecol Medit 2:3–86

Barbero M, Bonin G, Loisel R, Quézel P (1990) Changes and disturbances of forest ecosystems caused by human activities in the western part of the Mediterranean basin. Vegetatio 87:151–173

Bellarosa R, Simeone MC, Schirone B (2002) Country reports: Italy. In: Bozzano M and Turok J (eds) Mediterranean oaks network. Euforgen report of the second meeting, Gozo, Malta

Bennett KD, Tzedakis PC, Wi Llis KJ (1991) Quaternary refugia of North European trees. J Biogeo 18:103–115

Biondi E, Casavecchia S, Guerra V, Medagli P, Beccarisi L, Zuccarello V (2004) A contribution towards the knowledge of semideciduous and evergreen woods of Apulia (south-eastern Italy). Fitosociologia 41(1):3–28

Blasi C, Marignani M, Copiz R, Fipaldini M, Del Vico E (2010) ) Le Aree Importanti per le Piante nelle Regioni d'Italia: il presente e il futuro della conservazione del nostro patrimonio botanico. Progetto Artiser, Roma

Bond WJ, Woodward FI, Midgley GF (2005) The global distribution of ecosystems in a world without fire. New Phytol 165:525–538

Bonin G (1978) Contribution a la connaissance de la végétation des montagnes de l'Apennin centro-meridional. PhD thesis. Faculté des Sciences et Techniques St. Jérôme, Aix-Marseille III, France

Bozzano M and Turok J (2002) Mediterranean oaks network. Euforgen report of the second meeting, Gozo, Malta

Bucci G, Vendramin GG, Lelli L, Vicario F (1997) Assessing the genetic divergence of Pinus leucodermis Ant. Endangered populations: use of molecular markers for conservation purposes. Theor Appl Genet 7:1138–1146

Cabezudo B, Perez La Torre A, Nieto JM (1995) Regeneración de un alcornocal incendiado en el sur de España (Istan, Málaga). Acta Bot Malacitana 20:143–151

Camia A, San-Miguel-Ayanz J, Kucera J, Amatulli G, Boca R, Libertà G, Durrant T, Schmuck G, Schulte E, M Bucki (2008a) Forest Fires in Europe 2007EUR 23492 EN – Joint Research Centre – Institute for Environment and Sustainability Luxembourg: office for official publications of the European communities

Camia A, Amatulli G, San-Miguel-Ayanz J (2008b) Past and future trends of forest fire danger in Europe. EUR 23427 EN – Joint Research Centre – Institute for Environment and Sustainability EUR – Scientific and Technical Research series, Luxembourg: office for official publications of the European communities

Cardille JA, Ventura SJ, Turner MG (2001) Environmental and social factors influencing wildfires in the Upper Midwest. U S Ecol Applic 11:111–127

Carreira JA, López-Quintanilla JB, Linares JC (2008) Conservation and management adaptation options for the in-situ preservation of endemic mountain conifer forests: the *Abies pinsapo* case in Andalusia (Spain). In: Regato P, Salman R (eds) Mediterranean mountains in a changing world, guidelines for developing action plans. IUCN, Gland/Malaga

Castellari S, Artale V (2010) Climate change in Italy: evidence, impacts and vulnerability. Euro-Mediterranean Centre for Climate Change – CMCC – Bononia University Press, Rome

Ceballos L, Ruiz de la Torre J (1979) Árboles y Arbustos de la España Peninsular. Escuela Técnica Superior de Ingenieros de Montes, Madrid

Christopoulou A, Kokkoris I, Kazanis D, Arianoutsou M (2008) Post-fire seed dispersal of the Greek fir (*Abies cephalonica* Loudon) in the mountain Parnitha National Park: the importance of the population's unburned patches. In: Book of abstracts of the 4th conference of the Hellenic Ecological Society, Athens (in Greek)

Cincotta RP, Wisnewski J, Engelman R (2000) Human population in the biodiversity hotspots. Nature 404:990–992

Conservation International (2005) http://www.biodiversityhotspots.org/xp/hotspots/Documents/cihotspotmap.pdf. Accessed 11 May 2011

Conti F, Abbate G, Alessandrini A, Blasi C (eds) (2005) An annotated checklist of the Italian vascular flora. Palombi Editori, Roma, 420 pp

Corbetta F, Pirone GF (1981) Carta della vegetazione di M.te Alpi e zone contermini (Tav. "Latronico" della carta d'Italia), AQ/1/122, CNR, Roma

Corbetta F, Pirone GF (1996) La flora e le specie vegetali di interesse fito-geografico in Basilicata. Documenti Regione Basilicata: Risorsa Natura in Basilicata 5–6:127–142

Cowling R, Rundel P, Lamont B, Arroyo M, Arianoutsou M (1996) Plant diversity in Mediterranean climate regions. Trends Ecol Evol 11:362–366

Crosti R, Piotto B (2006) Soil seed bank restoration: the role of post-fire enhancing gents, such as smoke, in germination of Mediterranean native species. Società Italiana di Ecologia, XVI Congresso Nazionale, Urbino

Crosti R, Ladd PG, Dixon KW, Piotto B (2006) Post-fire germination: the effect of smoke on seeds of selected species from the central Mediterranean basin. For Ecol Manag 221:306–312

Dafis S (1986) Forest ecology. Ghiahoudi and Ghiapouli Publications, Thessaloniki (in Greek)

Dafis S (1988) Silvicultural treatment of firs' forests in Greece, vol LA. Scientific Annals of the Department of Forestry and Natural Environment, Aristotle University of Thessaloniki, Thessaloniki (in Greek)

Dale VH, Joyce LA, Mcnulty NRP, Ayres MP, Flannigan MD, Hanson PJ, Irland LC, Lugo AE, Peterson CJ, Simberloff D, Swanson FJ, Stocks BJ, Wotton BM (2001) Climate change and forest disturbances. Bioscience 9:723–734

Dech JP, Maun MA (2006) Adventitious root production and plastic resource allocation to biomass determine burial tolerance in woody plants from central Canadian Coastal Dunes. Ann Bot 98:1095–1105

Delipetrou P (2010) JUNICOAST (Actions for the conservation of coastal dunes with *Juniperus* spp. in Crete and the South Aegean (Greece)) Action A.8: Target habitat protection and restoration specifications. Deliverable A. 8 Habitat protection and restoration specifications, Chania, Greece

Deque M, Marquet P, Jones R (1998) Simulation of climate change over Europe using a global variable resolution general circulation model. Clim Dyn 14:173–189

Dury M, Hambuckers A, Warnant P, Henrot A, Favre E, Ouberdous M, François L (2011) Responses of European forest ecosystems to 21st century climate: assessing changes in inter-annual variability and fire intensity. iForest 4:82–99. doi:10.3832/ifor0572-004

Esteban LG, De Palacios P (2007) Pinsapo forests: past, present and future. Bois et forêts des tropiques 292:39–47

Esteban LG, De Palacios P, Rodriguez-Losada L (2010) *Abies pinsapo* forests in Spain and Morocco: threats and conservation. Fauna Flora Int Oryx 44:276–284. doi:10.1017/S0030605310000190

Esteve MA (2009) 9570 Bosques de *Tetraclinis articulata*. In: Ministerio de Medio Ambiente, y Medio Rural y Marino (ed) Bases ecológicas preliminares para la conservación de los tipos de hábitat de interés comunitario en España, Madrid

European Environmental Agency (EEA) (2007) European forest types. Categories and Types for sustainable forest management reporting and policy. EEA technical report no 9, 2nd edn. Copenhagen

Fady B, Arbez M, Ferrandéz P (1991) Variability of juvenile Greek firs (*Abies cephalonica* Loud.) and stability of characteristics with age. Silvae Genet 40:91–100

FAO (1995) Non-wood forest products from conifers. http://www.fao.org/docrep/x0453e/x0453e00.htm. Accessed 8 May 2011

Farjon A, Page CN (1999) Conifers. Status Survey and Conservation Action Plan. IUCN/SSC Conifer Specialist Group, IUCN, Switzerland & Cambridge, UK

Farjon A (2005) Monograph of Cupressaceae and Sciadopitys. Royal Botanic Gardens, Kew

Farjon A, Rushforth KD (1989) A classification of *Abies* Miller (Pinaceae). Notes Roy Bot Gard Edinburgh 46:59–79

Farjon A, Page CN, Schellevis N (1993) A preliminary world list of threatened conifer taxa. Biodiv Conserv 2:304–326

Farjon A, Pasta S, Troìa A (2006) *Abies nebrodensis*. In: IUCN red list of threatened species version 2010.4. http://www.iucnredlist.org. Accessed 1 May 2011

Fascetti S (2001) Aspetti botanici e forestali del Pino loricato (*Pinus leucodermis* Antoine). TI Parco Nazionale del Pollino. Basilicata Regione Notizie 99:111–118

Fernandéz-Gonzaléz F, Loidi J, Moreno JM (2005) Impacts on plant biodiversity. In: Moreno JM (ed) Impacts on climatic change in Spain. OCCE, Ministerio de Medio Ambiente, Madrid

Fillbrandt T (1999) Structure of planted and spatially separated tree collectives (Rotten) in high montane and subalpine zones of the Swiss Alps. In: Schoenenberger W, Frey W, Montero G, Piussi P (eds) Structure of mountain forests: assessment, impacts, management, modelling. Iufro Congress, Davos

Fiorentin R (2006) Habitat dunali del litorale Veneto. In: AA.VV Progetto LIFE Natura Azioni concertate perla salvaguardia del litorale veneto–Gestione degli habitat nei siti Natura 2000. Veneto Agricoltura; Servizio Forestale Regionale per le province di Padova e Rovigo; Servizio Forestale regionale per le province di Treviso e Venezia, Italy

Founda D, Giannakopoulos C (2009) The exceptionally hot summer of 2007 in Athens, Greece – A typical summer in the future climate? Glob Planet Chang 67:227–236

Furyaev VV, Wein RW, MacLean DA (1983) Fire Influences in *Abies*-dominated forests. In: Wein RW, MacLean DA (eds) The role of fire in Northern Circumpolar ecosystems. Wiley, Chichester

Gargano D, Bernardo L (2006) Defining population structure and environmental suitability for the conservation of *Pinus leucodermis* Antoine in central Mediterranean areas. Plant Biosyst 140(3):245–254

Génova-Fuster M (2007) El crecimiento de *Abies pinsapo* y el clima de Grazalema: aportaciones dendroecológicas. For Syst 16:145–157

Good P, Moriondo M, Giannakopoulos C, Bindi M (2008) The meteorological conditions associated with extreme fire risk in Italy and Greece: relevance to climate models studies. Int J Wildland Fire 17:1–11

Greuter W (1991) Botanical diversity, endemism, rarity, and extinction in the Mediterranean area: an analysis based on the published volumes of med-checklist. Bot Chron 10:63–79

Greuter W (1994) Extinctions in Mediterranean areas. Philos Trans R Soc Lond B 344:41–46

Guerrieri MR, Todaro L, Carraro V, De Stefano S, Lapolla A, Saracino A (2008) Risposte ecofisiologiche di *Pinus leucodermis* ad alta quota in ambiente mediterraneo. Forest 5:28–38

Habrouk A, Retana J, Espelta JM (1999) Role of heat tolerance and cone protection of seeds in the response of three pine species to wildfires. Plant Ecol 145:91–99

Juan R, Pastor J, Fernández I, Diosdado JC (2003) Relationships between mature cone traits and seed viability in *Juniperus oxycedrus* L. subsp. *macrocarpa* (sm.) Ball (Cupressaceae). Acta Biol Cracov Ser Bot 45(2):69–78

Kazakis G, Ghosn D, Remoundou E, Vogiatzakis I, Delipetrou P (2010) JUNICOAST (Actions for the conservation of coastal dunes with *Juniperus* spp. in Crete and the South Aegean (Greece)) Action A.2: Determining the plant communities composition and structure Deliverable A.2.1 report on plant associations, community types, composition and structure of coastal dunes with *Juniperus* spp in Crete, Chania, Greece

Kazanis D, Arianoutsou M (2004) Long-term post-fire vegetation dynamics in *Pinus halepensis* forests of central Greece: a functional group approach. Plant Ecol 171:101–121

Klimko M, Boratyńska K, Boratyński A, Marcysiak K (2004) Morphological variation of *Juniperus oxycedrus* subsp. *macrocarpa* (Cupressaceae) in three italian localities. Acta Soc Bot Pol 73:113–119

Lavalle C, Micale F, Houston TD, Camia A, Hiederer R, Lazar C, Conte C, Amatulli G, Genovese G (2009) Climate change in Europe. 3. Impact on agriculture and forestry. Rev Agron Sustain Dev 29:433–446. doi:10.1051/agro/2008068

Lavorel S, Canadell J, Rambal S, Terradas J (1998) Mediterranean terrestrial ecosystems: research priorities on global change effects. Glob Ecol Biogeogr Lett 7:157–166

Leone V, Lovreglio R (2004) Conservation of Mediterranean pine woodlands: scenarios and legislative tools. Plant Ecol 171:221–235

Leone V, Koutsias N, Martinez J, Vega-Garcia C, Allgöwer B, Lovreglio R (2003) The human factor in fire danger assessment. In: Chuvieco E (ed) Wildland fire danger estimation and mapping, the role of remote sensing data. World Scientific Publishing, Singapore

Liétor J, García-Ruiz R, Viñegla B, Ochoa V, Linares JC, Hinojosa B, Salido T, Carreira JA (2002) Variabilidad biogeoquímica en masas de pinsapar: efecto de la litología y el estado sucesional. Ecología 16:45–57

Linares JC, Carreira JA (2006) El pinsapo, abeto endémico andaluz. O, ¿Qué hace un tipo como tú en un sitio como éste. Ecosistemas 3:170–190

Linares JC, Camarero JJ, Carreira JA (2009a) Plastic responses of *Abies pinsapo* xylogenesis to drought and competition. Tree Physiol 29:1525–1536

Linares JC, Camarero JJ, Carreira JA (2009b) Interacting effects of changes in climate and forest cover on mortality and growth of the southernmost European fir forests. Global Ecol Biogeog 18:485–497

Linares JC, Camarero JJ, Carreira JA (2010) Competition modulates the adaptation capacity of forests to climatic stress: insights from recent growth decline and death in relict stands of the Mediterranean fir *Abies pinsapo*. J Ecol 98:592–603

Lindner M, Garcia-Gonzalo J, Kolström M, Green T, Reguera R, Maroschek M, Seidl R, Lexer MJ, Netherer S, Schopf A, Kremer A, Delzon S, Barbati A, Marchetti M, Corona P (2008) Impacts of climate change on European forests and options for adaptation. AGRI-2007-G4-06 report to the European commission directorate-general for agriculture and rural development. http://ec.europa.eu/agriculture/analisis/external/euro_forests/full_report_en.pdf. Accessed 12 May 2011

Longhi G (1956) Alcune osservazioni fitogeografiche e biologiche sul pino loricato. Italia Forestale e Montana 5:227–228

López-Hernández (2000) Respuesta ambiental de las principales especies arbustivas en sistemas áridos y semiáridos mediterráneos: Modelos y Aplicaciones. PhD Thesis, Universidad de Murcia, Spain

López-Hernández J, Calvo J, Esteve-Selma M, Ramírez-Díaz L (1995) Respuesta de Tetraclinis articulata (Vahl) Masters al fuego. Ecología 9:213–221

Macchia F, Cavallaro V, Forte L, Terzi M (2000) Vegetazione e clima della Puglia. Options Medit 53:33–49

Malcolm JR, Liu C, Ronald P, Neilson RP, Hansen L, Hannah L (2006) Global warming and extinctions of endemic species from biodiversity hotspots. Conserv Biol 20:538–548

Mañez M, Cobo D, Jimenez J (1997) *Tetraclinis articulata* (Vahl) Masters en la provincia de Huelva. An Jard Bot Mad 55(2):462

Martín-Albertos S, Gonzalez-Martinez SC (2000) Conservación de recursos genéticos de coníferas en España. For Syst 2:151–184

Martínez J, Vega-Garcia C, Chuvieco E (2009) Human-caused wildfire risk rating for prevention planning in Spain. J Env Manag 90:1241–1252

Maselli VG (1940) Contributo alla conoscenza delle querce d'Italia. Il Fragno, Riv Forest Ital 2:20–35

Massei G, Watkins R, Hartley SE (2006) Sex-related growth and secondary compounds in *Juniperus oxycedrus macrocarpa*. Acta Oecol 29:135–140. doi:10.1016/j.actao.2005.08.004

Médail F, Quézel P (1999) Biodiversity hotspots in the Mediterranean Basin: setting global conservation priorities. Conserv Biol 13:1510–1513

Menzel A, Sparks TH, Estrella N, Koch E, Aasa A, Ahas R, Alm-Kubler K, Bissolli P, Braslavska O, Briede A, Chmielewski FM, Crepinsek Z, Curnel Y, Dahl A, Defila C, Donnelly A, Filella Y, Jatczak K, Mage F, Mestre A, Nordli O, Penuelas J, Pirinen P, Remisova V, Scheifinger H, Striz M, Susnik A, Van Viet AJH, Wielgolaski F, Zach S, Zust A (2006) European phenological response to climate change matches the warming pattern. Global Change Biol 12:1969–1976

Misano G, Di Pietro R (2007) L'Habitat 9250 "Boschi a *Quercus trojana*" in Italia. Fitosociologia 44:235–238

Mitsopoulos DJ, Panetsos CP (1987) Origin of variation in fir forests in Greece. Silvae Genetica 36:1–15

Morandini R (1950) Tra i boschi di M. Pollino. Monti e Boschi 8:361–365

Morandini R, Ducci F, Menguzzato G (1994) *Abies nebrodensis* (Lojac.) Mattei. Inventario 1992. Ann Ist Sper Selv 22:5–51

Moreno JM (2010) Climate change, wildland fires and biodiversity in Europe. Convention on the conservation of European wildlife and natural habitats standing committee, first draft, T-PVS/Inf, Strasbourg, France

Morgante M, Rossi P, Vendramin GG, Boscherini G (1994) Low levels of outcrossing in Pinus leucodermis: further evidence in artificial stands. Can J Bot 72:1289–1293

Moriondo M, Good P, Durao R, Bindi M, Giannakopoulos C, Corte-Real J (2006) Potential impact of climate change on fire risk in the Mediterranean area. Clim Res 31:85–95

Mugnaini S, Nepi M, Artese D, Gaggi C, Pacini E (2004) Il particolato aerodisperso può ridurre l'efficienza riproduttiva dei ginepri? In: Ecologia. Atti del XIV Congresso Nazionale della Società Italiana di Ecologia, (Siena, 4-6 ottobre 2004) a cura di Carlo Gaggi, Valentina Nicolardi e Stefania Santoni

Muñoz–Reinoso JC (2003) *Juniperus oxycedrus* ssp. *macrocarpa* in SW Spain: ecology and conservation problems. J Coast Conserv 9:113–122

Muñoz–Reinoso JC (2004) Diversity of maritime juniper woodlands. For Ecol Manag 192:267–276

Myers N (1988) Threatened biotas: "hotspots" in tropical forests. Environmentalist 8:187–208

Nicolás MJ, Esteve MA, Palazón JA, López Hernández J (2004) Modelo sobre las preferencias de hábitat a escala local de *Tetraclinis articulata* (Vahl) Masters en una población del límite septentrional de su área de distribución. An Biol 26:157–167

Nyland RD (1996) Silviculture concepts and applications. McGraw-Hill Co, New York

Oliver CD, Larson BC (1996) Forest stand dynamics. Wiley, New York

Pacini E, Piotto B (2004) I ginepri come specie forestali pioniere: efficienza riproduttiva e vulnerabilità. APAT Rapporti 40/2004

Palutikof J, Wigley T (1996) Developing climate change scenarios for the Mediterranean region. In: Jeftic L, Keckes S, Pernetta JC (eds) Climate change and the Mediterranean, vol 2. Arnold, London

Panetsos KP (1975) Monograph of *Abies cephalonica* Loudon. Ann For Zagreb 7:1–22

Parducci L (2000) Genetics and evolution of the Mediterranean Abies species. Dissertation, Swedish University of Agricultural Sciences

Parducci L, Szmidt AE, Madaghiele A, Anzidei M, Vendramin GG (2001) Genetic variation at chloroplast microsatellites (cpSSRs) in *Abies nebroidensis* (Lojac.) Mattei and three neighboring *Abies* species. Theor Appl Genet 102:733–740

Parmesan C, Yohe G (2003) A globally coherent fingerprint of climate change impacts across natural systems. Nature 421:37–42

Parry ML, Canziani OF, Palutikof JP, van der Linden PJ, Hanson CE (2007) Contribution of working group II to the fourth assessment report of the intergovernmental panel on climate change, 2007. Cambridge University Press, Cambridge/New York

Pausas JG (2004) Changes in fire and climate in the Eastern Iberian Peninsula (Mediterranean Basin). Glob Chang 63:337–350

Pausas JG, Llovet J, Rodrigo A, Vallejo R (2008) Are wildfires a disaster in the Mediterranean basin?- A review. Int J Wildland Fire 17:713–723

Picchi S (2008) Management of Natura 2000 habitats. 2250* Coastal dunes with *Juniperus* spp. European Commission. Technical report 2008 06/24

Pignatti S (1998) Boschi d'Italia. Sinecologia e biodiversità, UTET, Torino

Piñol J, Terradas J, Lloret F (1998) Climate warming- wildfire hazard and wildfire occurrence in coastal eastern Spain. Clim Chang 38:345–357

Piotto B, Di Noi A (2001) Propagazione per seme di alberi e arbusti della flora mediterranea. Manuale ANPA, Roma

Polemio M, Casarano D, Limoni PP (2009) Karstic aquifer vulnerability assessment methods and results at a test site (Apulia, southern Italy). Nat Hazard Earth Syst Sci 9:1461–1470

Politi PI, Arianoutsou M, Stamou GP (2009) Patterns of *Abies cephalonica* seedling recruitment in Mount Aenos National Park, Cephalonia, Greece. For Ecol Manage 258:1129–1136

Politi PI, Georghiou K, Arianoutsou M (2011) Reproductive biology of *Abies cephalonica* Loudon in Mount Aenos National Park, Cephalonia, Greece. Trees. Struct Funct 25:655–668. doi:10.1007/s00468-011-0542-1

Quézel P (1985) Definition of the Mediterranean region and the origin of its flora. In: Gómez-Campo C (ed) Plant conservation in the Mediterranean area, Dr. W. Junk Publishers, Dordrecht

Quézel P, Barbéro M (1990) Les forêts méditerranéeennes. Problèmes posés par leur signification historique, écologique et leur conservation. Acta Bot Malacitana 15:145–178

Raftoyannis Y, Radoglou K (2001) Crown condition of a fir forest in Karpenisi, Central Greece. In: Proceedings of the international conference on forest research: "a challenge for an integrated European approach"; NAGREF, Forest Research Institute, Thessaloniki

Richardson DM, Rundel PW (1998) Ecology and biogeography of Pinus: an introduction. In: Richardson DM (ed) Ecology and biogeography of Pinus. Cambridge University Press, Cambridge

Rivas-Martinez S, Penas A, Diaz TE (2004) Bioclimatic map of Europe–thermoclimatic belts. Cartographic Service, University of León. http://www.globalbioclimatics.org/form/tb_med. htm. Accessed 7 May 2011

Rodríguez-Silva F (1999) Los usos tradicionales del monte y sus implicaciones en la aparición de los incendios forestales: una perspectiva desde los pinsapares andaluces. In: Araque E (ed) Incendios Históricos: una Aproximación Multidisciplinar. Universidad Internacional de Andalucia, Jaén

Root TL, Price JT, Hall KR, Schneider SH, Rosenzweig C, Pounds JA (2003) Fingerprints of global warming on wild animals and plants. Nature 421:57–60

Rosenzweig C, Casassa G, Karoly DJ, Imeson A, Liu C, Menzel A, Rawlins S, Root TL, Seguin B, Tryjanowski P (2007) Assessment of observed changes and responses in natural and managed systems. In: Parry ML, Canziani OF, Palutikof JP, van der Linden JP, Hanson CE (eds) Europe. Climate change 2007: impacts, adaptation and vulnerability. Cambridge University Press, Cambridge

Rundel PW (1991) Fire as an ecological factor. In: Lange OL, Nobel PS, Osmond CB, Ziegler H (eds) Encyclopedia of plant physiology. Vol 2A. Physiological plant ecology. Springer, Berlin/Heidelberg/New York

Schirone B, Spada F (1995) Anomalies in reproductive phenology and vegetation history: the case of SE Italy. Colloques Phytosociol 24:847–857

Schönenberger W (2001) Cluster afforestation for creating diverse mountain forest structures-a review. For Ecol Manag 145:121–128

Sicoli G, de Gioia T, Luisi N, Lerario P (1998) Multiple factors associated with oak decline in southern Italy. Phytopathol Mediterr 37:1–8

Simon JA (1996) Manual de la flora para la restauración de áreas críticas y diversificación en masas forestales. Junta de Andalucía

Souček J, Špulák O (2011) Cluster reforestation near the timber line. J For Sci 57:16–23

Syphard AD, Radeloff VC, Keeley JE, Hawbaker TJ, Clayton MK, Stewart SI, Hammer RB (2007) Human influence on California fire regimes. Ecol Appl 17:1388–1402

Thanos CA, Kaltsis A, Koutsovoulou K, Skourti E, Sarris D (2010) JUNICOAST (Actions for the conservation of coastal dunes with *Juniperus* spp. in Crete and the South Aegean (Greece)) Action A.3. Composition and structure of *Juniperus* populations. Deliverable A.3 Composition and structure of *Juniperus macrocarpa* subpopulations, Chania, Greece

Todaro L, Andreu L, D'Alessandro CM, Gutiérrez E, Cherubini P, Saracino A (2007) Response of *Pinus leucodermis* to climate and anthropogenic activity in the National Park of Pollino (Basilicata, Southern Italy). Biol Conserv 137:507–519

Tolika K, Maheras P, Tegoulias I (2009) Extreme temperatures in Greece during 2007: could this be a "return to the future"? Geophys Res Lett 36:L10813. doi:10.1029/2009GL038538

Tsopelas P, Angelopoulos A, Economou A, Voulala M, Xanthopoulou E (2001) Monitoring crown defoliation and tree mortality in the fir forest of Mount Parnis, Greece. In: Proceedings of the international conference on forest research: "a challenge for an integrated European approach". NAGREF, Forest Research Institute, Thessaloniki

Ucler AO, Yucesan Z, Demirci A, Yavuz H, Oktan E (2007) Natural tree collectives of pure oriental spruce [*Picea orientalis* (L.) Link] on mountain forests in Turkey. J Environ Biol 28:295–302

Valentini G, Maggi F, Bellomaria B, Manzi A (2003) The leaf and female cone oils of Juniperus oxycedrus L. ssp. oxycedrus and J. oxycedrus ssp. macrocarpa (Sibth. et Sm.) Ball. from Abruzzo. J Essent Oil Res 15:418–421

Valladares A (2009) 9520: Abetales de *Abies pinsapo* Boiss. In: VV.AA. Bases ecológicas preliminares para la conservación de los tipos de hábitat de interés comunitario en España. Dirección General de Medio Natural y Política Forestal, Ministerio de Medio Ambiente y Medio Rural y Marino, Madrid

Vega JA (1999) Historia del fuego de *Pinus pinaster* en la cara norte de Sierra Bermeja (Malaga). In: Araque E (ed) Incendios Históricos: una Aproximación Multidisciplinar. Universidad Internacional de Andalucia, Jaén

Velasco-Negueruela A, Pérez-Alonso MJ, Palá-Paúl J, Íñigo A, López G (2005) The volatile oil composition of the berries of *Juniperus macrocarpa* Sibth. and Sm., gathered in Spain. J Essent Oil Res 17:61–63

Vendramin GG, Fineschi S, Fady B (2008) EUFORGEN Technical Guidelines for genetic conservation and use for Bosnian pine (*Pinus heldreichii*). Bioversity International, Rome

Weber MG, Flannigan MD (1997) Canadian boreal forest ecosystem structure and function in a changing climate: impact on fire regimes. Environ Rev 5:145–166

Wein RW, Moore JM (1977) Fire history and rotations in the New Brunswick Acadian Forest. Can J For Res 7:285–294

Wein RW, Moore JM (1979) Fire history and recent fire rotation periods in the Nova Scotian Acadian Forest. Can J For Res 9:166–178

Zohary M (1973) Geobotanical foundations of the middle East, vol 1, 2. Gustav Fischer Verlag, Stuttgart

# Chapter 12
# Post-Fire Management of Shrublands

**Leonor Calvo, Jaime Baeza, Elena Marcos, Victor Santana, and Vasilios P. Papanastasis**

## 12.1 Ecological Context

### 12.1.1 Origin and Historical Uses of Shrublands in the Mediterranean Basin

Shrublands are natural or semi-natural ecosystems dominated by shrubs, although herbaceous species and, occasionally, trees are also present. In the whole Mediterranean Basin shrublands have been coevolving for millennia with human societies, which have used them as sources of, mainly, fuel for burning and fodder for livestock (Perevolotsky and Seligmann 1998; Wessel et al. 2004). Currently, they occupy extensive areas of the Mediterranean region (Fig. 12.1) covering a total surface of 110,854 km$^2$.

There is no commonly accepted classification of Mediterranean shrublands, but three main categories can be distinguished (Papanastasis 2000): (1) *Maquis* or *matorral*: sclerophyllous evergreen shrublands, growing on relatively deep, mainly siliceous, soils (Le Houérou 1992). (2) *Garrigues* or *pseudomaquis*: plant communities of sclerophyllous evergreen or deciduous shrubs growing mainly on calcareous

L. Calvo (✉) • E. Marcos
Area of Ecology, Faculty of Biological and Environmental Sciences, University of León, León, Spain
e-mail: leonor.calvo@unileon.es

J. Baeza • V. Santana
Fundacion CEAM, Parque Tecnológico, Paterna, Spain

V.P. Papanastasis
Laboratory of Rangeland Ecology, School of Forestry and Natural Environment, Aristotle University of Thessaloniki, Thessaloniki, Greece

F. Moreira et al. (eds.), *Post-Fire Management and Restoration of Southern European Forests*, Managing Forest Ecosystems 24, DOI 10.1007/978-94-007-2208-8_12, © Springer Science+Business Media B.V. 2012

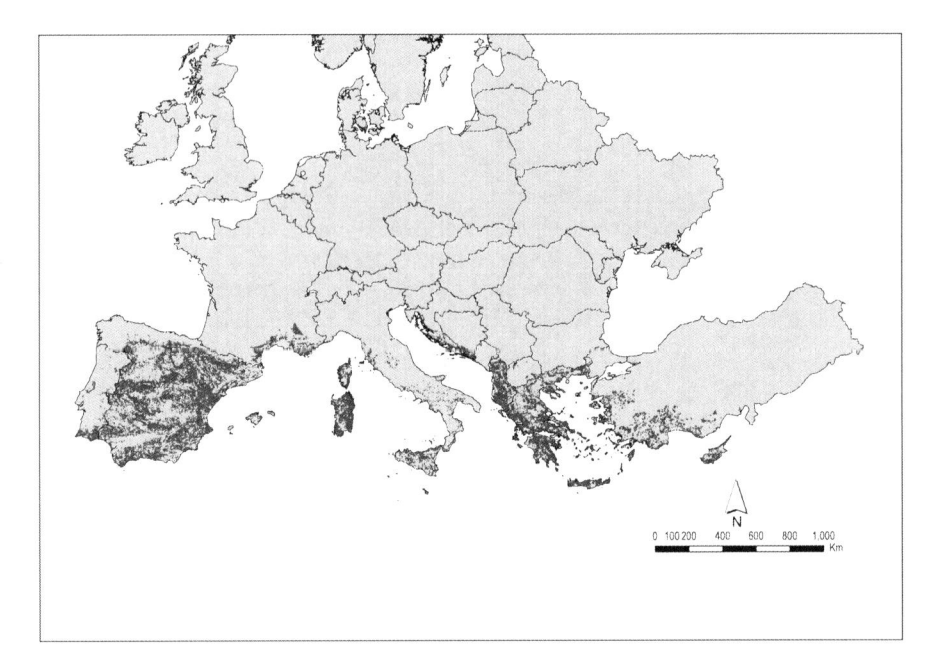

**Fig. 12.1** Sclerophyllous evergreen shrublands in the Mediterranean region (source: European Environment Agency/CORINE landcover 2000)

soils (Le Houérou 1992, 1993); (3) *Phrygana*: open communities of seasonal dimorphic dwarf shrubs growing mainly on rocky soils. The latter are distributed in the eastern Mediterranean and are known as *batha* in Palestine (Zohary 1962). Their equivalent type in the western Mediterranean is *tomillar*, the small brushwoods found in semiarid zones of Spain (Alcaraz and Delgado 1998).

The majority of shrublands have resulted from intense anthropogenic disturbance in forest ecosystems, particularly wildfires, cutting and grazing (Calvo et al. 1998; Casal 1987; Casal et al. 1990; Le Houérou 1993; Trabaud 1981, 1987). These long-term disturbances have led to the extensive destruction of native forests in large areas of the Mediterranean Basin (Naveh and Dan 1974) which were progressively replaced by shrublands (Calvo et al. 2002a, 2005; Luis-Calabuig et al. 2000; Naveh and Lieberman 1994; Pausas 1999; Trabaud 2000). However, in several regions of Southern Europe, shrublands also originate from the absence of human disturbance. This is the case of scrub encroachment in abandoned agricultural fields where shrublands constitute a stage in the secondary vegetation succession. In the Aegean islands, for example, several old cereal fields, especially on terraces, have been invaded by phrygana communities (Arianoutsou-Faraggitaki and Margaris 1982; Giourga et al. 1998). The same has happened in several regions of the Iberian Peninsula (Moreira et al. 2001). Nevertheless, there are Mediterranean shrublands that represent the potential vegetation (climax), usually located in sites with low precipitation as well as in sites in which soil and parent rock material conditions lead to high water deficiency during summer (Arianoutsou-Faraggitaki and Margaris 1982).

In some parts of the Mediterranean Basin, shrublands have been considered marginal lands as they have low productivity and economic value (Calvo et al. 1998). For this reason, they have been often destroyed and replaced by forests through extensive afforestation programs in order to protect soils from erosion and increase land use profitability, especially in Spain (Calvo et al. 2005) and Portugal (Moreira et al. 2001). Other forms of disturbance concern the use of fire by pastoralists in order to suppress the woody species and create open areas for grazing (Arianoutsou-Faraggitaki 1985, 2001; Papanastasis 1977, 2004; Perevolostky and Seligman 1998), cutting for use as a domestic fuel or fertilizers (Calvo et al. 1989, 2002c; Casal 1987), and ploughing up in order to create arable land or to plant olive trees (Papanastasis and Kazaklis 1998). With changes in the current agricultural policies, several of these areas cleared for grazing or farming may be abandoned again in the future and may turn back to shrublands. So, current vegetation patterns in many parts of the Mediterranean Basin dominated by shrublands cannot be understood without taking into account past and current priorities in land-use (Baeza et al. 2007; Giourga et al. 1998; Lloret et al. 2003).

In spite of the low value modern societies attribute to shrublands, they represent distinctive European habitats for their biodiversity, as well as their aesthetic and cultural values (Wessel et al. 2004). Overall, they have significant properties as ecosystems and provide several services (Riera et al. 2007). In the future, shrubland ecosystems could be threatened as a consequence of large-scale human activities, like use of artificial fertilizers, rural abandonment and wildfires (Harrison 2010).

### 12.1.2 Ecosystem Services of Shrublands in the Mediterranean Basin

Ecosystem services are the benefits that people obtain from ecosystems. They support, directly or indirectly, our survival and quality of life. Mediterranean shrublands have significant functions as ecosystems and provide provisioning, regulatory, and cultural services (Harrison et al. 2010). One of the most important services associated with shrublands is to provide forage to grazing animals. Such animals may be wild, including game that can be hunted. Hunting is an important economic activity in Spain (Wessel et al. 2004) and elsewhere in Europe. Mediterranean shrublands are also an important source of forage for domestic animals such as sheep, goats, cattle and horses (Perevolotsky and Seligman 1998; Henkin et al. 2005, Papanastasis et al. 2008). Woody species constitute an indispensable animal food, particularly in areas with dry to semi-arid Mediterranean climates. Such species can alleviate feed shortages, or even fill feed gaps in the winter and especially in the summer, when grassland growth is limited or dormant due to unfavorable weather conditions. Among domestic animals, goats have a predilection for woody forage. Therefore, grazing is an important activity in these ecosystems, often keeping them in the same successional stage. In fact, reduction of grazing intensity often causes successional changes to forests (de Bello et al. 2005; Papanastasis and Kazaklis 1998). This reduction is due to the modernization of livestock production activities that result in a decline of use of extensive rangelands and grasslands (Papanastasis 2004).

Shrub vegetation can also be used as a source of fuel, but this practice has declined in the last decades due to its replacement by commercial fossil fuels (Papanastasis 2004). There is less evidence for other provisioning services. However, Mulas and Mulas (2005) report on an investigation related to screening of Sardinian populations of *Rosmarinus officinalis* (rosemary) for their potential as new cultivars for biomass quality and chemical composition in essential oils. Rosemary, like many other Mediterranean species, is a culinary plant and its oils are used in beauty products. Other examples of provisioning services are related with honey production associated with the numerous flowers during spring, especially from aromatic plants. Maquis shrubs are also used for ornamental purposes, for the production of fruits or liquors (e.g. *Arbutus unedo, Ceratonia siliqua, Myrtus communis*) (Papanastasis 2004).

About 30 different types of Mediterranean shrublands are protected under EU designation (Habitats Directive-92/43/EEC). They include several types with a priority conservation status such as *Cistus palhinhae* formations on maritime wet heaths, arborescent matorral with *Zyziphus* ssp. and *Laurus nobilis*, and several others sub-mediterranean, temperate scrub, mediterranenean arborescent matorral, thermo-mediterranean and phrygana types. Some animals with threatened status are also associated to shrublands. For example, the wildcat is a very rare and charismatic mammal with high conservation value in the Mediterranean area (Lozano et al. 2003). Other species with a high conservation value include birds like *Sylvia sarda, Sylvia rueppelli, Sylvia undata, Sylvia conspicillata* and mammals like *Hystrix cristata*, all of them exclusively associated with shrublands. Hence, their conservation requires the maintenance of these types of communities (Moreira and Russo 2007).

There is evidence of shrub ecosystems regulating climate, air, water and erosion, and probably other services, such as pollination, seed dispersal by birds and pest regulation. Both, shrubs and herbaceous species have been shown to reduce water runoff and hence reduce soil erosion thus helping to avoid desertification (Boeken and Orenstein 2001; Scott et al. 1998). However, shrublands may have a negative role in fire management through their flammability (Potts et al. 2006). On the other hand, they affect nutrient cycling, as above and below ground carbon sequestration can be enhanced by the presence of woody plants (Wessel et al. 2004). However, all of these services could be altered in the future. We expect that livestock production and pollination in shrublands will decline due to modernization leading to the declining use of extensive rangelands (Harrison et al. 2010), while urbanization will lead to their transformation into croplands and urban areas. The status of other services, such as wood fuel and erosion regulation, has also been degraded as Europe is estimated to have lost about 90% of its shrublands. Shrubland restoration and recreation in the 1990s, partly to fulfill conservation objectives (e.g. EU Habitats Directive), has meant that some services, especially cultural, have been enhanced in a few regions of Europe (Harrison 2010).

## 12.2 Fire in Mediterranean Shrubland Ecosystems

Fire has been an important management tool used by people since prehistoric times (Moreno et al. 1998). On the other hand, Homer reports in Iliad that naturally started wildfires were a common phenomenon in prehistoric Greece (Liacos 1973). At first, it was a natural component that appeared more or less regularly in the natural cycle of vegetation succession (Trabaud 1994). Lightning and volcanic eruptions were natural causes of vegetation fires. However, the appearance of humans disrupted this balance of nature. The first evidence of its presence in the maquis belt of the Mediterranean Region was during the Mesolithic period in Israel and Greece, where fire was used to facilitate hunting and food gathering. Since then, fire became an important tool for pasture improvement (Naveh 1994). As human population increased, an increase in the extent and intensity of burning has been also observed.

The Mediterranean Basin has a marked prevalence of human induced fires, amounting to about 95% of the total number of fires (FAO 2007). According to the European Commission (2009), around 50% of the burned surface in Southern Europe corresponds to non-wooded areas in the last decade (data not applicable to France). Specifically for Spain, this percentage is much higher (Fig. 12.2). During the 60s, when the great depopulation of rural areas occurred, the percentage of burned non-wooded surface was about 50% in Spain. Since then it has increased and reached 66% in the 90s and 67% in the last decade.

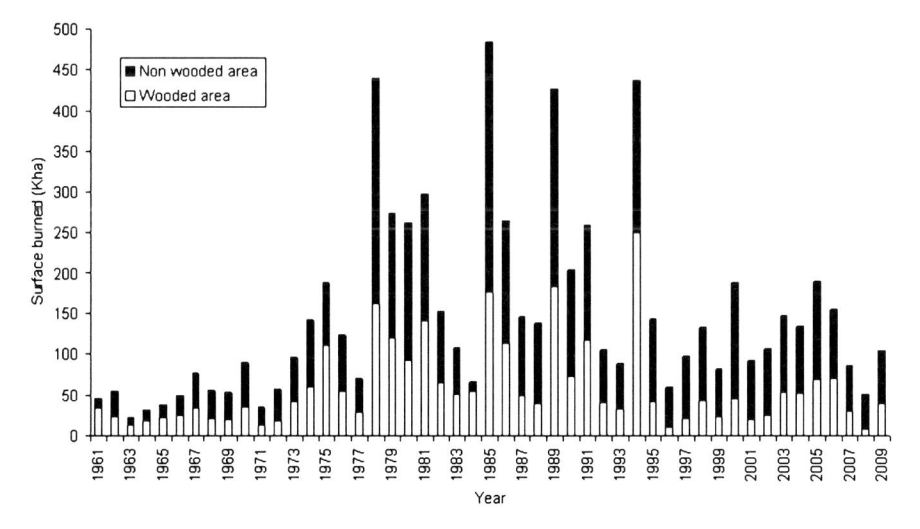

**Fig. 12.2** Yearly wooded surface burnt (*open bars*) and non wooded surface burnt (*close bars*) in Spain for the period 1961–2009. Elaborated from MARM (Ministerio de Medio Ambiente y Medio Rural y Marino)

Also in Greece, 63% of the number of fires occurred in non-wooded areas in the period 1980–1996, which accounted for 61% of the total burned lands (Dimitrakopoulos 2001). This increase in the proportion of burned area composed of shrublands may be related to the difficulty of changing the behaviour deeply rooted in rural society (APAS 2004). In Spain, for example, rural people continue to burn shrublands in order to increase pasture availability, a traditional practice, regardless of the decrease in the number of livestock. In Greece, at least 10% of the fire incidents are attributed to shepherds who burn shrublands for pasture improvement (Dimitrakopoulos 2003). The main problem is that fuel accumulation in shrublands across Southern Europe is currently higher than in the past and there is a high risk for pasture burning to initiate a devastating wildfire.

Not only the behavior of rural population is "responsible" for this increase, but other causes linked to depopulation of the rural areas have also contributed to the big changes in the traditional land use (Etienne et al. 1998; Maheras 2002). These changes have led to a loss of landscape heterogeneity and, consequently, to an increase of unmanaged shrubs and biomass (Rego 1992) with a high risk of fires (as in Portugal and Spain) and a loss of inhabitants with a sense of responsibility for the forest (Le Houérou 1987).

Changes in the landscapes over the recent past (Mazzoleni et al. 2004) in interaction with the more frequent heat waves during the summer due to climate change will alter the fire regime by increasing the frequency (Arianoutsou-Faraggitaki 2001; Pausas and Vallejo 1999), the wildfire hazard (Piñol et al. 1998) and the presence of very large fires (Moreno et al. 1998) thus affecting large shrublands areas.

## 12.3  Post-Fire Responses in Shrublands

### 12.3.1  Vegetation Response Strategies

Regardless of the different rate of post-fire regeneration between the plant species showing different response strategies, Mediterranean-shrublands are generally recognized as highly resilient to fire as a consequence of the ability of their plant species to recover from fire by means of resprouting from fire-resistant structures or from fire-protected seeds (Calvo et al. 2002c; Lloret 1998). However, some studies suggest that the resilience of Mediterranean shrublands is also related to the fire frequency (Díaz-Delgado et al. 2002). So, high fire recurrence could be a serious threat in shrublands dominated by obligate hard-seeder species, which need sufficient time to produce enough viable seeds to survive subsequent fires (Vallejo and Alloza 1998). The dynamics of communities dominated by resprouting species are more independent of the fire interval (Calvo et al. 2002a) and in general, more resilient under high fire recurrence than shrublands dominated by non-sprouting species (Vallejo and Alloza 1998).

Most Mediterranean shrublands are composed of obligate resprouters and obligate seeders as well as facultative resprouters that possess both regeneration strategies. For example, *Erica arborea*, a dominant species in maquis shrublands and *Sarcopoterium spinosum*, a dominant species in phrygana shrublands, regenerate after fire by both resprouting and seeds (Arianoutsou and Margaris 1981; Papanastasis 1978a; Seligman and Henkin 2000). For this reason, they are resilient to relatively short fire intervals. Phrygana ecosystems are well adapted to recurring wildfires every 3–5 years (Papanastasis 1980) while maquis and garrigues may tolerate a fire cycle of 5–10 years (Papanastasis 1988a, b).

## 12.3.2  Post-Fire Dynamics in Shrublands Dominated by Resprouters

Resprouting species are able to regenerate from dormant buds provided they have carbohydrates stored in their root system adequate to support their regeneration till they will develop new photosynthetic tissues. In some species like *Erica australis* this storage material is found in the lignotuber (Canadell and López-Soria 1998). The formation of a lignotuber has been associated with recurring perturbations that eliminate all the aboveground biomass (Cruz et al. 2003). This starch is degraded and freed to supply the sprout (Canadell and López-Soria 1998). However, in addition to resprouting vigorously, it is also capable of germinating from seeds stored in the soil seed bank.

One example of high resilience after fire is provided by a shrubland dominated by *Erica australis*, a typical resprouting species in the northwest of Iberian Peninsula. Soils are classified as humic cambisols; they are not easily erodible and have very low nutrient content and acidic pH. Calvo et al. (2002a) found that fire induces an increase in the nutrients of the soil (mainly nitrogen and phosphorus) that can be used by resprouting species. This is because, although the aboveground biomass is eliminated after fire, the sprouting organs are kept (Calvo et al. 2002c). As a result, there is a very fast recovery of *Erica* one year after burning (Fig. 12.3).

Woody species as a whole increase significantly their cover values over time until year 9 (Fig. 12.4), but thereafter the community is stabilized without further changes. The highest cover values for herbaceous species, mainly perennial, are recorded in the first 8 years after burning. As a whole, herbaceous species show an inverse relationship with woody species (Fig. 12.4), as their values increase when the woody cover is not very high and then start to decrease.

Similar patterns of secondary succession after fire have been found in a phryganic ecosystem in Greece (Arianoutsou-Faraggitaki and Margaris 1982) dominated by resprouting species such as *Phlomis fruticosa, Euphorbia acanthothamnos* and with the presence of seeding species like *Cistus* spp. During the first post-fire years, there was an abundance of herbaceous plants that did not last more that 5–8 years. After that, their numbers decreased and there was a clear dominance of woody species.

**Fig. 12.3** One year (*upper image*) and 15 years (*lower image*) after burning in a shrubland dominated by *Erica australis* in NW of Spain

In general, the abundance of herbaceous species after fire depends very much on the structure of the original community. In maquis, where herbaceous species have limited contribution to the plant cover, a flush of annual herbaceous species appears the first year after the fire, coming from the seed bank or the adjacent unburned areas, mainly consisting of legumes that disappear the following years (Papanastassis 1988a) (Fig. 12.5).

In garrigues, on the contrary, where herbaceous species make a significant proportion of plant cover and consist of mainly perennials, their cover increases only

**Fig. 12.4** Mean percentage cover and standard error of annual herbaceous, perennial herbaceous and woody species after burning in a shrubland dominated by *Erica australis* in NW of Spain. (Modified from Calvo et al. 2002c)

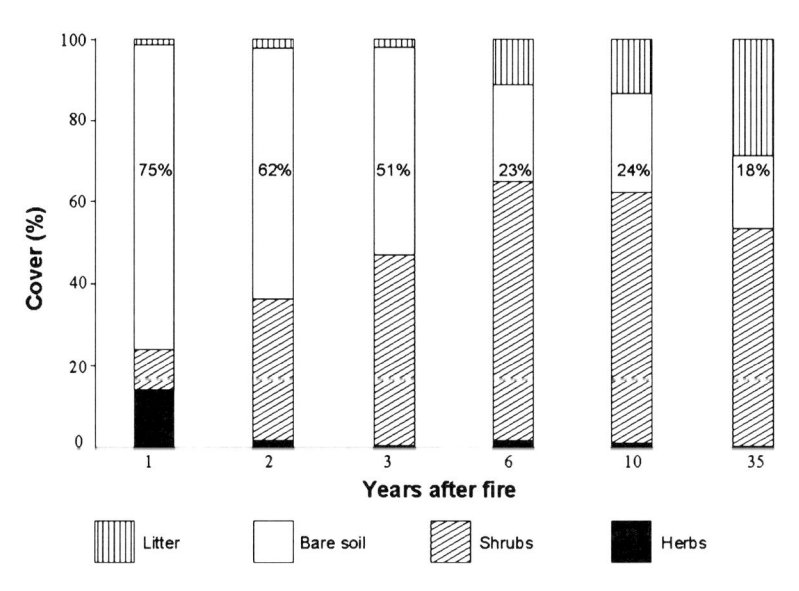

**Fig. 12.5** Evolution of plant cover 10 years after a wildfire in a maquis shrubland of northern Greece compared with the control (35 years old) (Papanastasis 1988a)

slightly the first years after the fire with annuals being favoured by burning but they do not disappear as fast as in maquis (Fig. 12.6).

So, in shrublands dominated by ericaceous resprouting species, the possibility of replacement by a different vegetation type is very low. The strong resprouting capacity of the component species confers a high degree of resilience in the spatial structure of shrub communities during recovering from perturbations in the short

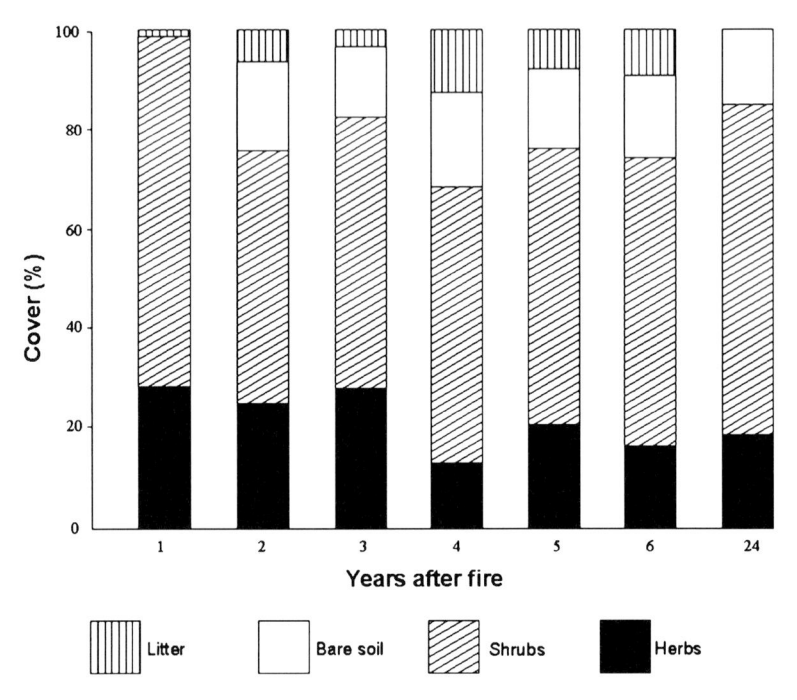

**Fig. 12.6** Evolution of plant cover 6 years after a wildfire in a *Quercus coccifera* garrigue of northern Greece as compared with the control (24 years old) (Papanastasis 1988b)

term (Calvo et al. 1998; Clemente et al. 1996). This type of post-fire regeneration has been traditionally described as an autosuccessional process where vegetation generally recovers without any great changes in the species composition (Hanes 1971; Arianoutsou and Margaris 1981; Trabaud 1992; Calvo et al. 2002b). This is also supported by long term studies carried out by Kadmon and Harari-Kremer (1999) in maquis communities.

### 12.3.3 Post-Fire Dynamics in Shrublands Dominated by Obligate Seeders

In some areas of the Mediterranean Basin where resprouters are scarce or absent, shrublands are dominated by seeding species (Pausas 2003; Vallejo and Alloza 1998). Among the species that appear in this type of shrublands, also known as gorse shrublands, *Ulex parviflorus*, *Rosmarinus officinalis* and *Cistus* spp. are included. Gorse shrublands are found on soils usually developed over marls. These soils are generally deep, highly calcareous and without rocky outcrops. Most of these soils have been cultivated and terraced, even in mountainous areas. Currently, farming has been abandoned and vegetation is maintained by recurrent fire.

These areas require a long fire-free period to be colonized by resprouting shrubs and trees (Santana et al. 2010).

One of the main characteristics of these obligate seeders is that they are very sensitive to high temperatures and do not survive a fire. These species generally depend on their soil seed banks to persist. Temperature is probably the most important environmental factor regulating the softening of hard seeds (Baskin and Baskin 1998) in these types of shrublands (Baeza and Roy 2008).

The regeneration of populations of these species from seeds implies that in the first years after a fire the vegetation cover will be low. This in turn implies that post-fire soil degradation due to water erosion is likely to be higher (Vallejo 1997). The combination of easily eroded marly soils and a vegetation type requiring a long period of time to reestablish its cover after fire makes these environments highly vulnerable to degradation processes. In fact, several studies have documented low resilience in Mediterranean shrublands dominated by seeders. Vallejo and Alloza (1998) carried out an extensive analysis of the capacity of this vegetation type to respond to fires that occurred in the Valencia region of Spain in 1991. They found that the mean vegetation cover registered was only 40.5% during the first year after fire and only 54.7% 3 years after. It should be noted that a high proportion of this cover (27%) corresponded to the herbaceous resprouting species, *Brachypodium retusum*; in contrast, the most abundant obligate seeding species, *Ulex parviflorus*, contributed only by 6.5% 3 years after fire. Similar results were reported by Baeza and Vallejo (2008), who carried out fuel reduction treatments in a 9-year-old *Ulex parviflorus* shrubland and found a very low regeneration capacity, with values around 60% 4 years after treatment application. However, the capacity of regeneration in shrublands dominated by obligated seeders also depends on the type of environmental conditions as in the case of Mediterranean shrubland grown in more humid areas, dominated by *Cistus ladanifer* and *Cistus laurifolius*, in which germination after fire is very fast and confers high resilience to the ecosystems (Tárrega et al. 1995, 1997).

Some conceptual models of post-fire vegetation dynamics in Mediterranean shrublands dominated by seeders (Fig 12.7) (Baeza et al. 2007) showed that when the dominant species in the initial stage of succession, for example *Ulex parviflorus* and *Cistus albidus*, become structurally disorganized and die (senescent phase), *Rosmarinus officinalis* becomes dominant, occupying the spaces opened by dead plants of *Cistus* after 10–15 years and of *Ulex* after 20–25 years.

As there is no transition towards later successional stages dominated by resprouters, *Rosmarinus officinalis* dominated shrublands could constitute an alternative state in which succession is arrested (Santana et al. 2010).

When fire occurs at short intervals in gorse shrubland (i.e. 2–3 fires within 20 years), the vegetation becomes dominated by open shrubland. *Rosmarinus officinalis* suffers an important decline after the second fire event– a decline similar to the one reported for Catalonia (Eugenio and Lloret 2006) and after the third fire plant community becomes dominated by nanophanerophytes (e.g. *Cistus albidus*), sub-shrubs (e.g. *Thymus vulgaris, Dorycnium pentaphyllum*) and, especially, the rhizomatous perennial grass *Brachypodium retusum*.

Vegetation dynamics in plant communities dominated by seeding species

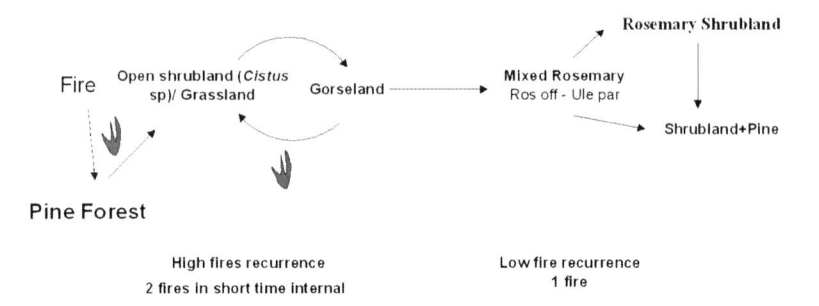

**Fig. 12.7** Vegetation dynamics after fire in shrublands dominated by seeder species with different fire recurrence (*1* and *2 fires*) on the Ayora site (Valencia Region, Spain). (Modified from Baeza et al. 2007)

## 12.4 Post-Fire Restoration in Mediterranean Shrublands

### 12.4.1 Natural Restoration Without Any Intervention

Often, the main objective of post-fire management of Mediterranean shrublands is to restore them to their original pre-fire status at the earliest time possible. This can be achieved through natural processes by allowing the burned species to recover through resprouting or seed regeneration. Therefore, just protecting the burned shrublands from any human intervention would allow them to recover and get restored to the pre-fire conditions.

The speed with which the burned ecosystems will recover depends on several factors: the fire regime and especially fire history and fire intensity, the climatic and soil conditions and the relative contribution of resprouters and seeders. In general, shrublands composed mainly of resprouting species on humid environments with deep soils will show a faster post-fire recovery than similar shrublands found on drier areas and poor soils. Even the post-fire regeneration mode of the individual species may be defined by the specific climatic conditions. For example, the phryganic species *Sarcopoterium spinosum*, regenerates mainly through seed germination by in the semi-arid environment of Attica, Greece (Arianoutsou–Faraggitaki and Margaris 1982) while in the subhumid environment of northern Greece, sprouting is its main regeneration mode (Papanastasis 1980).

### 12.4.2 Emergency Revegetation with Herbaceous Species

Despite the relatively rapid regeneration capacity of the constituent species, there is a great risk of soil erosion in burned shrublands, especially on steep slopes and in the first year after the fire. In the gorse (Baeza et al. 2007) or maquis shrubland (Papanastasis et al. 1988a), approximately 75% of the ground (Fig. 12.5) was bare at the end of the first growing season after the fire, being susceptible to high erosion risk. In the following years, this risk is reduced as the vegetative cover is gradually restored but there is still danger for soil erosion, at least until the third year after the fire. To cope with this problem, the practice of emergency revegetation of burned shrublands with herbaceous species is being applied in California since the 1940's (Keeler-Wolf 1995), but, this process is not an established practice in the Mediterranean region. However, several experiments have been conducted in various countries with interesting results. In the burned maquis shrubland composed by *Arbutus unedo, Erica arborea, Quercus coccifera, Phyllirea latifolia, Pistacia lentiscus, Olea oleaster* and *Cistus* spp., a mixture of grasses and legumes such as *Dactylis glomerata, Phalaris aquatica* and *Trifolium hirtum* was sown in November, about 3 months after the wildfire. At the end of first growing season, plant cover in the seeded plots was 62% compared with 51% in the non-seeded ones. In the following 2 years, seeded species declined except *Dactylis glomerata* which was the best to establish in the burned area and retained a 10% contribution to the plant cover at the end of the third season after the fire (Papanastasis 1978b). Better results were achieved in a *Quercus coccifera* garrigue, which was seeded with a mixture of annual and perennial grasses, namely *Bromus mollis, Lolium rigidum* and *Phalaris aquatica*, immediately after a wildfire. At the end of the first growing season, plant cover in the seeded plots was 69% of which 28% was contributed by the seeded species while in the non-seeded ones plant cover was only 45%. *Bromus mollis* was the most successful species, followed by *Lolium rigidum* while *Phalaris aquatica* failed again as in the previous experiment. By the end of the 3rd year, the seeded species contributed only 9% to the cover but there were no statistical differences in plant cover between the seeded and non-seeded plots (Papanastasis and Platis 1990).

Experiments to promote the presence of herbaceous species have also been also performed in eastern Iberian Peninsula as emergency post-fire treatments. Badia and Martí (2000) as well as Serrasolses et al. (2004) observed that sowing commercial pasture and forage species did not have a significant effect on the regeneration of the community: the herbaceous species were only slightly enhanced during the first post-treatment year in combination with mulch treatments. This fact put into relevance the importance of using autochthonous species collected in the area. Fernández-Abascal et al. (2003, 2004) used commercial varieties of autochthonous species in northwest Iberian Peninsula to restore a burned shrubland with *Erica australis*. The results obtained were different according to the meteorological conditions prevailing after sowing and the time of sowing. So, when the restoration was carried out 8 months after fire, using a mixture of three herbaceous species: *Agrostis capillaris, Festuca rubra* and *Lotus corniculatus* the results showed that no significant differences in the herbs cover between sown and unsown plots at any sampling period was observed. In particular, *Agrostis capillaris* and *Festuca rubra* presented higher mean cover in the sown plots than in the control, whereas the cover of *Lotus corniculatus* was very low in all cases. The low success of the sown herbaceous

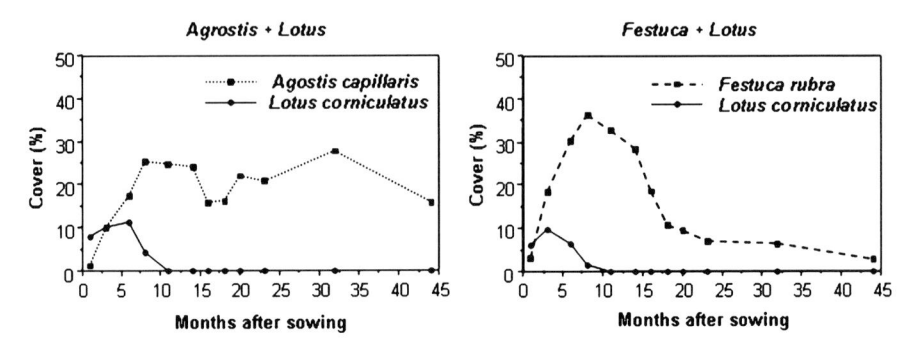

**Fig. 12.8** Mean cover of sown species in different experimental plots fixed in a burned shrubland co-dominated by *Erica australis* (resprouter) and *Erica umbellata* (obligate seeder) in NW of Spain. (Modified from Fernández-Abascal et al. 2003)

species was due to the adverse meteorological conditions (very cold spring and dry summer) after revegetation (Fernández-Abascal et al. 2004). However, another experiment carried out in a burned shrubland co-dominated by *E. australis* (resprouter) and *Erica umbellata* (obligate seeder) in NW Spain (Fernández-Abascal unpublished), where they used the combinations of the same herbaceous species (*Festuca rubra, Lotus corniculatus, Agrostis capillaris*) and was done very soon after fire (1 month) showed better results for post-fire restoration. Total cover was found significantly higher in the sown plots than in the control during the first few months, however, after 18 months no differences were found between treatments. One interesting aspect is the different behavior showed by the species used in the experiment (Fig. 12.8). *Lotus corniculatus* appeared only in the first year (values above 10%) and had no effect on the total cover. *Festuca rubra* appeared earlier than *Agrostis capillaris* and reached a 40% of cover at 8 months after sowing but decreased significantly in cover after 18 months. *Agrostis capillaris* cover values were lower at the beginning of the experiment but then maintained a constant cover value (25–30%) throughout the study period.

These differences can be a key aspect to select the more suitable species in relation to the objective of restoration. *Festuca rubra* is more preferable for soil protection and *Agrostis capillaris* for maintaining a good herb cover in time. Sowing did not affect the natural recovery of the woody species in the first stages, but in the last stages woody cover was significantly lower in the plots sown with *Agrostis capillaris* than in the unsown and these differences remained in time. The interference of sown species, particularly *Agrostis capillaris*, seemed to be related to the regeneration strategy of the woody species, since the greatest differences from the unsown control plots were in the cover of *Erica umbellata*, which could only recover from seed in these areas (Fernández-Abascal et al. 2003).

When climatic conditions are favourable, sowing with herbaceous species is not necessary (Calvo et al. 2005, 2007). However, fertilisation with nitrogen together as well as burning or cutting favoured the increase of perennial herbaceous species (Calvo et al. 2002b).

It appears, therefore, that the success of the seeded herbaceous species on burned shrublands depends on several factors. One factor is the particular species to be seeded which should be well adapted to the area under restoration. Annual species, especially grasses, are more preferred than perennials because they grow faster during the winter months after the fire. Another factor is the time of seeding. Seeds should be sown right after the fire on the ash which can serve both as a cover and as nutrient provider (fertilizer). If the seeds are seeded after the ash has been washed down and they fall on bare ground then their chances to germinate are limited. Finally, another factor is the weather conditions after fire and especially the intensity of the first rains. If they are too heavy, then the seeds may be drifted down slope, particularly in steep terrains.

### 12.4.3  Livestock Grazing of Burned Shrublands

Livestock grazing is not a management practice to restore burned shrublands but to reduce fuels aiming at decreasing the intensity and frequency of potential fires. In the Mediterranean Basin, fire and grazing have shaped the landscape over millennia, since indigenous agriculture and livestock farming have been practised in combination with deforestation and fire management for more than 10,000 years (Blondel and Aronson 1999). Fire has been traditionally used by shepherds to manage shrublands for the benefit of their livestock. Nowadays, the practice is still common in several parts including the Mediterranean islands such as Corsica, Crete and Sardinia. Pastoral burning is done in order to suppress the unpalatable woody species to animals in favor of the herbaceous plants that appear in abundance after the fire and are very palatable to sheep and cattle. However, both pastoral burning and grazing are practiced in an uncontrolled way thus resulting in land degradation (Papanastasis 1977). The phrygana shrublands are subjected to periodic pastoral wildfires and overgrazing by sheep. This practice can be illustrated with a conceptual model showed in Fig. 12.9.

Farmers set the fires on the very hot days during summer in order to suppress unpalatable phryganic shrubs. After the first autumn rains, a flush of herbaceous species, mainly legumes, appears in the burned areas that are allowed to be overgrazed by sheep. In the absence of competition, phryganic seedlings grow up fast and in a period of 3–5 years they overgrow. Then, shepherds set another fire to control them again and the vicious cycle goes on. On the long run, the fire cycle gets shorter and shorter and the phryganic ecosystem is degraded due to soil erosion after the fires, the elimination of several herbaceous species and the predominance of woody species. Overgrazing combined with wildfires is considered as the main cause of degradation and desertification of rangelands in the Mediterranean basin (Papanastasis 1977; Arianoutsou-Faraggitaki 1985).

In order for livestock grazing to serve the sustained use of burned shrublands, the fire-grazing system must be rationalized. Specifically for fire, instead of using it in an uncontrolled way as shepherds do, prescribed burning may be applied.

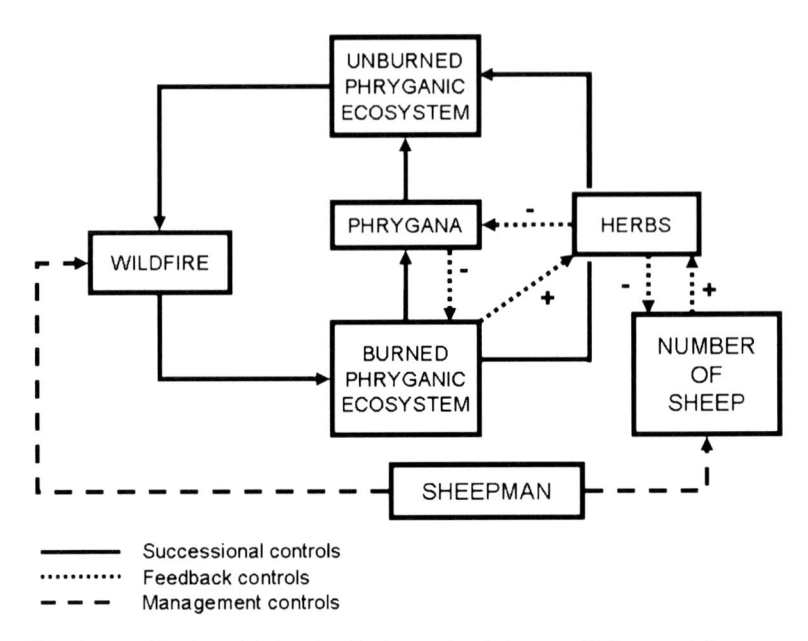

**Fig. 12.9** A generalized model showing the interaction between wildfire, vegetation succession and sheep grazing in phryganic rangelands where overgrazing is combined with pastoral wildfires (Papanastasis 1977)

For example, prescribed fire was used in Greece in order to convert a tall (2 m in height) and dense *Quercus coccifera* garrigue to grassland for the benefit of goats (Liacos et al. 1980). For this reason, the burned site was seeded right after the fire on the ash with a mixture of grasses and legumes such as *Dactylis glomerata, Bromus inermis, Phalaris aquatica, Lolium multiflorum* and *Lotus corniculatus*. At the same time, an adjacent shrubland was improved by thinning and cutting the shrubs to a height of 0.80 m so that they can be easily reached by goats. After a protection period of one growing season both treatments were grazed by goats for 3 consecutive years. Although the productivity of the converted treatment declined from the first to the third year due to the gradual decline of the seeded species, live weight gains of goats in burned plus sowed plots were on the average 60% higher than the ones in the improved treatment (Liacos et al. 1980).

As far as the grazing in the fire-grazing system is concerned, the problem does not usually lie on wildfires themselves but on the grazing management applied in burned shrublands. The question that often arises is how soon after the fire should livestock be allowed to enter the burned areas. Shepherds let their animals to graze during the first autumn months after the fire when seedlings and sprouts are too tender and weak, but very palatable, and a large part of the area is bare. Such a practice is detrimental to the burned ecosystem. The proper approach is to delay grazing until vegetation is well established after the fire. The grazing management to be applied should be adapted to the particular area. Plant cover in burned

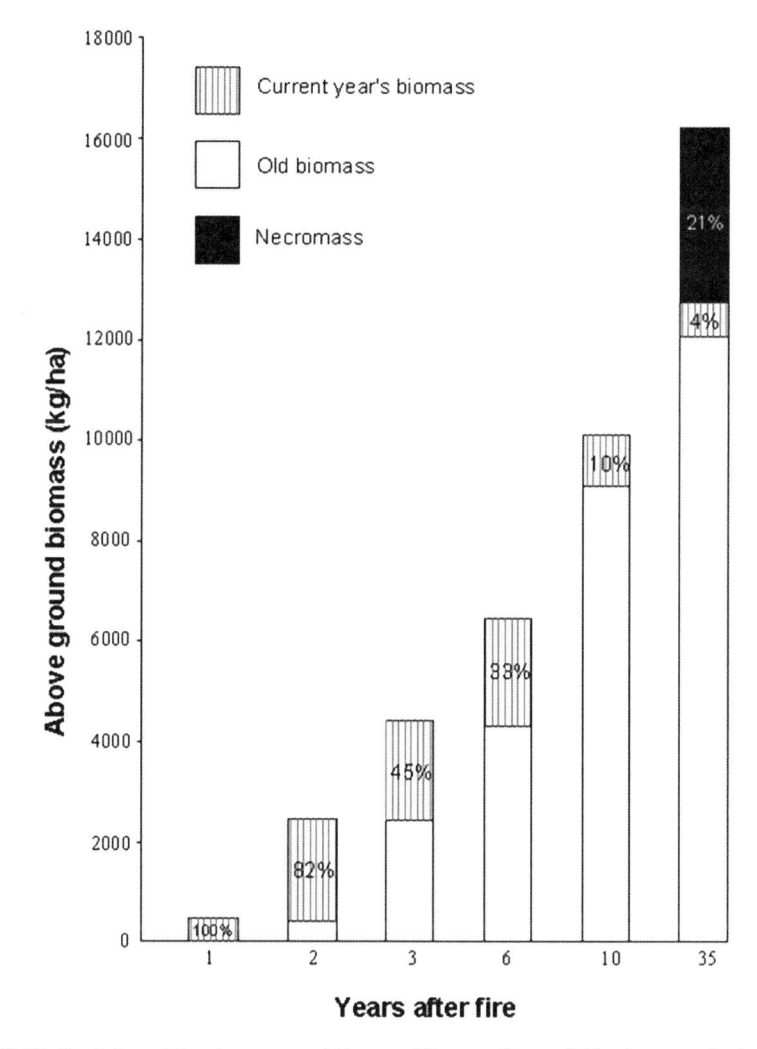

**Fig. 12.10** Evolution of the above-ground biomass 10 years after a wildfire in a maquis shrubland of northern Greece as compared with the control (35 years old) (Papanastasis 1988a)

shrublands is restored faster after fire than the above ground biomass. In a maquis shrubland in Greece, vegetation cover was almost established 6 years after the fire (Fig. 12.5) but the above ground biomass was only 62% in the 10th year after the fire compared with the unburned shrubland (Papanastasis 1988a). On the other hand, grazable (current year's growth) biomass was 100% in the first year after the fire but reduced to only 10% of the above ground biomass 10 years after (Fig. 12.10). This means that grazing could start when plant cover is restored but it should be of moderate intensity so that the ecosystem is not prevented from restoration due to overgrazing.

Grazing impact upon vegetation depends on the density of animals as well as the duration of grazing. Henkin et al. (2005) found that under heavy grazing (175–206 cow grazing days per hectare) vegetation was maintained open in a scrub woodland of *Quercus calliprinos* in Israel while under more moderate grazing (122–148 cow grazing days per hectare) the same vegetation tended to return to a dense thicket. Perevolotsky et al. (2002) observed that shrub removal by mechanical procedures together with grazing during the green season for 4 years was the most effective firebreak treatment in shrublands from Israel. Reduction of shrub cover in favour of herbaceous vegetation results in improved foraging conditions for grazing animals such as sheep and goats. For this reason, a mosaic of treated areas with prescribed fire or mechanical means to reduce shrub cover and non-treated ones should be pursued in shrubland management in order to develop semi-open, highly heterogeneous and diversified landscapes. Such landscapes will ensure the optimum use of shrublands by livestock, reduce the fire risk and serve hydrological purposes as well as aesthetic values.

It is important, therefore, for land managers to be able to estimate the thresholds under which livestock grazing can be applied without causing vegetation degradation. For example, in open maquis of south France the carrying capacity range is 0.6–1.4 animals (sheep equivalents) per hectare and year, whereas in woodlands it can reach 1.65 animals per hectare and year (Etienne et al. 1993). In drier areas of the south-east of the Iberian Peninsula, the carrying capacity is close to 1 animal per hectare and year, or even less (Correal et al. 1992; Baeza 2004). Also, maquis produce generally less usable forage for livestock than garrigue (Papanastasis 2000). Excessive livestock pressure can destroy soil structure, accelerating erosion processes and ecosystem productivity losses (Papanastasis 2000). On the other hand, light grazing may even facilitate shrub invasion by creating niches for woody species establishment through the trampling and browsing of the competing herbaceous vegetation (Noy-Meir et al. 1989).

In summary, further work on developing management tools is needed to fully understand the variables influencing fuel removal. It has been proposed that the integration of different treatments could provide the best strategy in these cases (Nader et al. 2007). For example, livestock cannot effectively control mature brush plants that either grow taller than the animals can effectively graze or have large diameter limbs (Liacos et al. 1980). In these cases, mechanical clearing, rotary slashing, prescribed burning control and hand-cutting can be used to manipulate the large diameter fuels, and grazing can be used as a follow-up treatment for controlling resprouting species or shifting the species composition to herbaceous plant fuel material (Liacos et al. 1980).

## *12.4.4* *Plantation with Fire-Resistant Resprouting Species*

Recent studies (Baeza et al. 2011) have shown that typical early-successional shrub species such as *Ulex parviflorus, Cistus albidus, Cistus clusii* and *Erica multiflora* present much higher percentages of dead fuel than trees or shrubs from later-successional stages such as *Quercus ilex, Juniperus oxycedrus* or *Pinus halepensis*, suggesting that vegetation communities dominated by the former shrub species are more likely to burn than the ones dominated by the latter. Several studies have pointed out that under less extreme weather conditions small fires seem to be particularly selective and show a strong preference for early successional shrublands (Díaz-Delgado et al. 2002; Nunes et al. 2005; Moreira et al. 2009). This could explain the process where forest communities degrade into shrub formations, favouring the phenomenon called *matorralización* (Zamora et al. 2004), where the conditions for new fires in short periods of time can increase the spatial aggregation of the fire.

The Clementsian succession model assumes that the successional process follows a linear trajectory in which facilitation between species allows more mature successional states to be reached by replacing some biological types with others. This model views succession as a deterministic process, whereby each stage in the successional series generates less favourable environmental conditions for the community already installed and more favourable ones for the following stage. In contrast, state-and-transition models recognise that multiple successional trajectories are possible, and that alternative meta-stable states can exist under the same environmental conditions (Hobbs and Norton 1996). An example of the latter state- and- transition model could be found in the Mediterranean gorse (*U. parviflorus*) shrubland, where inter-species tolerance mechanisms predominate since most species are present in the initial phase of colonization after fire.

Gorse shrublands are widespread in eastern Spain, colonizing old-fields as well as other naturally or anthropogenically disturbed areas. The large amount of dead fine fuel fractions in these shrublands constitutes an extreme fire risk during the summer dry period (Baeza et al. 2006). Moreover, the accompanying species are mainly obligate seeders typical of the first stages of ecological succession (i.e. *Rosmarinus officinalis, Cistus albidus, Cistus clusii*). The scarce presence of resprouting species, together with the dominance of highly flammable seeders, increases the fire hazard, slows down the recovery rate after disturbance and reduces the resilience of the ecosystem, facilitating the generation of a fire-degradation loop.

Restoration is frequently regarded as accelerated succession where planting of resprouting species can result in changes in aggradations trajectories towards reference ecosystems. The main aim of creating a more suitable environment for late successional stages is to convert Mediterranean shrublands into more resilient communities with reduced fire hazard and increased ecosystem quality. Several studies in the Mediterranean Basin have shown that late successional hardwood species may be able to establish in degraded areas during the restoration process. Valdecantos et al. (2009) used two different silvicultural techniques to plant *Quercus ilex* and *Rhamnus alaternus* in shrublands: one was designed to cause the least alteration in the vegetation structure of the ecosystem while the other involved clearing of the standing vegetation (brush-chipping) and preserving desirable species. Five years later they observed that the combination of planting and managing natural vegetation was successful in reintroducing native resprouting species. Both species showed survival percentages above 90% in all treatments, which represents very high values in comparison with other experiments under similar environmental and climatic conditions (Valdecantos et al. 2006). Both species showed higher survival rates within the undisturbed shrubland and faster growth rates on the cleared plots where the light and water limitations were less extreme and the slash generated by clearing had increased the soil water content. The net positive effect of the presence of shrubs on the survival and growth of 16 tree seedlings has been also shown in restoration experiments in south-eastern Spain (Gomez-Aparicio et al. 2004). The canopy of the nurse shrubs improves both the microclimatic conditions and the physicochemical properties of the soil, thus reducing abiotic stress and improving resource acquisition This facilitation effect allows a fraction of seedlings of the species planted to survive the first year of outplanting, resulting in the establishment of proportionally more seedlings than in other microhabitats devoid of protective vegetation cover (Maestre et al. 2004; Castro et al. 2002; Gomez-Aparicio et al. 2004).

An alternative to establishment of a hardwood forest in fire-prone shrublands is to create agro-forestry systems by planting trees way apart so that wood production is combined with forage production to be used by domestic and wild animals. For the western part of the Mediterranean, the main dehesa species *Q ilex* and *Q. suber* may be used while for the eastern part *Q. ithaburensis* and *Q. coccifera* could be promoted.

## 12.5 Post-Fire Management in a Global Warming Scenario: Implications

The last IPCC report predicts warmer and drier conditions, in particular, in the southern part of Europe with important influence in Mediterranean area. The new climate conditions that are predicted to occur in the future will very likely cause an increase in fire danger conditions and in fire risk in the Mediterranean area (Moreno et al. 2010; Moriondo et al. 2006).

In the Mediterranean Basin, shrubland ecosystems represent an important vegetation component that will be most affected by fire. These effects will be conditioned by the type and timing of the fire, the intrinsic regeneration capacities of different species (from obligate resprouters to obligate seeders) and the pre-fire status of the vegetation. For example, the response of resprouting species to fire is regulated by a positive-feedback mechanism of vegetative regeneration that allows for quick space occupation, often before seeding species have germinated (Riera et al. 2007). Therefore, in order to predict the post-fire regeneration in the scenario of climate change, a good knowledge about plant regeneration strategies is needed.

The differences in the regeneration modes: seedlings vs vegetative sprouts (Lloret 1998, 2004) condition the ability to regenerate in the framework of the climate change. Seedlings are new genets with increased response plasticity to environmental constraints. They also provide units of dispersal, which can be crucial for the survival of species under changing climates or highly disturbed environments. Finally, although seedlings develop their own above-ground structure and root system, sprouts share anatomical structures and resources in other parts of the same plant (Bond and Midgley 2001). Also, sprouts obtain benefit from a root system well developed by a previously established individual, while seedlings may be more vulnerable to water or nutrient stress, due to their lower ability to uptake resources from soil (Savé et al. 1999). Thus, lower survival and growth rates are expected in seedlings compared to sprouts. However, this different response to the availability of below-ground resources would diminish in older individuals (Lloret 2004). Such a different behaviour could be the sign of a possible change in species composition with possible consequences on the evolution of the community. A new community structure could imply different processes at the soil (Hawkes et al. 2005) and plant level (Armas and Pugnaire 2005).

Another factor that could be considered is the possible increase in the recurrence of the fires. Mediterranean vegetation may not be adapted to an even-more-frequent fire regime and negative consequences may occur as a result (Diaz- Delgado et al. 2002). More frequent fires may contribute to a further reduction in the productivity of the vegetation and the organic content of the soil. Reductions in soil organic matter will diminish soil aggregate size and stability and together with a decrease in vegetation cover, it will reduce the infiltration of water into the soil and increase surface flow. All these lead to soil erosion and further deterioration of the hydrological characteristics of the soil (Garcia et al. 2002). Reduction of plant cover, increased fire and erosion risk are three of the most pronounced and socially worrying potential effects of climatic change in Mediterranean shrublands (Riera et al. 2007).

Therefore, since an increase in the frequency of fires is expected, in shrublands dominated by seeders, an appropriated management strategy would be to transform them in more resilient communities, i.e. dominated by resprouting species. Also, another good management strategy would be to use shrubland communities for grazing by domestic and wild animals so that the fuel loads are controlled and economic returns are ensured for the rural people.

## 12.6 Key Messages

- Some Mediterranean shrublands can recover through natural processes after fire. Therefore, just protecting the burned shrublands from any human intervention would allow them to recover and get restored to the pre-fire conditions.
- In shrubland, with great risk of soil erosion after fire, the practice of emergency planting with herbaceous species may be applied with positive results.
- In shrublands dominated by obligate seeders, an appropriate management strategy would be to transform them into more resilient communities by favouring the introducing resprouting species.
- Another effective management strategy would be to use shrubland communities for grazing by domestic and wild animals so that the fuel loads are controlled minimizing the risk of future fires and decreasing their intensity while economic returns are ensured for rural people.

**Acknowledgments** Research data reported here were obtained in the research projects: GRACCIE (Consolider-Ingenio 2010, CSD2007-00067), FIREMED (AGL2008-04522/FOR). CEAM is supported by the Generalitat Valenciana and the Fundación Bancaja. JCYL LE021A08 and MCYT CGL2006-10998-C02-01/BOS. We would like to thank to Margarita Arianoutsou and Francisco Moreira for their valuable comments to help to improve the first versions of this manuscript.

# References

Alcaraz F, Delgado MJ (1998) Thyme-brushwood communities ("tomillares") of semiarid South-eastern Spain. Phytocoenologia 28:427–453

APAS (2004) Estado del conocimiento sobre las causas de los incendios forestales en España. Publicación electrónica: www.idem21.com/descargas/pdf/CAUSAS_IF.pdf

Arianoutsou- Faraggitaki M (2001) Landscape changes in Mediterranean ecosystems of Greece: implications for fire and biodiversity issues. J Medit Ecol 2:165–178

Arianoutsou-Faraggitaki M (1985) Desertification by overgrazing in Greece: the case of Lesvos island. J Arid Environ 9:237–242

Arianoutsou-Faraggitaki M, Margaris NS (1981) Producers and the fire cycle in a phryganic eco-system. In: Margaris NS, Mooney NA (eds) Components of productivity of Mediterranean climate regions: basic and applied aspects. Dr. W Junk Publishers, Hague

Arianoutsou-Faraggitaki M, Margaris NS (1982) Phryganic (East Mediterranean) ecosystems and fire. Ecol Medit 8:473–480

Armas C, Pugnaire FI (2005) Plant interactions govern population dynamics in a semi-arid plant community. J Ecol 93:978–989

Badía D, Martí C (2000) Seeding and mulching treatments as conservation measures of two burned soils in the central Ebro valley, NE Spain. Arid Soil Res Rehabil 14(3):219–232

Baeza MJ (2004) El manejo del matorral en la prevención de incendios forestales. In: Vallejo VR, Alloza JA (eds) Avances en el estudio de la gestión del monte Mediterráneo. Fundación Centro de Estudios Ambientales del Mediterráneo, Valencia

Baeza MJ, Roy J (2008) Germination of an obligate seeder (*Ulex parviflorus*) and consequences for wildfire management. For Ecol Manag 256:685–693

Baeza MJ, Vallejo R (2008) Vegetation recovery after fuel management in Mediterranean shru-blands. Appl Veg Sci 11:151–158

Baeza MJ, Raventós J, Escarré A, Vallejo VR (2006) Fire risk and vegetation structural dynamics in Mediterranean shrubland. Plant Ecol 187:189–201

Baeza MJ, Valdecantos A, Alloza JA, Vallejo R (2007) Human disturbance and environmental factors as drivers of long-term post-fire regeneration patterns in Mediterranean forests. J Veg Sci 18:243–252

Baeza MJ, Santana VM, Pausas JG, Vallejo VR (2011) Successional trends in standing dead biomass in Mediterranean Basin species. J Veg Sci 22:467–474

Baskin C, Baskin M (1998) Seeds: ecology, biogeography and evolution of dormancy and germi-nation. Academic, London

Blondel J, Aronson J (1999) Biology and wildlife of the Mediterranean region. Oxford University Press, New York

Boeken B, Orenstein D (2001) The effect of plant litter on ecosystem properties in a Mediterranean semi-arid shrubland. J Veg Sci 12(6):825–832

Bond WJ, Midgley JJ (2001) Ecology of sprouting in woody plants: the persistence niche. Trends Ecol Evol 16:45–51

Calvo L, Luis E, Tárrega R (1989) Regeneración de herbáceas en parcelas experimentales de matorral. Options Méditerranéennes – Série Séminaires 3:127–130

Calvo L, Tárrega R, Luis E (1998) Space–time distribution patterns of *Erica australis* L. subsp. *aragonensis* (Willk) after experimental burning, cutting, and ploughing. Plant Ecol 137:1–12

Calvo L, Tárrega R, Luis E (2002a) Secondary succession after perturbations in a shrubland community. Acta Oecol 23:393–404

Calvo L, Tárrega R, Luis E (2002b) Regeneration patterns in a *Calluna vulgaris* heathland in the Cantabrian mountains (NW Spain): effects of burning, cutting and ploughing. Acta Oecol 23:81–90

Calvo L, Tárrega R, Luis E (2002c) The dynamics of Mediterranean shrubs species over 12 years following perturbations. Plant Ecol 160:25–42

Calvo L, Alonso I, Fernández AJ, Luis-Calabuig E (2005) Short term study of effects of fertilisation and cutting treatments on the vegetation dynamics of mountain heathlands in Spain. Plant Ecol 179:181–191

Calvo L, Alonso I, Marcos E, De Luis E (2007) Effects of cutting and nitrogen deposition on biodiversity in Cantabrian heathlands. Appl Veg Sci 10:43–52

Canadell J, López-Soria L (1998) Lignotuber reserves support regrowth following clipping of two Mediterranean shrubs. Funct Ecol 12:31–38

Casal M (1987) Post-fire dynamics of shrublands dominated by Papilionaceae plants. Influence of fire on the stability of Mediterranean forest ecosystems. Ecol Medit 13:87–98

Casal M, Basanta M, González F, Montero R, Pereiras J, Puentes A (1990) Post-fire dynamics in experimental plots of shrubland ecosystems in Galicia (NW Spain). In: Goldamer JG, Jenkins MJ (eds) Fire in ecosystem dynamics. SPB Acedemic Publishing, Hague

Castro J, Zamora R, Hódar JA, Gómez JM (2002) Use of shrubs as nurse plants: a new technique for reforestation in Mediterranean mountains. Restor Ecol 10:297–305

Clemente AS, Rego FC, Correia OA (1996) Demographic patterns and productivity of post-fire regeneration in Portuguese Mediterranean maquis. Int J Wildland Fire 6:5–12

Correal E, Robledo A, Ríos S (1992) Recursos forrajeros herbáceos y leñosos de zonas áridas y semiáridas. 43 Reunión de la Federación Europea de Zootecnia, Madrid, Spain

Cruz A, Pérez B, Moreno JM (2003) Resprouting of the Mediterranean-type shrub *Erica australis* with modified lignotuber carbohydrate content. J Ecol 91:348–356

De Bello F, Lepš J, Sebastià MT (2005) Predictive value of plant traits to grazing along a climatic gradient in the Mediterranean. J Appl Ecol 42:824–833

Diaz-Delgado R, Lloret F, Pons X, Terradas J (2002) Satellite evidence of decreasing resilience in Mediterranean plant communities after recurrent wildfires. Ecology 83:2293–2303

Dimitrakopoulos AP (2001) Fuel models of the Mediterranean vegetation types of Greece. Geotech Sci Issues 12:192–206 (In Greek with an English summary)

Dimitrakopoulos AP (2003) Analysis of forest fire causes, during the period 1956–1997. For Res 16:17–28 (In Greek with an English summary)

Etienne M, Armand D, Julian P, Napoleone M (1993) Un contrat d'entretien de pare-feu par des moutons. Bilan 1987–1992. INRA, Avignon, France

Etienne M, Aronson J, Le Floch E (1998) Abandoned lands and land use conflicts in southern France. In: Rundel PW, Montenegro G, Jaksic FM (eds) Landscape disturbance and biodiversity in Mediterranean type ecosystems. Springer, New York

Eugenio M, Lloret F (2006) Effects of repeated burning on Mediterranean communities of the northeastern Iberian Peninsula. J Veg Sci 17:755–764

European Commission (2009) Forest fires in Europe 2008, EUR 23492 EN, office for official publications of the European communities, Luxembourg

FAO (2007) Fire management – global assessment 2006. FAO forestry paper 151. Available at: http://www.fao.org/docrep/009/a0969e/a0969e00.htm. Accessed Oct 2010)

Fernández-Abascal I, Tárrega R, Luis-Calabuig E, Marcos E (2003) Effects of sowing native herbaceous species on the post-fire recovery in a heathland. Acta Oecol 24:131–138

Fernández-Abascal I, Tárrega R, Luis-Calabuig E (2004) Ten years of recovery after experimental fire in a heathland: effects of sowing native species. For Ecol Manag 203:147–156

Garcia C, Hernández T, Roldan A, Martin A (2002) Effect of plant cover decline on chemical and microbiological parameters under Mediterranean climate. Soil Biol Biochem 34:635–642

Giourga H, Margaris NS, Vokou D (1998) Effects of grazing pressure on succession process and productivity of old fields on Mediterranean islands. Environ Manag 2:589–596

Gómez-Aparicio L, Zamora R, Gómez JM, Hódar JA, Castro J, Baraza E (2004) Applying plant facilitation to forest restoration in Mediterranean ecosystems: a meta-analysis of the use of shrub as nurse plants. Ecol Appl 14:1128–1138

Hanes TL (1971) Succession after fire in the chaparral of southern California. Ecol Monogr 41:27–52

Harrison PA (2010) Ecosystem Services and biodiversity conservation: an introduction to the RUBICODE project. Biodivers Conserv 19(10):2767–2772

Harrison PA, Vandewalle M, Sykes MT, Berry PM et al (2010) Identifying and prioritising services in European terrestrial and freshwater ecosystems. Biodivers Conserv 19:2791–2821

Hawkes CV, Wren IF, Herman DJ, Firestone MK (2005) On the separation of net ecosystem exchange into assimilation and ecosystem respiration: review and improved algorithm. Glob Change Biol 11:1424–1439

Henkin Z, Gutman M, Aharon H, Perevolotsky A, Ungar ED, Seligman NG (2005) Suitability of Mediterranean oak woodland for beef herd husbandry. Agric Ecosyst Environ 109:255–261

Hobbs RJ, Norton DA (1996) Towards a conceptual framework for restoration ecology. Restor Ecol 4:93–110

Kadmon R, Harari-Kremer R (1999) Landscape-scale regeneration dynamics of disturbed Mediterranean maquis. J Veg Sci 10:393–402

Keeler-Wolf T (1995) Post-fire emergency seeding and conservation in southern California shrublands. In: Keeley JF, Scott T (eds) Brushfires in California wildlands: ecology and resource management. International Association of Wildland Fire, Fairfield

Le Houérou HN (1987) Vegetation wildfires in the Mediterranean Basin: evolution and trends. Ecol Medit 13:13–24

Le Houérou HN (1992) Grazing land of the Mediterranean basin. In: Coupland RT (ed) Ecosystems of the world, natural grasslands, vol 8B. Elsevier Scientific Publications, Amsterdam

Le Houérou HN (1993) Land degradation in Mediterranean Europe: can agroforestry be a part of the solution? A prospective review. Agrofor Sys 21:43–61

Liacos LG (1973) Present studies and history of burning in Greece. In: Proceedings of the 13th annual tall timbers fire ecology conference, Tallahassee

Liacos LG, Papanastasis VP, Tsiouvaras CN (1980) Contribution to the conversion of kermes oak (*Quercus coccifera* L.) brushlands to grasslands and comparison of their productivity with improved brushlands in Greece. For Res 2:97–142 (In Greek with an English summary)

Lloret F (1998) Fire, canopy cover and seedling dynamics in Mediterranean shrubland of northeastern Spain. J Veg Sci 9:417–430

Lloret F (2004) Régimen de incendios y regeneración. In: Valladares F (ed) Ecología del bosque mediterráneo en un mundo cambiante. Ministerio de Medio Ambiente. EGRAF, S.A, Madrid

Lloret F, Pausas JG, Vilà M (2003) Responses of Mediterranean plant species to different fire frequencies in Garraf Natural Park (Catalonia, Spain): fields observations and modelling predictions. Plant Ecol 167:223–235

Lozano J, Virgos E, Malo AF, Huertas DL, Casanovas JG (2003) Importance of scrub-pastureland mosaics for wild-living cats occurrence in a Mediterranean area: implications for the conservation of the wildcat (*Felis silvestris*). Biodivers Conserv 12:921–935

Luis-Calabuig L, Tárrega R, Calvo L, Marcos E, Valbuena L (2000) History of landscape changes in northwest Spain according to land use and management. In: Trabaud L (ed) Life and environment in the Mediterranean. WIT Press, Southampton

Maestre FT, Cortina J, Bautista S (2004) Mechanisms underlying between *Pinus halepensis* and the native late-successional shrub *Pistacea lentiscus* in a semi-arid plantation. Ecography 27:776–786

Maheras G (2002) Forest fires in Greece. The analysis of the phenomenon affecting both natural and human environment. The role of sustainable development in controlling fire effects. M. Sc. thesis, Lund University Sweden

Mazzoleni S, di Pasquale G, Mulligan M, di Martino P, Rego F (eds) (2004) Recent dynamics of the Mediterranean vegetation and landscape. Wiley, West Sussex

Moreira F, Russo D (2007) Modelling the impact of agricultural abandonment and wildfires on vertebrate diversity in Mediterranean Europe. Landsc Ecol 22:1461–1476

Moreira F, Rego F, Ferreira P (2001) Temporal (1958–1995) pattern of change in a cultural landscape of northwestern Portugal: implications for fire occurrence. Landsc Ecol 16:557–567

Moreira F, Vaz P, Catry F, Silva JS (2009) Regional variations in wildfire susceptibility of landcover types in Portugal: implications for landscape management to minimize fire hazard. Int J Wildland Fire 18:563–574

Moreno JM, Vázquez A, Vélez R et al (1998) Recent history of forest fires in Spain. In: Large forest fires. Backhuys Publishers, Leiden

Moreno JM, Zavala G, Martín M, Millán A (2010) Forest fire risk in Spain under future climate change. In: Josef Settele LP, Teodor Georgiev RG, Vesna G, Volker H, Stefan K, Mladen K, Ingolf K (eds) Atlas of biodiversity risk. Pensoft Publishers, Sofia

Moriondo M, Good P, Durao R, Bindi M, Giannakopoulos C, Corte-Real J (2006) Potential impact of climate change on forest fire risk in Mediterranean area. Clim Res 31:85–95

Mulas M, Mulas G (2005) Cultivar selection from rosemary (*Rosmarinus officinalis* L.) spontaneous populations in the Mediterranean area. Acta Hortic 676:127–133

Nader G, Henkin Z, Smith E, Ingram R, Narvaez N (2007) Planned herbivory in the management of wildfire fuels. Rangelands 29:18–24

Naveh Z (1994) The role of fire in the conservation of Mediterranean ecosystems. In: Moreno JM, Oechel WC (eds) The role of fire in Mediterranean-type ecosystems. Springer- Verlag, New York

Naveh Z, Dan J (1974) Effects of fire in Mediterranean region. In: Kozlowski TT, Ahlgren CE (eds) Fire and ecosystems. Academic, New York

Naveh Z, Lieberman AS (1994) Landscape ecology. Theory and application. Springer, New York

Noy-Meir I, Gutman M, Kaplan Y (1989) Responses of Mediterranean grassland plants to grazing and protection. J Ecol 77:290–310

Nunes MC, Vasconcelos MJ, Pereira JM, Dasgupta N, Alldredge RJ, Rego FC (2005) Land cover type and fire in Portugal: do fires burn land cover selectively? Landsc Ecol 20:661–673

Papanastasis VP (1977) Fire ecology and management of phrygana communities in Greece. In: Mooney HA, Conrad CE, Coors (eds) Proceedings of the Symposium. Environmental consequences of fire and fuel management in Mediterranean Ecosystems. USDA Forest Service, General Technical Report. WO-3, Washington DC

Papanastasis VP (1978a) Early succession after fire in a maquis-type brushland of Northern Greece. Forest 79–80:19–26 (In Greek with an English summary)

Papanastasis VP (1978b) Potential of certain range species for improvement of burned brushlends in Greece. In: Hyder DN (ed) Proceedings of the 1st international Rangeland congress. Society for range management, Colorado

Papanastasis VP (1980) Effects of season and frequency of burning on a phryganic rangeland in Greece. J Range Manag 33:251–255

Papanastasis VP (1988a) Rehabilitation and management of vegetation after wildfires in maquis-type brushlands. Forest Research 2(IX): 77–90 (In Greek with English summary)

Papanastasis VP (1988b) Evolution of vegetation in a burned kermes oak brushland seeded with range grasses. Scientific annals of the department of forestry and natural environment, LA: 255–270 (in Greek with English summary)

Papanastasis VP (2000) Shrubland management and shrub plantations in southern Europe. In: Gintzburger G, Bounejmate M, Nefzaoui N (eds) Fodder shrub development in arid and semi-arid zones, vol 1. ICARDA, Aleppo, Syria

Papanastasis VP (2004) Traditional *vs* contemporary management of Mediterranean vegetation: the case of the island of Crete. J Biol Res 1:39–46

Papanastasis VP, Kazaklis A (1998) Land use changes and conflicts in the Mediterranean-type ecosystems of western Crete. In: Rundel PW, Montenegro G, Jaksik FM (eds) Landscape disturbance and biodiversity in Mediterranean-type ecosystems ecological studies 136. Springer, Berlin

Papanastasis VP, Platis P (1990) Effects of range grass seeding on a wildburned kermes oak shrubland in Greece. In: 6th meeting of the FAO subnetwork on Mediterranean pastures and fodder crops, Bari

Papanastasis VP, Yiakoulaki MD, Decandia M, Dini-Papanastasi O (2008) Integrating woody species into livestock feeding in the Mediterranean areas of Europe. Anim Feed Sci Tech 140:1–17

Pausas J (1999) Mediterranean vegetation dynamics: modeling problems and functional types. Plant Ecol 140:27–39

Pausas J (2003) The effect of landscape pattern on Mediterranean vegetation dynamics- A modelling approach using functional types. J Veg Sci 14:365–374

Pausas J, Vallejo VR (1999) The role of fire in European Mediterranean Ecosystems. In: Chuvieco E (ed) Remote sensing of large wildfires in the European Mediterranean Basin. Springer-Verlag, Berlin

Perevolotsky A, Seligman NG (1998) Role of grazing in Mediterranean rangeland ecosystems – Inversion of a paradigm. Bioscience 48:1007–1017

Perevolotsky A, Schwartz-Tzachor R, Yonatan R (2002) Management of fuel breaks in the Israeli Mediterranean ecosystem. J Medit Ecol 3:13–22

Piñol J, Terradas J, Lloret F (1998) Climate warning, wildfire hazard, a wildfire occurrence in coastal eastern Spain. Clim Chang 38:345–357

Potts SG, Petanidou T, Roberts S, O'Toole C, Hulbert A, Willmer P (2006) Plant pollinator biodiversity and pollination services in a complex Mediterranean landscape. Biol Conserv 129:519–529

Rego F (1992) Land use changes and wildfires. In: Teller A, Mathy P, Jeffers JNR (eds) Response of forest fires to environmental change. Elsevier, London

Riera P, Peñuelas J, Farreras V, Estiarte M (2007) Valuation of climate-change effects on Mediterranean shrublands. Ecol Appl 17:91–100

Santana VM, Baeza MJ, Mars RH, Vallejo VR (2010) Old-field secondary succession in SE Spain: can fire divert it? Plant Ecol 211:337–349

Savé R, Castell C, Terradas J (1999) Gas exchange and water relations. In: Roda F, Retana J, Gracia CA, Bellot J (eds) Ecology of Mediterranean evergreen oak forest. Springer-Verlag, Berlin

Scott MJ, Bilyard GR, Link SO, Ulibarri CA, Westerdahl HE, Ricci PF, Seely HE (1998) Evaluation of ecological resources and functions. Environ Manag 22:49–68

Seligman N, Henkin Z (2000) Regeneration of a dominant Mediterranean dwarf-shrub after fire. J Veg Sci 11:893–902

Serrasolses I, Llovet J, Bautista S (2004) Degradación y restauración de suelos forestales Mediterráneos. In: Vallejo VR, Alloza JA (eds) Avances en el estudio de la gestión del monte Mediterráneo. Fundación Centro de Estudios Ambientales del Mediterráneo, Valencia

Tárrega R, Luis Calabuig E, Alonso I (1995) Comparison of the regeneration after burning, cutting and ploughing in a *Cistus ladanifer* shrubland. Vegetatio 120:59–67

Tárrega R, Luis Calabuig E, Alonso I (1997) Space-time heterogeneity in the recovery after experimental burning and cutting in a *Cistus laurifolius* shrubland. Plant Ecol 129:179–187

Trabaud L (1981) Man and fire: impacts on Mediterranean vegetation. In: Di Castri F, Goodall DW, Specht RL (eds) Ecosystems of the world 11: Mediterranean-type shrublands. Elsevier, Amsterdam

Trabaud L (1987) Natural and prescribed fire: survival strategies of plants and equilibrium in mediterranean ecosystems. In: Tenhunen JD, Catarino FM, Lange OL, Oechel WC (eds) Plant response to stress. Springer-Verlag, Berlin

Trabaud L (1992) Community dynamics after fire disturbance: short-term change and longterm stability. Ekistics 356:287–292

Trabaud L (1994) Post-fire plant community dynamics in the Mediterranean Basin. In: Moreno JM, Oechel WC (eds) The role of fire in Mediterranean-type ecosystems. Springer- Verlag, New York

Trabaud L (2000) Seeds: their soil bank and their role in post-fire recovery of ecosystems of the Mediterranean Basin. In: Trabaud L (ed) Life and environment in the Mediterranean. WIT Press, Southampton

Valdecantos A, Cortina J, Vallejo VR (2006) Nutrient status and field performance of tree seedlings planted in Mediterranean degraded areas. Ann Forest Sci 63:249–256

Valdecantos A, Baeza MJ, Vallejo VR (2009) Vegetation management for promoting ecosystem resilience in fire-prone Mediterranean shrublands. Restor Ecol 17:414–421

Vallejo VR (1997) La restauración de la cubierta vegetal en la Comunidad Valenciana. CEAM, Valencia

Vallejo VR, Alloza JA (1998) The restoration of burned lands: the case of Eastern Spain. In: Moreno JM (ed) Large forest fires. Backhuy Publishers, Leiden

Wessel WW, Tietema A, Beier C, Emmett BA, Peñuelas J, Riis-Nielsen T (2004) A qualitative ecosystem assessment for different shrublands in western Europe under impact of climate change. Ecosystems 7(6):662–671

Zamora R, Gacía-Fayos P, Gómez-Aparicio L (2004) Las interacciones planta-planta y planta animal en el contexto de la sucesión ecológica. In: Valladares F (ed) Ecología del bosque mediterráneo en un mundo cambiante.. Ministerio de Medio Ambiente. EGRAF, S.A, Madrid

Zohary M (1962) Plant life in Palestine. Ronald Press, New York

# Author Index

# Subject Index

F. Moreira et al. (eds.), *Post-Fire Management and Restoration of Southern European Forests*, Managing Forest Ecosystems 24, DOI 10.1007/978-94-007-2208-8,
© Springer Science+Business Media B.V. 2012